Achieving net zero dairy farming

It is widely recognised that agriculture is a significant contributor to global warming and climate change. Agriculture needs to reduce its environmental impact and adapt to current climate change whilst still feeding a growing population, i.e. become more 'climate-smart'. Burleigh Dodds Science Publishing is playing its part in achieving this by bringing together key research on making the production of the world's most important crops and livestock products more sustainable.

Based on extensive research, our publications specifically target the challenge of climate-smart agriculture. In this way we are using 'smart publishing' to help achieve climate-smart agriculture.

Burleigh Dodds Science Publishing is an independent and innovative publisher delivering high quality customer-focused agricultural science content in both print and online formats for the academic and research communities. Our aim is to build a foundation of knowledge on which researchers can build to meet the challenge of climate-smart agriculture.

For more information about Burleigh Dodds Science Publishing simply call us on +44 (0) 1223 839365, email info@bdspublishing.com or alternatively please visit our website at www.bdspublishing.com.

Related titles:

Achieving sustainable production of milk Volume 2
Print (ISBN 978-1-78676-048-7); Online (ISBN 978-1-78676-050-0, 978-1-78676-051-7)

Advances in sustainable dairy cattle nutrition
Print (ISBN 978-1-80146-205-1); Online (ISBN 978-1-80146-207-5, 978-1-80146-208-2)

Assessing the environmental impact of agriculture
Print (ISBN 978-1-78676-228-3); Online (ISBN 978-1-78676-230-6, 978-1-78676-231-3)

Improving rumen function
Print (ISBN 978-1-78676-332-7); Online (ISBN 978-1-78676-334-1, 978-1-78676-335-8)

Reducing greenhouse gas emissions from livestock production
Print (ISBN 978-1-78676-439-3); Online (ISBN 978-1-78676-441-6, 978-1-78676-442-3)

Chapters are available individually from our online bookshop:
https://www.bdspublishing.com/webshop/

climate•SMART•publishing
IN AGRICULTURAL SCIENCE

BURLEIGH DODDS SERIES IN AGRICULTURAL SCIENCE
NUMBER 171

Achieving net zero dairy farming

Edited by Emeritus Professor John Webster, University of Bristol, UK

burleigh dodds
SCIENCE PUBLISHING

Published by Burleigh Dodds Science Publishing Limited
82 High Street, Sawston, Cambridge CB22 3HJ, UK
www.bdspublishing.com

Burleigh Dodds Science Publishing, 1518 Walnut Street, Suite 900, Philadelphia, PA 19102-3406, USA

First published 2026 by Burleigh Dodds Science Publishing Limited

Library of Congress Control Number: 2026930443

British Library Cataloguing in Publication Data
A catalogue record for this book is available from the British Library

ISBN 978-1-83545-018-5 (print)
ISBN 978-1-83545-019-2 (ePub)
ISBN 978-1-83545-020-8 (PDF)
ISSN 2059-6936 (print)
ISSN 2059-6944 (online)

DOI: 10.19103/AS.2025.0157

Typeset by Deanta Global Publishing Services, Dublin, Ireland

Contents

Part 2 Strategies to achieve net zero: supporting the dairy cow

Part 4 Conclusion

Series list

Title	Series number
Achieving sustainable cultivation of maize - Vol 1 From improved varieties to local applications Edited by: Dr Dave Watson, CGIAR Maize Research Program Manager, CIMMYT, Mexico	001
Achieving sustainable cultivation of maize - Vol 2 Cultivation techniques, pest and disease control Edited by: Dr Dave Watson, CGIAR Maize Research Program Manager, CIMMYT, Mexico	002
Achieving sustainable cultivation of rice - Vol 1 Breeding for higher yield and quality Edited by: Prof. Takuji Sasaki, Tokyo University of Agriculture, Japan	003
Achieving sustainable cultivation of rice - Vol 2 Cultivation, pest and disease management Edited by: Prof. Takuji Sasaki, Tokyo University of Agriculture, Japan	004
Achieving sustainable cultivation of wheat - Vol 1 Breeding, quality traits, pests and diseases Edited by: Prof. Peter Langridge, The University of Adelaide, Australia	005
Achieving sustainable cultivation of wheat - Vol 2 Cultivation techniques Edited by: Prof. Peter Langridge, The University of Adelaide, Australia	006
Achieving sustainable cultivation of tomatoes Edited by: Dr Autar Mattoo, USDA-ARS, USA and Prof. Avtar Handa, Purdue University, USA	007
Achieving sustainable production of milk - Vol 1 Milk composition, genetics and breeding Edited by: Dr Nico van Belzen, International Dairy Federation (IDF), Belgium	008
Achieving sustainable production of milk - Vol 2 Safety, quality and sustainability Edited by: Dr Nico van Belzen, International Dairy Federation (IDF), Belgium	009
Achieving sustainable production of milk - Vol 3 Dairy herd management and welfare Edited by: Prof. John Webster, University of Bristol, UK	010
Ensuring safety and quality in the production of beef - Vol 1 Safety Edited by: Prof. Gary Acuff, Texas A&M University, USA and Prof. James Dickson, Iowa State University, USA	011
Ensuring safety and quality in the production of beef - Vol 2 Quality Edited by: Prof. Michael Dikeman, Kansas State University, USA	012
Achieving sustainable production of poultry meat - Vol 1 Safety, quality and sustainability Edited by: Prof. Steven C. Ricke, University of Arkansas, USA	013
Achieving sustainable production of poultry meat - Vol 2 Breeding and nutrition Edited by: Prof. Todd Applegate, University of Georgia, USA	014

Introduction

In order to help reduce the greenhouse gas (GHG) emissions that contribute to the current climate crisis, many countries are aiming to be 'climate-neutral' by the year 2050. Achieving 'net-zero', according to the United Nations, can be defined as 'cutting greenhouse gas emissions to as close to zero as possible, with any remaining emissions re-absorbed from the atmosphere'.

The livestock sector, primarily dairy cattle, is estimated to contribute approximately 11-17% of global GHG emissions. The dairy sector is therefore a prime target for reducing emissions from the agricultural sector. This volume focuses on how and how far dairy farming can achieve net-zero. Chapters focus on fundamental aspects of the dairy cow and how emissions can be measured. The book also discusses particular strategies to reduce emission to achieve net zero.

Part 1 Fundamentals and measurement on the dairy farm

Chapter 1 helps set the scene for the book. The first part of the chapter begins by introducing some basic concepts involved in defining sustainability. The chapter then goes on to provide a simple introduction to the principles of biochemistry and physiology that determine energy and protein exchanges in the dairy cow. This provides context for more detailed discussions that explore the ways these fundamental processes can be leveraged to optimise nutrient use efficiency in the dairy cow, reduce unwanted by-products such as methane and thus help achieve the goal of net zero.

Chapter 2 discusses the production and mitigation of methane emissions from dairy cattle. After reviewing what we know about the processes driving methanogenesis, it reviews the evidence on how certain feed additives, particularly 3-NOP and *Asparagopsis taxiformis*, can help in reducing methane production. It then goes on to analyze different factors that may affect their effectiveness, including potential rumen adaptation to counteract their effects. Following this, the chapter then reviews models that can predict methane emissions in dairy cows, as well different systems of emissions measurement such as the GreenFeed system which can be used to assess the effectiveness of interventions such as feed additives.

As Chapter 3 points out, quantifying GHG emissions at the whole farm level is challenging. The primary GHG gases emitted from dairy farms are methane (CH_4), nitrous oxide (N_2O) and carbon dioxide (CO_2). Ammonia (NH_3) is also an important pollutant from dairy farms. Whilst direct measurement is important, a key tool is the use of models to estimate emissions of differing GHGs from particular sources on dairy farms. Chapter 3 explores how individual model sub-components can be integrated into a comprehensive modelling framework to ascertain whole farm emissions. Life cycle assessment is a modelling approach used to account for resource use, emissions and other environmental impacts. The chapter focuses on the important use of dairy farm models to quantify and compare carbon footprints to determine best management practices. Mitigation strategies can then be evaluated alone and in combination to refine production systems toward the goal of net zero milk production.

This next chapter focuses on the water footprint in dairy farming. Chapter 4 evaluates the findings from recent research and case studies, focusing on the latest developments in measuring the water footprint of dairy farming. The chapter begins by exploring the concept and components of a water footprint, followed by the exploration of water management technologies. Recent studies focus on the variability of water consumption across, particularly in feed production. Studies that focus on the strategies used for calculating water footprints under varying conditions of water scarcity are also highlighted. It also discusses the effectiveness of various calculation methods in enhancing region-specific approaches, water productivity and local water challenges.

Chapter 5 focuses on various methods for measuring methane emissions in particular at both the animal and farm levels. It highlights the need for accurate, reliable, cost-effective and practical approaches, which is crucial for the sustainability of dairy operations. This chapter lays out the sources of methane in dairy farming, as well as comparing the pros and cons for different methods for measuring these emissions. There is also discussion of strategies for mitigating methane emissions, and how integrating these technologies with precision agriculture offers new opportunities for reducing the environmental impact of dairy farming while maintaining productivity.

Part 2 Strategies to achieve net zero: supporting the dairy cow

Focusing on the integration of methane traits into dairy cattle breeding programs, Chapter 6 begins by exploring various methods for recording methane emissions from individual cows, in order to make informed breeding decisions. Thereafter, the chapter discusses how data collected by different recording techniques can be related to promising traits to include in breeding programmes. The chapter then reviews genetic parameters estimated for these traits and considers their implications for designing recording schemes for genetic evaluations. Finally, the chapter addresses the implementation of selection strategies aimed at methane mitigation within existing or new breeding programs.

Chapter 7 delves into the importance of improving feed conversion ratios for advancing sustainability in the dairy industry. Traditional nutritional and genetic technologies have advanced feed efficiency and productivity in the dairy industry over the past several decades, and the movement toward individual animal feeding is expected to advance this trend. The chapter also discusses how precision livestock monitoring and management technologies will be critical to advancing the individual feeding of animals. However, numerous improvements in the functionality and usability of these systems will be important for optimizing the utility of these technologies.

This next chapter explores key housing strategies to mitigate GHG emissions while maintaining animal welfare. Chapter 8 discusses measures such as managing pasture access and stocking density, improving animal surveillance, enhancing diet quality and optimising manure management in housed systems. It points out that while restricting pasture can reduce methane and nitrous oxide emissions, there must be a balance with welfare concerns. This chapter also addresses precision livestock farming technologies that enhance herd monitoring, improve health, reproduction, feed efficiency and reduce emissions in housed systems.

Chapter 9 draws attention to the fact that, despite decades of research, the availability of low-inclusion additives with precise, specific and significant effects on methanogenesis has still not resulted in their widespread use. The chapter explores the various potential target mechanisms involved: reduce the production of hydrogen during fermentation; provide alternatives to methanogenesis as pathways for hydrogen utilisation; inhibit methanogenesis (or methanogenic archaea) directly; or metabolise methane before it leaves the rumen. It reviews the evidence supporting potential use of particular products, including those based on species of red seaweed in the genus *Asparagopsis* containing bromoform ($CHBr_3$), which directly inhibits methanogenesis, as well as the evidence underpinning use of the methane inhibitor 3-NOP, As the chapter points out, these products must satisfy a challenging range of criteria. The chapter explores opportunities and hurdles relating to product development, regulatory approval, accounting for secondary effects, supply chain issues and social acceptance.

Part 3 Strategies to achieve net zero: system-level improvements

The subject of Chapter 10 is manure management and processing. The beginning of the chapter discusses how manure management in the United States is considered one of the key hotspots of greenhouse gas emissions. The dairy industry has begun implementing a series of interventions throughout the manure management system to address emissions, helping to reduce the volatile fraction of manure through solid separation, improve nutrient management, and capture biogas in anaerobic digesters.

Chapter 11 explores the issues and advances in using dairy manure, especially slurry, to replace commercial fertilisers and improve soils. It touches upon the issues of preserving nutrients and providing the best crop response and carbon balance, with minimal impact on farms and their surroundings. The advances in using dairy slurry employ short and long term field studies, integrated farm management systems, and farmer based approaches. This chapter also uses Canada as an example to show how dairy manure can be better utilised and managed.

The next chapter explains how anaerobic digestion for biogas and biomethane production and methane utilisation can be implemented on dairy farms and how this contributes to achieving net-zero greenhouse gas emissions. Chapter 12 discusses how enteric fermentation in ruminants and the handling of manure and slurry cause unwanted GHG emissions on dairy farms. In contrast, anaerobic digestion through engineered systems yields biogas or biomethane as a renewable energy source. The production and utilisation of methane-rich renewable gas represents an important pathway to achieve net-zero greenhouse gas emissions for dairy farms, as demonstrated by different case studies.

The focus of Chapter 13 is the exploration of integrating dairy farming and crop production. It emphasises the multifunctionality of different types of integrated crop-livestock systems in accounting for GHG emissions across farm boundaries, assessing the mitigation potential of crops as well the potential of integrated systems to reduce overall emissions associated with livestock production. It also assesses the drivers for developing integrated crop- livestock systems to achieve net zero and climate resilience.

Part 4 Conclusion

The final chapter of the book takes stock of the key insights provided by individual chapters in the book. Chapter 14 reviews the evidence relating to the emission of GHG gases and other pollutants, approaches to their mitigation, and the potential for carbon sequestration. Drawing from a wealth of expert analysis, the chapter provides an incisive assessment of the nature of the challenge facing dairy production in contributing to sustainability goals. It also provides a pragmatic assessment of options for mitigation and what they can achieve.

Part 1

Fundamentals and measurement on the dairy farm

Chapter 1

Sustainability, nutrient and energy flows in the dairy cow: an overview

John Webster, University of Bristol, UK

1 Introduction

The poison is in the dose

Paracelsus

'Net zero' has become the fashionable term to describe a concept as old as agriculture, namely sustainability or, to use an even more old-fashioned phrase, 'the balance of nature'. Chapters in this book examine critical issues of sustainability within the context of dairy farming in the light of new scientific understanding with particular emphasis on reducing or ameliorating the effects of greenhouse gases (GHG or GHGs), carbon dioxide (CO_2), methane (CH_4) and nitrous oxide (N_2O). This is a very specific objective but one that generates a very wide range of questions. The chapters that follow this introduction review the state of the art in respect to the many factors in play within the cow, on the land, in the water and in the air. They have been written by scientists well equipped to review advances in understanding within their specialist fields of expertise. The aim of this introductory chapter is not to provide answers but to set out the questions within the context of established biological principles that determine the production of GHG and other potential pollutants. Subsequent

http://dx.doi.org/10.19103/AS.2025.0157.01

chapters explore the scope for manipulating these things in the pursuit of sustainability in dairy farming.

There are two simple truths that need to be stated at the outset. The first is that the accumulation of GHG leading to problems of global warming and climate change is a very recent phenomenon. It has been calculated that the concentration of CO_2 in the atmosphere remained stable at about 280 ppm for about 10 000 years and only began to rise to its current (but still rising) concentration at the beginning of the twentieth century[1] (Callendar 1958).

The second relates to ruminant numbers. In 2024, the total cattle population in the USA (cows, calves and finishing beef cattle) was approximately 87 million of which about 9.3 million were adult dairy cows and 28 million beef cows. However, in the seventeenth century, the bison population in the USA was approximately 60 million (Kelliher and Clark 2010). The current production of GHG from ruminants in the USA has been estimated as no more than 14% greater than that when bison roamed the plains – *sustainably* (Hristov 2012).

Nevertheless, we cannot escape the fact that we have a problem. It is trite but true to claim that agricultural pollution is simply a case of too much fertiliser in the wrong place. This aphorism has special relevance to dairy farming. While it is undeniable that intensification leading to increased milk production per cow has increased milk yield relative to GHG production (by the cows), it is also the case that pastoral or agro-forestry systems, though less productive, can be more compatible with sustainability because of their greater capacity to recycle waste products and sequester carbon.

I prefaced this chapter with the words of Paracelsus, *The poison is in the dose*. This is especially applicable to the business of dairy farming. On its current scale and in its current forms it has been estimated to make a significant contribution to the global production of GHG and scientists have a duty to contribute to the resolution of this problem through the application of new understanding.

2 Sustainable dairy farming: some basic concepts

This introductory chapter comes in two sections. I shall first set out the rules of engagement: the complex set of interactions that determine the sustainability of dairy farming. In particular, I review the requirements, release, recapture and recycling of the main compounds involved in the cow and on the farm. I start with some basic elements/concepts involved in sustainability.

1 The most recent ice age, the period of 'quaternary glaciation', was preceded by a period when atmospheric CO_2 concentration was less than 200ppm. There is debate as to the extent to which these two things are related (Eggleton 2013).

2.1 Energy

Sources include 'free energy', i.e. solar energy converted into plant energy by photosynthesis, and 'fuel energy', used for everything else in the production, fertilisation, harvesting and processing of feed, manure disposal, construction and maintenance of resources etc. (The list is long.) Feed energy for dairy cattle may be further subdivided into *competitive*: feed energy (e.g. in cereals that could be fed to and digested by humans) and *complimentary* (e.g. grasses that provide energy for human consumption after processing by animals, in this case as milk).

2.2 Protein

Competitive sources of protein for dairy cows include high protein beans (e.g. soya, rape). Complementary sources of dietary nitrogen include proteins available as a by-product of food and drink production for humans (e.g. brewers' grains) and non-protein nitrogen (e.g. urea) available for protein synthesis by microorganisms in the rumen.

2.3 Water

Analysis of water management in dairy farming recognises *green water*, essentially free water from rainfall, *blue water*, ground water from reservoirs, which can become a problem in times of water shortage and *grey water*, polluted as a consequence of farming activities.

2.4 Disposal of wastes

This includes the costs of handling and dispersal of manure products, recycling as natural fertiliser, use to generate biogas or disposal of in ways that contribute to pollution of water and the air. It also must include the production of GHG directly from the cows and from other farm activities (e.g. manure management).

2.5 Quality of life

This must incorporate not only the provision of nutritious, healthy food for humans, the health and welfare of the farm animals but also the sustained health of the natural environment: soils, water, trees, wildlife etc.

3 Measuring sustainability: life cycle assessment

The various and complex elements involved in assessments of sustainability can form the basis for a life cycle assessment (LCA), which is an accounting framework used to inventory the material and energy inputs and emissions associated with each stage of a production life cycle. I shall explore this approach in more detail in the final chapter when I attempt to bring together the insights emerging from our authors contributing their specialist knowledge, to present an overview of the disparate issues and set out a list of priorities for action and further research. At this stage, I shall simply draw attention to two approaches to LCA that have special relevance to questions of GHG production and sustainability in dairy farming.

The first approach is to conduct an inventory of CO_2, CH_4 and nitrogen (N) cycles in systems that range widely in the proportion of time cattle are housed or at pasture. Studies such as those by Pelletier et al. (2010), Soussana et al. (2007, 2010) include measures of the generation, recapture and sequestration of carbon (C) and N in plants and soils so offer an approach to the full accounting of the GHG (CO_2, N_2O, CH_4) budget.

The other supremely elegant though fiendishly complex approach to LCA for farming systems is 'emergy' analysis (Brown and Ulgiati 1997). This concept expresses all the work processes and resources (sunlight, water, fossils fuels, minerals etc.) used in the generation of a product in terms of a common unit of measurement, namely embedded solar energy (E_m). It is particularly well suited to the assessment of the efficiency and sustainability of farming because it can quantify the yield of food energy for human consumption (Y) in terms of renewable (R) resources of sun, soil and water embedded in farmland, non-renewable sources (NR), loss of energy from (e.g. soil degradation), and (F), purchased goods and services such as fuel, fertiliser, labour and imported feeds.

Examples of the application of emergy analysis to livestock systems are given in Table 1 which compares yields and sustainability within agricultural systems as described by three ratios:

- EYR: emergy yield ratio;
- ELR: environmental load ratio; and
- ESI: emergy sustainable index.

In more detail:

- EYR = (R+ NR + F)/F and describes the contribution of sun and local resources (land) to product;
- ELR = (NR + F)F and describes the ratio of non-renewables to renewables in product; and
- ESI = (EYR/ELR) is a measure of yield relative to environmental compatibility.

These ratios are dimensionless. The data in Table 1 are taken from Pereira et al. 2013, Rotolo et al. 2007, and Vigne et al. 2013.

Table 1 A comparison of yields and sustainability within agricultural systems assessed in terms emergy as described by three ratios, emergy yield ratio (EYR), environmental load ratio (ELR) and emergy sustainable index (ESI). For further details see text.

	EYR	ELR	ESI
Corn (USA)[a]	1.07	18.8	0.06
Conventional pig (Sweden)[a]	1.04	22.3	0.05
Organic pig (Sweden)[a]	1.13	7.80	0.15
Intensive dairy (Brittany)[b]	1.35	3.25	0.42
Extensive dairy (Mali)[b]	1.89	1.25	1.57
Extensive cattle (Argentina)[c]	3.73	0.55	6.80

a Pereira et al. 2013
b Rotolo et al et al. 2007
c Vigne et al et al. 2013

Values for EYR show that the relative contribution of renewable resources do not differ greatly between corn production, conventional and organic pig production. The contribution of local resources was greater for dairy production, especially low-intensity dairy production in South Mali. The grazing of beef cattle was by far the most efficient in terms of the contribution of renewable resources. Intensive pig production imposed by far the greatest environmental load ratio (ELR). However, the most striking differences between the systems are revealed by ESI, the measure of yield in relation to environmental compatibility. By this measure, small scale dairy production was more sustainable than intensive production even in Brittany where a large proportion of feed comes from pasture, and extensive beef production on the Argentinian pampas outstrips all others. This may come as a surprise to urbanised critics of livestock production in general and beef production in particular, but it would appear as an overcomplicated proof of the obvious to the gauchos of the pampas or the indigenous races of North America living in perfect symbiosis with the bison.

4 Nutrient and energy flows within the dairy cow

This section addresses the physiological and biochemical processes that determine the digestion and metabolism of the energy and nutrients involved in maintenance and lactation in dairy cows with special reference to the production of GHG. This is a necessary precursor to subsequent specialist chapters that consider different approaches to manipulating these things within the cow and on the farm to improve sustainability in ways that are compatible with cow productivity, health and welfare. In the specific context of GHG, we need to be clear that we should not see our aim as simply to reduce GHG production. We could achieve that by giving our cows less to eat. The aim must be to reduce GHG production *relative to milk yield*.

4.1 Energy metabolism

Metabolizable energy, conventionally expressed in kilojoules, is that made available to an animal by the digestion of feed. Thus:

Metabolizable energy = Gross energy (GE) in feed – energy lost (in faeces + urine + methane)

In cattle, ME will range from about 82% to 92% digestible energy (DE), and this is largely due to variation in methane production. Strategies to reduce methane production will form a major theme of this book. At this stage, I wish to consider the rules that determine the efficiency of conversion of ME into milk in relation to the production of GHG.

Dietary ME must first meet the demands of the animal for maintenance: the metabolic processes necessary to sustain life at a constant body weight. The energy used for these essential processes is all dissipated as heat. Thus, at maintenance ME intake = heat production. The metabolic processes involved in the manufacture of adenosine triphosphate (ATP) (the primary energy source for cells) incur an energy cost. For cattle, the *net* efficiency of utilisation for maintenance (k_m) is approximately 0.7, the remainder (0.3) is defined as the heat increment of feeding. This value (k_m) is unaffected by the composition of the diet.

ME fed to a lactating cow in excess of maintenance is available for milk production and utilised at a net efficiency (k_l) (Blaxter 1967). For a limited range of nutritious feed for productive dairy cows, this figure will also be of the order 0.7. For less digestible feeds offered low yielding cows, it will be lower. High-performance cows at peak lactation will contribute reserves of fat and protein (if they have reserves to contribute). Most dairy cows in late lactation will use some ME to restore body reserves so the apparent value for (k_l) will be lower.

The primary determinant of the *gross* efficiency of utilisation of ME for lactation (RE_l)/ME will be determined by the extent to which milk production and the ME intake required for milk production exceed the need for maintenance, when the efficiency of production is zero. Compare the utilisation of ME (MJ/day) for lactation in high and low yield cows (50 v. 10 l/day):

	Holstein	'Unimproved' cow
Milk yield (l/day)	50	10
Energy flow (MJ/day), ME intake	320	100
Energy in milk	135	25
Heat production	185	120
Gross efficiency (RE_l)/ME intake)	0.42	0.25
Milk yield (l/day): energy loss as heat (MJ)	0.27	0.08

Since the production of carbon dioxide is a direct function of heat production in metabolism, these figures (adapted from Webster 2020) show that cows giving 50 l milk/day produce approximately 3.4 times as much milk per unit of GHG emission than cows producing 10 l/day.

However, this is only part of the story. When high yielding dairy cows are kept in barns, the production of GHG arising from processes involved in manure disposal will be much greater and can exceed 50% (Ellis et al. 2010). Moreover, in pastoral systems manure becomes a fertiliser not a cost. In silvo-pastoral systems (agroforestry) carbon sequestration can absorb a high proportion of GHG production and dairy farming can approach the goal of net zero. Evidence for this is provided by Soussana et al. (2007, 2010) who compared the balance of GHG, C and N in European beef and dairy systems with differing inputs from grassland. By their own admission, these calculations carry a high degree of uncertainty but they confirm that of sequestration in grassland to mitigate the GHG balance of ruminant production systems.

5 Fermentation and digestion in the rumen

Digestion in the dairy cow is a two-stage process:

- Anaerobic fermentation by microbes in the reticulo-rumen; and
- Acid digestion in the abomasum.

Carboniferous feeds are primarily digested by anaerobic fermentation. Rumen microbes make energy and protein available to their host, the cow, by fermenting what they can (principally carbohydrates and fibre) and incorporating organic nitrogen into microbial protein that passes out of the rumen for digestion downstream.

It has become conventional to consider feed energy and protein supply to ruminants within four categories (Webster, Dewhurst and Waters 1988):

- Material that is quickly completely fermented or degraded. Quickly fermented energy (QFE) describes carbohydrates (e.g. starch) that are quickly fermented to volatile fatty acids (VFA). Quickly degradable nitrogen (QDN) describes organic nitrogenous material rapidly degraded to ammonia;
- Material that is slowly and incompletely fermented/degraded is termed SFE and SFN;
- Material that is unfermented or degraded in the rumen is termed unfermentable digestible energy (UDE) and undegradable dietary nitrogen (UDN); and

- Material that is not fermented or degraded in the rumen, nor digested downstream. Since this does not contribute to energy or protein supply it requires no code.

Fermentation, described by Pasteur as 'life without oxygen', describes the process of anaerobic metabolism of energy sources by microorganisms to provide fuel in the form of ATP to support their own needs for maintenance, growth and multiplication. The population (microbiome) includes bacteria, ciliated protozoa, anaerobic fungi and archaea, which are of special relevance to methane production (Ungerfeld 2023).

Carboniferous substrates for fermentation purposes include hexoses (sugars and starches in cell contents) and cell walls: cellulose, hemicellulose, pectin and lignin. Sugars and starches ferment rapidly and form the main constituents of QFE. Cellulose and hemicellulose are essentially complex polymers of hexose and classed as SFE. Pectin is not strictly a polysaccharide but rapidly fermented. Lignin is an indigestible phenolic polymer.

The capacity of the microbiota to capture energy to produce ATP from carbohydrates in the anaerobic environment of the rumen is low: most of their substrate energy is excreted in the form of simple VFA: acetate, propionate and butyrate. These serve as the main source of digestible energy for the host animal. In a healthy rumen environment (pH 6–7) the VFA are present as sodium salts, the sodium arising from saliva and direct secretion of fluid across the rumen wall. These salts of the VFA are absorbed across the rumen wall. The *extent* of fermentation of complex polysaccharides (SFE) is determined both by fermentation rate and retention time in the rumen. Cell wall material that escapes fermentation in the rumen cannot be broken down by acid digestion, but some further fermentation can take place in the hind gut (normally less than 10%).

The final steps involved in anaerobic fermentation are critical in the context of feed efficiency, milk quality and environmental impact. To keep it simple: we may assume that all carbohydrates are fermented via monosaccharides to pyruvic acid:

$$C_6H_{12}O_6 \rightarrow 2CH_3COCOOH + 4H\ (+2ADP \rightarrow 2ATP)$$

At this stage, anaerobic fermentation has generated two ATP and four protons (4H). Pyruvic acid is metabolised further to a mixture of VFA, principally acetic, propionic and butyric acids. The yields of ATP to the microbiota from fermentation to acetate, propionate and butyrate are, respectively, 4.3, 4.0 and 3.0 moles ATP per mole glucose. So far as the rumen microbiota are concerned, acetate production is the most efficient process:

Acetate production: $CH_3COCOOH + H_20 \rightarrow CH_3COOH + CO_2 + 2H$ (+4.3 ATP)

Propionate production: $CH_3COCOOH + 4H \rightarrow CH_3CH_2COOH + H20$ (+4 ATP)

Butyrate production: $2CH_3COCOOH \rightarrow CH_3CH_3CH_2COOH + CO_2$ (+3 ATP)

Fermentation of pyruvate to acetate generates a further excess of two protons (2H), propionate production requires four protons and butyrate production is neutral. In normal rumen fermentation (i.e. without additives) excess protons are harvested by *archaea* and used for methanogenesis.

Table 2 (from Webster 2020) illustrates the fate of 1000 kJ of energy fermented in the rumen of a cow fed with different combinations of forage (mainly SFE from cell walls) and cereals (mainly QFE from starch). In all cases the great majority of fermented energy (FE) is unavailable to the microbiota and absorbed by the cow as VFAs (780–900 J/kJ FE). The amount of FE captured by the microbiota for their own use is very small (50–70 J/kJ FE). The ratio of acetate to propionate production is highly dependent on the ratio of forage to concentrate (SFE:QFE). This determines proton balance and thereby methane production. On a diet consisting entirely of forage (SFE), energy loss as methane is 150 J/kJ FE, on an 80% cereal diet, it is only 50 J/kJ FE.

This table (though theoretical rather than drawn from measurements of practical diets) carries the clear message that the most effective way to reduce methane production is to increase the proportion of cereal-based, propionate forming feeds in the diet. Propionate, being a 3C compound, carries a further benefit in that it is glucogenic and glucose synthesis can be a limiting factor in ruminant metabolism. Acetate and butyrate are ketogenic, effective substrates for ATP synthesis in a balanced diet but if present in excess can lead to ketosis.

All this is, of course, entirely consistent with the practice of incorporating cereals (up to a point) into diets for dairy cattle to promote high milk yields.

Table 2 Fate of 1000 kJ of energy fermented in the rumen (from Webster, 2020)

Diet	Forage 100%	Forage:Cereal 50:50	Forage:Cereal 20:80
Energy yield to cow (kJ)			
Acetate	448	395	322
Propionate	150	260	393
Butyrate	182	185	185
Total VFA	780	840	900
Energy yield to microbiota	70	60	50
Energy loss as methane	150	100	50

The 80:20 cereal:forage mix illustrated in Table 1 would be too rich in QFE and likely to generate problems of rumen acidosis. However, the formulation of nutritious, healthy *Total Mixed Rations* (TMR) with an optimal QFE:SFE balance for barn-fed dairy cows has not only made it possible for high genetic merit cows to express their potential to achieve lactation yields in the order of 10 000 l, it has significantly contributed to the aim of reducing GHG production relative to milk yield by increasing the gross energy efficiency of milk yield relative to ME intake and reducing the proportion of FE lost as methane.

The third source of ME for dairy cows is UDE, feed that escapes fermentation in the reticulo-rumen but is available for acid digestion in the abomasum. This will include undegradable dietary proteins and can include protected fats. Incorporation of these into TMRs will further reduce methane production relative to ME intake.

The formulation of diets for high-performance dairy cows is a very well researched topic. These diets work not only because they are highly nutritious but because they support healthy digestion and therefore the appetite necessary to sustain the high intake of ME necessary to support the work of lactation in high genetic merit cows. A Holstein cow producing 50 l milk/day will (day after day) consume ME at a rate 3–4 times maintenance requirement and 'work' (i.e. produce heat) at a rate twice that of a dry cow at rest. They are by some measure, among the hardest working of all animals (Webster 2013). I suggest that it may not be necessary to review ration formulation in further detail within the context of exploring ways to reduce GHG production because it is already a fortunate, unplanned consequence of the strictly economic aim to maximise milk yield.

This suggests that there is limited scope for further research designed to reduce GHG produced directly from high performance, barn-fed cows through manipulation of diet. There is, of course, considerable scope to reduce GHG production from other farm activities in intensive systems (e.g. manure handling) and to exploit the potential for manure recycling and carbon sequestration in pastoral and silvo-pastoral systems but these are beyond the scope of this opening chapter.

What is important in this context is to consider the possible effects of feed additives designed to reduce methane production from fermentation. Table 1 showed that, *so far as the microbiota are concerned*, fermentation to acetate yields the most ATP to support their maintenance, growth and multiplication. It is reasonable to conclude from this that a relatively high production of acetate is consistent with the maintenance of a large, stable microbiome, a high, stable rate of fermentation and, in consequence, a high, stable appetite for fibrous feed. Thus, any study of the use of feed additives to reduce methane production should incorporate measures of feed intake and fibre digestibility.

For dairy cows, as always, the aim must be to reduce GHG production relative to milk yield.[2]

6 Digestion of protein and nitrogenous compounds

The specific role of dietary protein is to provide amino acids for protein synthesis. The main source of amino acids to the cow comes from the population of microorganisms that pass out of the reticulo-rumen and are digested in the acid medium of the abomasum by pepsin, or in the neutral medium of the duodenum by pancreatic proteases. The rumen microbiota synthesise microbial protein from dietary protein and simpler nitrogenous compounds, principally ammonia.

The totality of nitrogenous (N) compounds in a ruminant diet is conventionally described as *crude protein*, or 6.25N. However, this term does not differentiate between true protein and non-protein nitrogen (NPN) sources such as urea, which is a significant substrate for microbial protein synthesis. It also does not distinguish nitrogenous compounds that are bound into the lignified portion of the cell wall, thus unavailable for degradation in the rumen or subsequent acid digestion. This fraction is described as acid-detergent insoluble nitrogen (ADIN). To understand protein digestion in the ruminant, it is best to avoid the term crude protein and describe all process in terms of dietary N. To sustain the link with digestion of energy sources, NPN equates to QDN, true protein to SDN and ADIN to UDN. The supply of QDN from dietary sources is augmented by UDN (principally urea) recycled from the blood stream via saliva and direct secretion across the rumen wall (Webster, Dewhurst and Waters 1988).

The rate of synthesis of microbial protein is linked to the rate of fermentation of organic matter (principally carbohydrates). The maximum yield of microbial protein N is approximately 50 g/kg fermentable organic matter (FOM). The cow has access only to the microbial protein that passes out into the duodenum, whereas VFA are absorbed across the rumen wall. The ratio of microbial protein to ME supply is therefore determined by the proportion of microbial protein leaving the rumen, and this is determined by rumen outflow rate. For a cow at maintenance with a rumen outflow rate of 4%/hour, the yield of microbial protein will be about 20 g/kg FOM, 40% of total microbial protein synthesis. At an intake of three times maintenance, and an outflow rate of 8%/hour microbial protein yield will be 40 g/kg FOM (approximately).

The relative rates at which energy and protein sources are, respectively, fermented and degraded depend on the relative proportions of the quickly

2 I began my scientific career in the nutrition department of the Hannah Dairy Research Institute, Ayr, Scotland in 1963. The department was headed by Kenneth Blaxter (later Sir Kenneth) and had five respiration chambers for studies of the energy metabolism of ruminants. When I arrived, four of the five respiration chambers were being used to measure the effects of a methane inhibitor in sheep (Czerkawski, Blaxter and Wainman 1966) the aim then being to increase the capture of ME rather than save the planet. *Plus ca change. . .*).

and slowly fermentable and degradable fractions. A meal of hay alone will provide a slow, relatively balanced increase in the production of ammonia and VFA. A meal of grass silage, high in QDN but low in QFE, produces an imbalanced yield of ammonia and VFA. Ammonia that exceeds the capacity of the microbiota for protein synthesis passes across the rumen wall and enters the liver where it is converted to urea that will be recycled to the rumen and, up to a point, incorporated into microbial protein.

However, in order to maximise both appetite and the efficiency of capture of fermentable energy and degradable protein as Metabolizable Energy and Metabolizable Protein, diets should be formulated to ensure the best possible balance between the supply of QDE and QDN, SFE and SFN. It follows that this will also maximise the capture of energy and protein relative to excretion of CO_2, CH_4 and nitrogenous compounds.

7 Competitive and complementary sources of feed energy and protein

Any analysis of the sustainability of dairy farming relative to other foods of animal origin destined for human consumption such as eggs and meat, needs to consider the capacity of ruminants to derive energy and protein for maintenance and production from feeds that do not compete with food that could be used directly to feed us. Table 3 (from Webster 2013) compares the efficiency of conversion of energy and protein in animal feed to eggs, pork, milk and beef in terms of total and 'competitive' energy and protein in feed for animals. Overall efficiency relates food energy and protein for human

Table 3 Efficiency of conversion of competitive and complementary feeds into eggs, pork, milk and beef (from Webster 2013)

	Eggs	Pork	Milk	Beef
Production unit	1 hen	22 pigs	1 cow	1 calf
Support unit	0.05 hens	1 sow	0.33 heifers	1 cow
Output/year (kg food)	15	1300	8000	200
Food energy (MJ)	130	13 000	28 000	2500
Protein (kg)	1.65	208	264	32
Input/year (MJ ME), total	389	67 038	67 089	29 850
'competitive'	351	53 630	20 127	10 268
Input/year (kg protein), total	5.2	818	946	361
'competitive'	5.0	736	236	108
Efficiency				
Food energy/total feed ME	0.33	0.19	0.42	0.08
Food energy/'competitive' feed ME	0.35	0.24	1.39	0.24

consumption to total ME and protein consumed by the productive and support animals. Competitive efficiency relates food energy and protein for human consumption to animal consumption of ME and protein from 'competitive' feed, i.e. feeds such as cereals that could have been eaten directly by us as distinct from 'complementary' feeds (e.g. grazing, forages and by-products remaining after preparation of food and drink for human consumption).

The *overall efficiencies* of ME conversion into eggs, pork, milk and beef are 0.33, 0.19, 0.42 and 0.08 respectively; for protein conversion they are 0.32, 0.25, 0.28 and 0.09. The reason why the efficiency of energy conversion to milk is greater than that for egg production can be attributed to the fact that there has, to date, been no limit to the ability of breeders to select cows to produce more and more milk, whereas hens are still restricted to the production of one egg per day. When energy conversion is examined in terms of *competitive efficiency* the picture changes. Here beef becomes as efficient as pork (or no less inefficient) and dairy farming becomes very efficient. Indeed, in this example, based on a typical diet fed to cows in the pasture-rich Southwest of England approximately 65% ME is complementary and the output of food energy for human consumption is 39% greater than their demand for feed that we could eat ourselves. The ability of the dairy cow to produce more food for human consumption than she derives from competitive feeds is most marked in advanced pastoral systems as seen in New Zealand but can be approached in fully housed systems though proper selection of complementary feeds.

8 Conclusion

I repeat: the aim of this multi-authored book is to examine dairy farming and explore, in the light of new understanding, routes to net zero in particular and sustainability in general. The chapters that follow have been prepared by experts, active in their specialist fields, and augmented by comprehensive references to the scientific literature. My opening chapter is much simpler, designed to keep you always in mind of the big picture by setting out the big issues, which are timeless. The second section is an equally simple introduction to the principles of biochemistry and physiology that determine energy and protein exchanges in the dairy cow. This is intended as a prelude to later chapters that explore ways to manipulate these fundamentals in the direction of net zero.

9 References

Blaxter, K.B. (1967). *The Energy Metabolism of Ruminants*, London, Hutchison.

Brown, M.S. and Ulgiati, I. (1997). Emergy-based indices and ratios to evaluate sustainability: monitoring economies and technology toward environmentally sound innovation. *Ecological Engineering* 9, 51–69.

Callendar, G. S. (1958). On the amount of carbon dioxide in the atmosphere. *Tellus* 10(2), 243–248. https://doi.org/10.3402/tellusa.v10i2.9231

Czerkawski, J.W., Blaxter, K.L. and Wainman, F.W. (1966). The metabolism of oleic, linoleic and linolenic acids by sheep with reference to their effects on methane production. *British Journal of Nutrition* 20, 349–362.

Eggleton, T. (2013). *A Short Introduction to Climate Change*. Cambridge University Press, Cambridge, p. 52.

Ellis, J.L., Bannink, J., France, E., Kebrear, E. and Dijkstra, J. (2010). Evaluation of enteric methane prediction equations for dairy cows used in whole farm models. *Global Change Biology* 16, 3246–3256.

Hristov, A.N. (2012). Historic, pre-European settlement, and present-day contribution of wild 955 ruminants to enteric methane emissions in the United States. *Journal of Animal Science*, 90(4), 1371–1375. https://doi.org/10.2527/jas.2011-4539

Kelliher, F.M. and Clark, H. (2010). Methane emissions from bison–an historic herd estimate 997 for the North American Great Plains. *Agricultural and Forest Meteorology* 150(3), 473–477. https://doi.org/10.1016/j.agrformet.2009.11.019

Pelletier, N., Pirog, R. and Rasmussen, R. (2010). Comparative life cycle environmental impact of three beef production strategies in in the upper mid-western United States. *Agricultural Systems* 103, 380–389.

Pereira, L. and Ortega, E. (2013). A modified footprint method: the case study of Brazil. *Ecological Indicators* 16, 113–127.

Rotolo, G.C., Rydberg, T. and Liebline, G. (2007). Emergy evaluation of grazing cattle in Argentina's Pampas *Agriculture, Ecosystems and Environment* 119, 383–395.

Rotz, C.A., Montes, F. Chianese, D.S. (2010). The carbon footprint of dairy production systems through partial life cycle assessment. *Journal of Dairy Science: Home PageJournal of Dairy Science* 93, 1266–1282.

Soussana, J.F., Tallec, T. and Blanfort, V. (2010). Mitigating the greenhouse gas balance of ruminant production systems through carbon sequestration in grasslands. *Animal* 4, 334–340.

Soussana, J.F. et al. (2007). Full accounting of the greenhouse gas (CO_2, N_2O, CH_4) budget of nine European grassland sites. *Agriculture, Ecosystems & Environment* 121, 121–134.

Ungerfeld, E.M., Cancino-Padilla, N. and Vera-Aguilera, N. (2023). Fermentation in the Rumen. In *Microbial Fermentations in Nature and as Designed Processes*, ed. C.J. Hurst, Wiley. https://doi.org/10.1002/9781119850007.ch4

Vigne, M., Peyraud, J.L., Leconte, P. and Corson, M.S. (2013). Emergy evaluation of contrasting dairy systems at multiple levels. *Journal of Environmental Management* 129, 44–53.

Webster, A.J.F. (2013). *Animal Husbandry Regained: The Place of Farm Animals in Sustainable Agriculture*. Earthscan from Routledge.

Webster, A.J.F. (2020). *Understanding the Dairy Cow* (3rd ed.) Wiley Blackwell.

Webster, A.J.F., Dewhurst, R.J. and Waters, C.J. (1988). *Alternative Approaches to the Characterisation of Feeds for Ruminants*. Recent Advances in Animal Nutrition. Butterworths.

Chapter 2

Advances in understanding methane production in the dairy cow

Mohammad Ramin and Petra Fant, Swedish University of Agricultural Sciences, Sweden; and Miroslav Joch and Mariana Vadroňová, Czech University of Life Sciences, Czech Republic

1 Introduction

The world population is growing, and the demand for livestock production is increasing, leading to rises in greenhouse gases (GHG). As the demand for food grows, sustainable food production becomes increasingly crucial. Makkar (2018) emphasized that by 2050, there will be a surge in demand for animal production, expected to rise by 70%, with the most significant contributions deriving from developing countries. The primary contributor to the greenhouse effect is water vapor, followed by carbon dioxide (CO_2) and methane (CH_4), as reported by Kiehl and Trenberth (1997). Enteric CH_4 from feed digestion and nitrous oxide (N_2O) from manure are among the most important GHGs in the agriculture sector, primarily emitted by ruminants (Arndt et al., 2022). Methane is generated by various biological sources, including natural wetlands, landfills, rice paddies, livestock, termites, solid wastes, and biomass burning (Immig, 1996). Additionally, from coal mining and natural gas production leakages.

http://dx.doi.org/10.19103/AS.2025.0157.03

The livestock sector contributes to almost 14.5% of total anthropogenic GHG emissions (Kennady et al., 2023). Within the livestock sector, dairy cows and beef cattle contribute the most, with recent values showing contributions of up to 70% of total emissions of CH_4 (Caro et al., 2014). Methane formation in the forestomach (rumen) of ruminants is a byproduct of enteric fermentation, and the gas is released to the atmosphere through eructation. These CH_4 emissions are considered an energy loss for the animal, energy that could be used for milk production instead, raising concerns about identifying strategies to reduce emissions in ruminants (Hristov et al., 2015a). Numerous studies have looked into the dynamics of CH_4 emissions and GHG contributions from ruminant digestive processes. Methane, a natural byproduct, is produced during the digestion of feed in the rumen of ruminants, with high correlations observed between dietary feed intake and CH_4 emissions in dairy cows (Ramin and Huhtanen, 2013). This complex process is crucial for the animals' digestive continuity, necessitating the elimination of hydrogen gas simultaneously generated in the digestion process. Methanogenesis plays a key role in managing the excess hydrogen, combining it with CO_2 to generate CH_4, which is subsequently expelled from the animal's mouth through belching.

Research has uncovered crucial insights into the biological mechanisms within ruminants, shedding light on the complex balance that maintains digestive functionality while contributing to environmental pollutants such as GHG emissions. It is important to recognize the complexity of these processes, as they both support the health of ruminants and impact the environment. Efforts toward understanding and optimizing these complex digestive processes offer possibilities for mitigating environmental impacts while maintaining the welfare of the animals. In addressing the crucial environmental challenge of mitigating GHG emissions from livestock production, especially those originating from dairy cows, a comprehensive approach is essential. This involves developing inventories, identifying interventions related to diet and microbial activity, and integrating fundamental and applied scientific knowledge, policy measures, and agricultural practices.

2 Methane emissions in ruminants

Ruminants, such as cows, sheep, and goats, are herbivorous animals characterized by a distinctive digestive system enabling efficient nutrient extraction from plant material. The digestive process for ruminants occurs in the rumen, a large fermentation chamber within their four-chambered stomach, housing a diverse community of microorganisms. The initial digestion of feed in the rumen is primarily carried out by rumen microorganisms, including bacteria, fungi, and protozoa, known as primary fermenters. These microorganisms break down feed components into simple monomers, which are subsequently utilized by both primary and secondary fermenters to generate final end-products such as volatile fatty acids (VFA), hydrogen (H_2), and CO_2 (McAllister et al., 1996). Methane is produced in the final stage by methanogenic archaea, utilizing H_2 or formate (HCOO) along with CO_2 as their primary substrates. The general equation for methanogenesis in the rumen is: $CO_2 + 4H_2 \rightarrow CH_4 + 2H_2O$. Over 95% of the CH_4 generated through enteric fermentation in the rumen is expelled through the mouth and nostrils into the atmosphere, while rectal emissions contribute only 2–3% (Murray et al., 1976). Depending on factors such as dry matter intake (DMI), feed quality, digestibility, animal type, size, live weight, and production characteristics, CH_4 emissions range from 2% to 12% of the animal's gross energy intake (GEI) (Johnson and Johnson, 1995). The metabolizable energy, calculated by subtracting CH_4 energy and urinary energy from digestible energy, is not entirely available for maintenance and production, as a portion is lost as fermentation heat.

Hydrogen gas plays a crucial role in rumen fermentation, originating from the reduction of protons by hydrogenases associated with ferredoxin enzyme systems in rumen microbes (Hegarty, 1999a). The accumulation of H_2 in the rumen hinders the re-oxidation of NADH, leading to an increase in lactate and a subsequent decrease in pH. This accumulation of H_2 adversely impacts fiber digestibility, as pointed out by Joblin (1999). From a scientific standpoint, in the absence of methanogens, NADH is regenerated through alternative H_2-consuming processes like the formation of lactate, ethanol, or succinate. Subsequently, lactate and succinate can be converted to propionate (Schönhusen et al., 2003).

Methanogens play a vital role in converting H_2 to CH_4 to maintain low H_2 pressure (preventing accumulation) in the rumen. For continuous monomer fermentation, re-oxidation of NADH to NAD+ is essential. The majority of rumen microbes utilize the Embden-Meyerhof-Parnas pathway to oxidize sugar units (such as glucose) to pyruvate. Hydrogen is produced during the enzymatic oxidation of NADH formed in glycolysis to NAD+. It has been observed that immediately after feeding, the rate of CH_4 production increases, attributed to the elevated partial pressure of H_2 and the rapid fermentation rate (Baker, 2002). In the rumen of sheep, approximately 100 liters of H_2 gas are produced (Hegarty and Gerdes, 1999b). The partial pressure of H_2 significantly influences methanogenesis (Hegarty and Gerdes, 1999b). Feeding increases the partial pressure of H_2 in the rumen, subsequently increasing the headspace H_2 (Hegarty and Gerdes, 1999b). The primary process of H_2 removal is through the conversion of CO_2 and H_2 into CH_4 by rumen methanogens (Zabranska and Pokorna, 2018). Propionate, produced during rumen fermentation, is likely the alternative sink for H_2 after methanogenesis. An increase in propionate production is closely associated with a reduction in CH_4 formation (Janssen, 2010).

The mutual association between fermenting species and the utilization of H_2 through methanogenesis is referred to as interspecies H_2 transfer (Crutzen, 1995). Acetate production enhances CH_4 production, whereas propionate production suppresses CH_4 production, given that H_2 is utilized in the formation of propionate (Moss et al., 2000). Various H_2 sinks exist in the rumen, with the primary one being CH_4 production. Additional sinks include microbial cell synthesis (approximately 10%), biohydrogenation of fatty acids (1–2%), acetogenesis, and the production of propionic acid (Czerkawski, 1986). The recovery rate of metabolic H_2 ranges from 78% to 96% (Demeyer, 1991). Assuming an average H_2 recovery of 90%, CH_4 emission is expected to be 10% lower than stoichiometric fermentation suggests (Moss et al., 2000). Nitrate and sulfates also serve as H_2 sinks in the rumen (Van Soest, 1994).

Numerous dietary factors affect CH_4 emissions in dairy cows, with dry matter intake (DMI) or feeding level being the primary factor. Furthermore, diet digestibility and composition affect CH_4 emissions at a given intake level. Research consistently confirms the significant impact of DMI on total CH_4 emissions in dairy cows, identifying feed intake as the primary factor driving CH_4 emissions in ruminants (Axelsson, 1949; Yan et al., 2000; Ramin and Huhtanen 2013; Niu et al., 2018). Three mechanisms contribute to reducing CH_4 as a proportion of gross energy (CH_4-E/GE) as intake increases. Firstly, increased feeding level (FL) increases the passage rate of feed particles, decreasing diet digestibility and the amount of fermented substrate per unit of intake. Secondly, increased FL enhances the efficiency of microbial cell synthesis. Because microbial cells are more reduced than fermented carbohydrates, more H_2 is utilized for cell synthesis with higher FL. Lastly, the shift in rumen fermentation pattern is the third factor contributing to the reduction in CH_4-E/GE as FL increases (Ramin and Huhtanen 2013). As CH_4 is solely produced from fermentation of organic matter (OM) in the rumen, a positive correlation between CH_4 emissions and diet digestibility is expected. Research has also demonstrated that improved digestibility has a positive effect on CH_4 emissions. Models pioneered by Axelsson (1949) were likely among the earliest to highlight the significance of digestibility in predicting CH_4 emissions. Axelsson (1949) observed that CH_4 emission per unit of DMI increased with a higher content of digested carbohydrates. Jentsch et al. (2007) showed the significant role of digestibility in predicting CH_4 emission, revealing a more robust relationship between CH_4 emission and digestible nutrients compared to crude nutrients ($R^2 = 0.896$ vs. 0.859).

It is well established that increasing the fat content in dairy cow diets reduces enteric CH_4 emissions. Numerous studies (Panyakaew et al., 2013; Alstrup et al., 2015; Bayat et al., 2018) and meta-analyses (Moate et al., 2011) consistently demonstrate this inhibitory effect. Several mechanisms explain this outcome. First, adding fat reduces the availability of fermentable substrates, primarily carbohydrates, since only the glycerol portion of fat is fermentable, while fatty acids are not. Additionally, dietary fat alters rumen fermentation patterns, promoting the production of propionate over acetate or butyrate. Furthermore, the biohydrogenation of unsaturated fatty acids acts as a H_2 sink, further contributing to reduced CH_4 emissions. Fat can be incorporated into the diet through oilseeds or plant oils. For example, adding rapeseed oil, safflower oil, or linseed oil at 50 g/kg of diet dry matter reduces CH_4 emissions (g/day) by 21–23% and CH_4 intensity (g/kg ECM) by 33%, with no adverse effects on milk yield or ECM yield, though feed intake may decrease slightly (Bayat et al., 2018). Similarly, other fat sources, such as rapeseed cake or whole cracked rapeseed, also reduce CH_4 emissions and intensity (Brask et al., 2013). Specifically, CH_4 emissions (L/d) and intensity decreased by 6.7% and 6.9%, respectively, with the addition of rapeseed cake, while whole cracked rapeseed led to reductions of 16% and 17%, respectively. The dietary fat content in these diets was 5.5% and 6.2%, respectively, compared to 3.5% in the control diet.

It is commonly hypothesized that an increased proportion of concentrate in the diet results in a reduction of CH_4 emission in dairy cows. However, substantial reductions are primarily observed in diets with very high concentrate content, such as those in feedlot conditions (Johnson and Johnson, 1995). Within the typical ranges of concentrate inclusion, the effects on CH_4 emission are relatively minimal. Ferris et al. (1999) reported a marginal difference in CH_4-E/GE, ranging between 37% and 59% of concentrate proportion. Conversely, Lovett et al. (2003) did not observe a statistically significant reduction in CH_4-E/GE when the concentrate proportion increased from 35% to 60%. The meta-analysis conducted by Ramin and Huhtanen (2013), reported that the impact of dietary carbohydrate composition exhibited only minor effects on CH_4 emissions. The absence of an evident effect of carbohydrate composition on CH_4 emissions in the study by Ramin and Huhtanen (2013) is unexpected, as cell wall carbohydrates have been observed to increase CH_4 emissions more than non-fibrous carbohydrates. Moreover, there is evidence that the degree and nature of silage fermentation can affect CH_4 emissions through two distinct mechanisms: alterations in the rumen fermentation pattern and microbial cell synthesis. This was shown in the study by Ramin and Huhtanen (2013) where the concentration of total acids in silages forages had a negative effect on CH_4 emissions ($P = 0.02$).

3 Methanogenesis

Methanogenesis is an anaerobic respiration that generates CH_4 as the final product of metabolism. It is mediated by anaerobic microorganisms from the domain archaea, commonly called methanogens (Smith and Hungate, 1958). Rumen methanogenesis is a thermodynamic necessity for microbial conversion of feed to VFA, which are essential nutrients for ruminants. Despite significant advancements, our understanding of rumen methanogenesis remains incomplete, partly due to the complexity of the rumen fermentation system and partly because of the technical hurdles associated with working with strict anaerobes (Khairunisa et al., 2023). The rumen is a highly dynamic and heterogeneous environment where a multitude of microbial interactions occur, making it challenging to decipher the specific contributions and interactions of individual microbial groups within this ecosystem.

The complex rumen metabolism can be divided into three levels of chemical transformation of plant material macromolecules. In the first level, the main polymers of plant cell wall (cellulose and hemicellulose) are colonized by microorganisms (bacteria, protozoa, and fungi) and decomposed into soluble sugars. In the second level, the soluble sugars are fermented. The products of fermentation are mainly VFA, as well as various metabolites such as formate, CO_2, and H_2. In the last level, some of these excreted metabolites, such as CO_2 and H_2, are further converted into CH_4 (Janssen, 2010; Mizrahi et al., 2021). Hydrogen represents a key molecule in the rumen ecosystem, as its high partial pressure may impair fermentation (Kelly et al., 2022). Most of H_2 (50–80%; Morgavi et al., 2023) is removed through methanogenesis, during which methanogens fulfill their cardinal role as the main electron sink in the rumen (Mizrahi et al., 2021). Further important electron sinks are propionate and valerate production (20–30%) and approximately 14% of H_2 is used for butyrate production (Mackie et al., 2024; Morgavi et al., 2023). The process of electron transfer from H_2 producer to a methanogen via H_2 as a vehicle was the first recognized case of interspecies electron transfer (Bryant et al., 1967). However, the utilization of hydrogen to reduce CO_2 to CH_4 is only one of the three major pathways of methanogenesis:

1. Hydrogenotrophic pathway: in this pathway H_2 (or formate/ some simple alcohols) is utilized to reduce CO_2 to CH_4. This pathway is the most exergonic ($\Delta G0 = -135$ kJ/mol CH_4) of all rumen methanogenesis pathway and is thought to predominate in the rumen (about 78% of rumen CH_4 production; Liu and Whitman, 2008; Sanjorjo et al., 2023; Ungerfeld and Pitta, 2024).
2. Methylotrophic pathway: in this pathway methylated compounds such as methanol and methylamines serve as methyl group donors to form methyl-coenzyme M, which then joins the hydrogenotrophic pathway. Although this pathway is less prevalent (22%; Sanjorjo et al., 2023), recent results suggest higher importance than previously thought (Pitta et al., 2022; Söllinger et al., 2018). The $\Delta G0$ of -113 kJ/mol CH_4 makes it the second most exergonic methanogenesis pathway in the rumen (Liu and Whitman, 2008; Ungerfeld and Pitta, 2024).

3 Acetoclastic pathway: in this pathway acetate is used as a substrate for methanogenesis. The methyl group of acetate is transferred to tetrahydromethanpterin (H4MPT) for further processing, generating CH_4. Acetoclastic methanogenesis is most active and important in freshwater sediments and anaerobic digestors, where acetate contributes roughly two-thirds of total CH_4 formation. In the rumen, this pathway has minimal importance, because it is not as exergonic ($\Delta G0 = -33$ kJ/mol CH_4) and may not produce ATP quickly enough for methanogen cells to replicate rapidly in the fast-paced environment of the rumen (Ungerfeld and Pitta, 2024). Methanogens capable of producing CH_4 from acetate, such as *Methanosarcina* spp. and *Methanosaeta* spp., are thus remarkably rare (Henderson et al., 2015).

Rumen CH_4 production is a multistep process, where the microbes involved can be classified into two main groups: those that produce methanogenesis substrates (mainly bacteria and protozoa), and CH_4 producers (methanogenic archaea; Zhang et al., 2023). Almost all archaea are methanogenic (92.3%; Ungerfeld, 2022) and belong to the phylum *Euryarchaeota* (Huws et al., 2018). *Euryarchaeota* comprises seven orders and the following four of them are commonly found in the rumen: *Methanobacteriales*, *Methanomicrobiales*, *Methanomassiliicoccales*, and *Methanosacrinales*. The mainly hydrogenotrophic genus *Methanobrevibacter* of the *Methanobacteriales* order is the most dominant, accounting for up to 74% of the rumen archaeal community (Henderson et al., 2015; Mizrahi et al., 2021; Ungerfeld, 2022). The dominant archaeal groups in the rumens of animals are strikingly similar across all regions of the world (Henderson et al., 2015). Unlike bacteria, archaea represent about 0.3–3% of the rumen microbiota (Ungerfeld, 2020) and are much less diverse, reflecting the few simple substrates they utilize (Beauchemin et al., 2020). This high universality and limited diversity of rumen methanogens could enable the successful mitigation of CH_4 emissions by developing strategies that specifically target the few dominant methanogens (Henderson et al., 2015).

Although methanogens are the only known CH_4 producers in the rumen, the relationship between their abundance and CH_4 production is not always straightforward. In some studies, the abundance of methanogens was directly correlated to high CH_4-emitting ruminants, whereas in other studies no direct correlation was apparent (Shabat et al., 2016; Wallace et al., 2015; Zhou et al., 2009). Reports suggest that methanogenesis is influenced not only by the abundance, but also by the metabolic activity and diversity of methanogens, as well as their interaction with other microorganisms (Danielsson et al., 2017a; Pitta et al., 2022; Wang et al., 2023). Understanding the relationship between the abundance, diversity, and activity of methanogens and the level of CH_4 production helps us consider how rumen methanogenesis might be curtailed.

For example, Shi et al. (2014) reported that RNA-based metatranscriptomics effectively distinguish metabolically active microbes that are linked to CH_4 production through gene expression. Specifically, methanogen activity was estimated by quantifying expression of the gene encoding methyl-coenzyme M reductase (MCR). Methyl-coenzyme M reductase is an enzyme catalyzing the last step in CH_4 production, which is present in all methanogens and needed in all methanogenesis pathways (Pitta et al., 2022; Roehe et al., 2016). The findings by Shi et al. (2014) are further supported by the reduction of methanogenesis using 3-nitrooxypropanol (3-NOP), a CH_4 inhibitor targeting the activity of MCR, which reduced gene expression more than gene content (Pitta et al., 2022).

Numerous mitigation strategies, both direct and indirect, have been suggested and implemented to target methanogenesis (see reviews). Feed additives with anti-methanogenic properties are among the most practical interventions for reducing CH_4 emissions (Hristov, 2024). These additives may reduce methanogenesis by indirectly altering the ruminal microbiome or through their direct biochemical activity (McGurrin et al., 2023).

The availability of H_2 for methanogenesis may be decreased by limiting its production. Generally, most cellulolytic microorganisms produce H_2 during carbohydrate degradation (Wang et al., 2016), among them many cellulolytic bacteria (e.g. *Ruminococcaceae*; Zhang et al., 2023), protozoa, and fungi (Mitsumori and Sun, 2008). Therefore, H_2 production in the rumen may be inhibited by supplying antimicrobial agents; for example, tannins. Tannins inhibit cellulolytic bacteria as well as protozoa (Cobellis et al., 2016; Kholif and Olafadehan, 2021). Protozoal populations significantly contribute to CH_4 production in the rumen. About 10–20% of methanogenic archaea live associated with protozoa both as endo- and ectosymbionts on their outer surface. Protozoa produce substantial amounts of H_2 and their close association with methanogenic archaea allows for an efficient H_2 transfer. Consequently, these methanogens are most metabolically active (Sanjorjo et al., 2023). Together they contribute significantly to CH_4 yield (about 30% of total methanogenesis; Fregulia et al., 2024; Patra and Puchala, 2023). Another option is to provide a thermodynamically more favorable H_2 sink to the rumen environment; for example, nitrates (Kelly et al., 2022; Mackie et al., 2024). Nitrate addition to ruminant diets may effectively decrease methanogenesis; they function as a highly competitive alternative electron sink, as they are reduced to ammonia in the rumen (Fouts et al., 2022), and thereby redirect H_2 flow from CO_2 to nitrate reduction (Honan et al., 2021).

The impact of enteric CH_4 inhibitors, such as 3-NOP and the red seaweed *Asparagopsis* spp., is achieved through direct interference with methanogens and their metabolic pathways (Mizrahi et al., 2021; Roque et al., 2021). The 3-nitrooxypropanol binds competitively to enzymes essential for CH_4 production (McGurrin et al., 2023). Typically, during the final step of methanogenesis, the enzyme MCR interacts with methyl-coenzyme M, which results in CH_4 formation. When 3-NOP is supplemented, this reaction is prevented, as 3-NOP mimics the chemical structure of methyl-coenzyme M, which allows it to bind to MCR instead. Since 3-NOP oxidizes the nickel in MCR, MCR can no longer bind to methyl-coenzyme M and CH_4 can no longer be produced. This oxidation process also generates nitrate and nitrite and ultimately leads to the degradation of 3-NOP into 1,3-propanediol in the rumen (Hegarty et al., 2021). In *Asparagopsis* spp., the active compound reducing CH_4 emissions is bromoform (Hristov, 2024). Bromoform and other halogenated CH_4 analogues like bromochloromethane inhibit methanogenesis by binding to the prosthetic group needed by MCR, which is crucial for the final step of methanogenesis (Roque et al., 2021). Studies have shown that methanogens may vary in their sensitivity to halogenated CH_4 analogues (Pitta et al., 2022; Ungerfeld and Pitta, 2024). These compounds inhibit methanogenesis by essentially blocking a key step where the cobamide-dependent transfer of a methyl group is needed for methanogenesis (Ungerfeld and Pitta, 2024). Cobamides are vitamin B12 family of cofactors essential for a variety of microbial metabolisms and are also produced and used by archaea (Shelton et al., 2019). Cobamides have structural variations among methanogen species and orders, which may result in different binding affinities for CH_4 halogenated analogues and influence how effectively these analogues inhibit methanogenesis (Ungerfeld and Pitta, 2024).

Still, we need to better understand the mechanism of action of feed additives as it pertains to archaeal growth and methanogenesis. Research in this direction could significantly benefit the agricultural industry by reducing CH_4 production in ruminants, leading to more efficient and sustainable livestock management.

4 Feed additives to reduce methane emission

Feed additives can mitigate enteric CH_4 emissions, though their modes of action and effectiveness vary. These additives also influence dairy cow productivity, animal health, and product safety in different ways. Extensive research on feed additives has been conducted through both *in vitro* and *in vivo* studies. While *in vitro* studies are essential for screening feeds and diets before they are given to animals *in vivo*, the effects observed *in vitro* often fail to replicate *in vivo*. Therefore, this section will focus exclusively on evidence from *in vivo* studies regarding CH_4-mitigating additives.

To date, the most promising feed additive for reducing CH_4 emissions is the synthetic compound 3-NOP. This compound inhibits the enzyme responsible for the final step of methanogenesis in the rumen. A recent meta-analysis, which included data from 14 experiments, reported a 33%, 31%, and 33% reduction in CH_4 production (g/d), CH_4 yield (g/kg dry matter intake), and CH_4 intensity (g/kg energy-corrected milk yield), respectively, when 3-NOP was fed at an average dose of 70.5 mg/kg dry matter to dairy cows (Kebreab et al., 2023). Most studies have found no effect on milk yield at dosages between 40 and 200 mg 3-NOP/kg feed dry matter (Hristov et al., 2015a; Melgar et al., 2020a, 2020b), although a recent study noted a slight decrease in milk yield at a dose of 80 mg/kg dry matter (Maigaard et al., 2024b). Although long-term studies are scarce, van Gastelen et al. (2024) reported CH_4 production, CH_4 yield, and CH_4 intensity in dairy cows supplemented with 3-NOP throughout lactation over one year to be 21%, 20%, and 27% lower, respectively, than in non-supplemented cows. The efficacy of 3-NOP, however, is influenced by the diet's chemical composition, with higher dietary fiber and fat content decreasing efficacy and higher dietary starch content increasing efficacy (Kebreab et al., 2023). Currently, 3-NOP is marketed as Bovaer® by dsm-firmenich and is authorized for use in lactating dairy cows in regions including the EU, United Kingdom, Switzerland, Australia, Brazil, and Canada.

Nitrate (NO_3^-) has been studied as a feed additive to reduce CH_4 emissions. In the rumen, nitrate is reduced to nitrite and further to ammonium, acting as an alternative hydrogen sink and diverting hydrogen away from CH_4 production. Nitrate can be incorporated into animal feed either as a part of a concentrate or added separately to a total mixed ration, similar to a mineral supplement. Van Zijderveld et al. (2011a) reported a 16% reduction in CH_4 yield over 89 days with an inclusion rate of 21 g nitrate/kg dry matter, with no adverse effects on diet digestibility or milk yield. Similarly, Olijhoek et al. (2016) found that adding 5.3, 13.6, and 21.1 g of nitrate/kg of dry matter reduced CH_4 yield by 6%, 13%, and 23%, respectively, without affecting feed intake, diet digestibility, milk yield, or milk composition. It is crucial that animals are gradually adapted to dietary nitrate in order to avoid methemoglobinemia, a condition where the oxygen-carrying capacity of erythrocytes is reduced. Methemoglobinemia is caused when nitrite is absorbed from the rumen to the blood due to incomplete reduction of nitrate to ammonia by nitrite-reducing bacteria (Bruning-Fann and Kaneene, 1993). In the studies by van Zijderveld et al. (2011a) and Olijhoek et al. (2016), methemoglobin levels in blood increased with dietary nitrate despite gradual adaptation. However, levels remained well below the clinical threshold for methemoglobinemia.

Seaweed additives have also been explored for reducing CH_4 emissions, with various groups and species of seaweed under investigation. Among these, the red seaweed *Asparagopsis* spp. has shown the most promise. *Asparagopsis* spp. contains halogenated compounds, primarily bromoform (CHBr3), which inhibits the final step of methanogenesis in the rumen, thereby reducing CH_4 production. Research has particularly focused on two species of this red seaweed for CH_4 mitigation: *Asparagopsis taxiformis* and *Asparagopsis armata*. Roque et al. (2019) reported 20% and 43% lower CH_4 yield from dairy cows fed *A. armata* at inclusion levels of 0.5% and 1% of organic matter, respectively. Stefenoni et al. (2021) reported a 55% reduction in CH_4 yield from dairy cows fed *A. taxiformis* at an inclusion level of 0.5% of dry matter. However, supplementation with *Asparagopsis* spp. significantly decreases feed intake and milk yield (Roque et al., 2019; Stefenoni et al., 2021). Additionally, there are other concerns posing challenges for the practical implementation of *Asparagopsis* spp. as a feed additive. Muizelaar et al. (2021) found considerable damage to the rumen wall in the form of absence of rumen papillae, hemorrhages, ulcers, and signs of inflammation in cows fed *Asparagopsis* spp. Another concern relates to the transfer of bromoform into milk, as bromoform has toxic and carcinogenic properties (NTP, 1989). Although bromoform concentration in milk increases when cows are fed *Asparagopsis* spp., the observed levels are still below what is considered safe concentrations by both the European Union and the Environmental Protection Agency in the United States of America. The mass production of red seaweed also poses challenges, requiring either ocean harvesting and long-distance transport or local production in heated tanks. Other seaweeds such as *Chondrus crispus*, *Saccharina latissima*, *Fucus serratus*, *Fucus vesiculosus*, *Ascophyllum nodosum*, and *Sargassum muticum* have been investigated but were found ineffective in reducing enteric CH_4 emissions in dairy cows (Muizelaar et al., 2023; Thorsteinsson et al., 2023; Thorsteinsson et al., 2024).

Various essential oils have been tested for their antimethanogenic properties, as they are complex mixtures of volatile, lipophilic secondary metabolites with antimicrobial properties (Calsamiglia et al., 2007). Although *in vitro* studies have shown potential, few *in vivo* studies have replicated this efficacy. For instance, *in vivo* studies have evaluated the supplementation of dairy cow diets with cinnamon oil and its main component, cinnamaldehyde (Benchaar, 2016), as well as diallyl disulfide (a main component of garlic oil) and yucca extract (van Zijderveld et al., 2011b). None of these additives had any effect on enteric CH_4 production. Feeding garlic-citrus extracts has shown some CH_4 mitigation potential *in vivo*. Bayat et al. (2022) fed dairy cows 20 g/d of a combination of garlic-citrus extract and essential oils and reported only minor effects on CH_4 production. Conversely, Khurana et al. (2023) fed dairy cows 44 g/d of garlic and citrus extract and observed that CH_4 production and CH_4 intensity were 10.3% and 11.7% lower, respectively. Methane yield also tended to be 9.7% lower in cows fed the extract compared with the control. Notably, dry matter intake and milk production were not affected. Oregano (*Origanum vulgare* L.) contains essential oils with antimicrobial activity, such as carvacrol. *In vivo* results from studies where oregano leaf has been fed to dairy cows are inconclusive. Hristov et al. (2013b) reported a linear decrease in CH_4 yield with increasing dietary inclusion of oregano leaf material, noting a 25% reduction in CH_4 yield at the highest inclusion rate of 750 g/d compared with 0 g/d within 8h after feeding. A 24-h feeding cycle was not represented in the study. More recent studies have reported no effect of oregano leaf on enteric CH_4 production (Benchaar, 2020; Stefenoni et al., 2021). Martins et al. (2024) supplemented a dairy cow diet with Capsicum oleoresin alone and in combination with clove oil. The combination decreased CH_4 yield by 11% in primiparous cows but not in multiparous cows, and there was no effect on feed intake or milk yield. Silvestre et al. (2023) fed an essential oil blend containing eugenol and geranyl acetate to dairy cows but found no effect on CH_4 production.

A common observation when feeding CH_4-mitigating additives to dairy cows and other ruminants is an increase in H_2 production. Studies involving 3-NOP, *Asparagopsis* spp., and nitrate have documented this rise in H_2 alongside a reduction in CH_4 (Hristov et al., 2015a; Stefenoni et al., 2021; Van Zijderveld et al., 2011b). Figure 1 demonstrates the production of H_2 as a result of inhibiting CH_4 production in the rumen. When CH_4 production is partially inhibited, H_2 accumulates and must be either emitted or redirected through

Figure 1 Illustration representing the consequences of applying dietary additives intended to mitigate CH_4 emissions. Reprinted from: An Introduction to Rumen Studies by J.W. Czerkawski, page 172. Copyright © (1986).

alternative pathways. If H_2 is not redirected, fermentation might be negatively affected (McAllister and Newbold, 2008). A recent study investigated rumen pulse-dosing of H_2 acceptors while feeding 3-NOP or nitrate to dairy cows (Maigaard et al., 2024a). The authors reported that dosing with fumaric acid or acrylic acid increased the proportion of propionate in rumen fluid, but had no major effect on the amount of dissolved H_2, headspace H_2, or H_2 emissions. Furthermore, dosing with phloroglucinol together with 3-NOP had no effect on any of the H_2 measures. Further research is needed to explore the combination of CH_4-mitigating feed additives with H_2 acceptors.

5 Feed efficiency

Feed efficiency is a useful measure to assess how effectively dairy cows convert feed nutrients into milk. It can be defined in various ways, such as the amount of milk produced per kilogram of dry matter consumed. However, feed efficiency is a complex trait influenced by many factors, including the structure and function of the rumen microbiome (Fregulia et al., 2022). For example, more efficient cows tend to have lower microbiome gene richness (Shabat et al., 2016). Another measure of feed efficiency is residual feed intake (RFI), which is calculated by subtracting predicted feed intake, based on production level, body weight, and physiological state, from actual feed intake. Low RFI animals are considered efficient as they consume less feed than expected, while high RFI animals are considered inefficient due to consuming more than expected. Importantly, RFI is phenotypically independent of both production level and body weight, although significant variation in RFI exists among individual cows. This variation is influenced by several biological processes, including feed intake and digestion, as outlined by Olijhoek et al. (2018).

A key assumption in RFI measurements is that energy sinks, such as energy-corrected milk yield, metabolic body weight, and changes in body weight, do not explain between-cow variation. Instead, these variations are linked to differences in energy use efficiency (Olijhoek et al., 2020). Furthermore, Olijhoek et al. (2018) found that low RFI cows (considered efficient) had a higher CH_4 yield (L/kg DM) compared to high RFI cows, a difference associated with dietary fiber and increased fiber digestibility. This underscores the importance of evaluating the impact of diet on both efficient and inefficient cows. Recently, a novel approach has emerged for identifying efficient cows by measuring the gases they produce. Measuring residual CO_2 in dairy cows over several weeks has proven to be a more reliable indicator of efficiency. A study by Huhtanen et al. (2021) showed that cows with low residual CO_2 (the difference between actual and predicted CO_2 production) had lower feed intake, produced less CH_4, and showed greater efficiency in utilizing metabolizable energy for lactation compared to cows with high residual CO_2. Huhtanen et al. (2021)

also noted that when residual CO_2 is calculated for mid-lactation cows, the risk of incorrectly ranking cows that are still mobilizing body energy as efficient is minimized, as these cows are no longer in a negative or positive energy balance. However, further research is needed to validate this approach for ranking cows based on feed efficiency.

6 Models in predicting methane emission in dairy cows

Hristov et al. (2013a) demonstrated a robust correlation between DMI and CH_4 emissions (R^2 = 0.86, n = 377). Many CH_4 emission predicting models in the literature rely on intake variables (such as DM, GE, digestible energy, and metabolizable energy). Consequently, accurately modeling the effect of DMI on CH_4 emission and understanding the associated biological mechanisms is essential. Although DMI is often the primary predictor in CH_4 production models, modest enhancements may be achieved by incorporating additional variables like dietary composition (Ellis et al., 2007; Nielsen et al., 2013). However, on some occasions, no notable improvements have been observed by including other factors in the model (Mills et al., 2003), highlighting the crucial role of DMI in predicting CH_4 emission in dairy cows.

There are different type of models in predicting CH_4 emission in dairy cows. Empirical and mechanistic models, in which empirical models are used widely compared to mechanistic models. In empirical models, mainly fewer input variables are needed as they relate enteric CH_4 emission to DMI and other dietary composition (Schils et al., 2013; Hristov et al., 2013a; Ramin and Huhtanen 2013; Niu et al., 2018). Empirical models also do take into account the most significant factors that influence CH_4 emission, which is DMI, digestibility, fat, and carbohydrates. Empirical models typically require easily measurable or estimable input data. These models offer a preliminary estimation of CH_4 emission, utilizing limited information about the animal or dietary factors (Schils et al., 2013). Jentsch et al. (2007) developed an empirical model where digested nutrients showed a superior predictive capacity for CH_4 emission compared to crude nutrient intake, with R^2 values of 0.90 and 0.86, respectively. Non-linear modeling also demonstrates a smaller Root Mean Square Error (RMSE) than linear modeling, making it advisable for use, particularly when models extend beyond the intake range covered in the datasets used for model development (Mills et al., 2003).

Limited literature covers mechanistic models for predicting CH_4 emission. The key benefit of mechanistic modeling lies in its capacity to enhance the comprehension of system behavior. Requirements for effective mechanistic models include addressing complex biological pathways, determining the rate of passage in the rumen, assessing VFA production, and establishing the kinetic rate of digestibility for various feed fractions. These models are valuable for screening diverse diets, particularly in evaluating interactions between different dietary factors. Among mechanistic models, the Nordic Karoline model has showed great potential in predicating CH_4 emission in dairy cows. The Nordic cow model, Karoline, is a dynamic, mechanistic model that describes digestion and metabolism in dairy cows (Danfær et al., 2006). Its updated version, mainly the sub-models predicting CH_4 emission (Huhtanen et al., 2015a) has been validated by Ramin and Huhtanen (2015) as a valuable tool for predicting CH_4 emission in cattle. The Molly model is also a mechanistic model that describes digestion processes in the rumen and is able to predict CH_4 emission in dairy cows (Hanigan et al., 2013). In a recent investigation conducted by Kass et al. (2022), the Karoline and Molly models underwent evaluation using an extensive dataset derived exclusively from studies employing respiration chambers to quantify CH_4 emissions in dairy cows (observed).

Upon conducting mixed model regression analysis, a strong concordance emerged between the predicted (using either Karoline or Molly model) and observed values for CH_4 emission in both models, indicating their capacity to accurately estimate variations in these variables within a study. Notably, the Karoline model exhibited smaller prediction errors for CH_4 emission compared to the Molly model (Figure 2). It was concluded by Kass et al. (2022) that the Karoline and Molly models prove to be valuable tools for understanding model behavior, estimating CH_4 emission in ruminants, and formulating effective mitigation strategies. The Karoline model exhibited no mean or linear bias when compared to the Molly model, which displayed both mean and linear bias.

7 Techniques to measure methane emission

Various methods are used to measure CH_4 emissions in dairy cows, with respiration chambers being the most reliable. Respiration chambers are considered the gold standard for accuracy in CH_4 emission measurement in dairy cows (Johnson and Johnson, 1995). This method serves as a standard for assessing the precision of other experimental techniques, as well as the performance of mechanistic and empirical equations (Ramin and Huhtanen, 2013; Ramin and Huhtanen, 2015). The method's principle involves capturing all exhaled air from the animal's breath, along with gases emitted from the lower digestive tract (rectal emissions). The animal is placed within a chamber for approximately 72 hours, ensuring ventilation for the inflow and outflow of air. Additionally, fresh air is introduced using a recycling fan, following the methodology outlined by McGinn et al. (2006). Methane concentrations are adjusted for background gases, and the ultimate flux is computed by multiplying the airflow by the CH_4 concentration in grams per liter (g/L). Concerns persist regarding potential decreases in intake when animals are

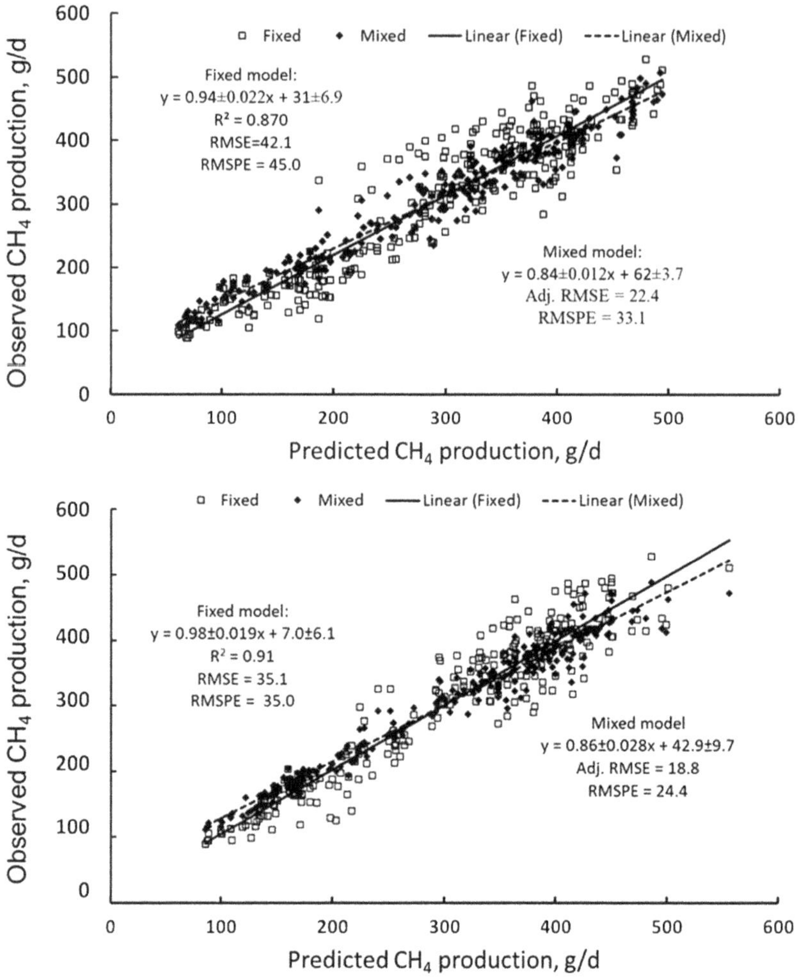

Figure 2 The relationship between predicted and observed CH_4 emission was examined through fixed and mixed model regression analyses using the Molly model (above) and Karoline model (below). From Kass et al., 2022.

confined within chambers. Therefore, it is vital to design chambers in a manner that allows animals to maintain visual contact to prevent any decline in intake. Hellwing et al. (2012) demonstrated that total intake remained unaffected when cows were housed in chambers equipped with transparent polycarbonate walls (Figure 3). The primary drawbacks of this technique are the construction costs and labor requirements involved.

Another method for measuring CH_4 emission is the use of the sulfur hexafluoride (SF_6) technique. The tracer technique, initially pioneered by

Figure 3 The respiration chamber technique often referred to as the Gold standard method'. Hellwing et al., 2012.

Johnson et al. (1994), involves the use of a known gas source (permeation tube) inserted into the rumen prior to the experiment's start. Subsequently, samples of the animal's breath are collected, and the gas concentrations are analyzed using gas chromatography. The ratio of CH_4 to SF_6 is then calculated. By knowing the release rate of SF_6, the release rate of CH_4 can be estimated. This method is primarily designed for grazing cattle due to the minimal background concentration of gases outdoors, but the technique can also be adapted for indoor experiments with adequate ventilation.

The use of SF_6 gas as a tracer for estimating CH_4 emission presents certain limitations, particularly when employed with cannulated animals. Leakage from the cannula could result in underestimation of CH_4 emission if the gas ratios in the leaked air differ from those in exhaled breath (Beauchemin et al., 2012). The release rate of SF_6 gas from permeation tubes is a critical consideration, as fluctuations in this rate could lead to variations in CH_4 emission measurements. Grainger et al. (2007) conducted a comparative study between the SF_6 method and the chamber method, finding similar CH_4 emission measurements from both techniques. However, they observed higher day-to-day variability within cows using the SF_6 method, with greater variability among cows compared to within-cow variation. Furthermore, the variability was more pronounced in the SF_6 technique compared to the respiration chamber method (Grainger et al., 2007). In summary, while the SF_6 method offers a means to estimate CH_4 emission, its application with cannulated animals requires caution due to potential leakage issues. Moreover, attention to the release rate of SF_6 gas is vital to ensure accurate CH_4 measurements, with considerations for variability in measurements when compared to other established techniques like respiration chambers.

GreenFeed, developed by Zimmerman (2011), is a relatively new technique in measuring gases from dairy cows. Referred to as the spot sampling technique, it measures the flux of gases once the animal visits the system and in return, the animal receives a concentrate bait. The frequency and duration of visits can be customized as needed. Each unit can be used for approximately 25–30 animals. A controlled amount of concentrate feed is dispensed from the feed bin to the manifold to tempt the animals to visit the unit. Once an animal enters the unit, its breath, along with the surrounding airflow, is directed through a system of pipes and mixed by a fan. Following this mixing process, a gas sample is extracted and later analyzed for CH_4, CO_2, H_2, and oxygen concentrations (Figure 4). Additionally, the system incorporates a head position sensor to filter out data when the cow's head is not properly positioned.

The GreenFeed device employs a comparable principle to measure gas emissions as seen in respiration chambers. It utilizes an active airflow to capture emitted air, integrating measurements of airflow, gas concentrations,

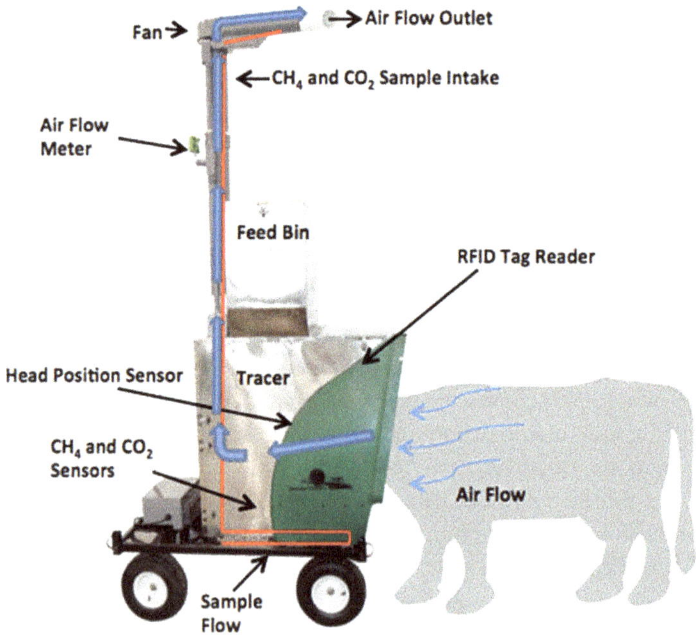

Figure 4 General layout of the GreenFeed system in the Stand-Alone Feeder. GreenFeed Stand-Alone Feeder instruction manual, C-Lock. Hristov et al., 2015b.

and detection of muzzle position (Zimmerman, 2011; Huhtanen et al., 2015b). Subsequently, the ideal gas law is applied to convert the data into mass fluxes, with adjustments made for head position relative to the airflow. However, training cows to visit the system regularly is crucial to ensure a sufficient number of visits, which is necessary for accurate estimations of gases produced.

In vitro gas production methods have been developed to assess factors affecting digestibility and rumen fermentation. These systems can also be employed to measure CH_4 emission from specific feeds or diets. Additionally, *in vitro* methods enable the screening of a large number of samples. However, conducting *in vivo* studies can be costly and may have significant effects on animals. To ease these factors, various *in vitro* systems have been developed. Continuous culture experiments, as described early by Czerkawski and Breckenridge (1977), are commonly utilized to evaluate the effects of diets and additives on enteric CH_4 emissions. The study by Ramin and Huhtanen (2012) introduced an *in vitro* method for estimating CH_4 emission in cows' rumen. This method utilizes kinetic parameters derived from an automated *in vitro* gas production system within a 2-compartment rumen model. By incorporating rumen dynamics, particularly digestion kinetics, this approach may offer advantages over single time-point batch culture systems. In an evaluation conducted by Danielsson et al. (2017b), it was observed that predicted CH_4 values using *in vitro* gas production system exhibited strong correlation with observed CH_4 values obtained from studies utilizing respiration chambers for CH_4 emission measurement. The mixed model regression analysis demonstrated a high R_2 value of 0.96 and a low root mean square prediction error of 40.1 (Figure 5). However, the primary drawback of the *in vitro* gas production system seems to be its lesser suitability for assessing diets with higher concentrate proportions. Its primary utility lies in ingredient screening and evaluating the impact of feed additives on CH_4 emissions, digestibility, and digestion kinetics. Nonetheless, any findings from *in vitro* mitigation studies need validation through *in vivo* experimentation before practical implementation.

In a recent study conducted by Fant and Ramin (2024), it was observed that the diet composition of rumen inoculum donor animals did not affect the comparison of diets concerning CH_4 emissions. Furthermore, the study demonstrated a robust correlation between the predicted *in vivo* CH_4 emissions, obtained through the *in vitro* system, and the observed CH_4 emissions measured using the GreenFeed system. This correlation was observed in dairy cows fed a diet primarily consisting of grass silage, with barley, oats, or dehulled oats serving as concentrate supplements.

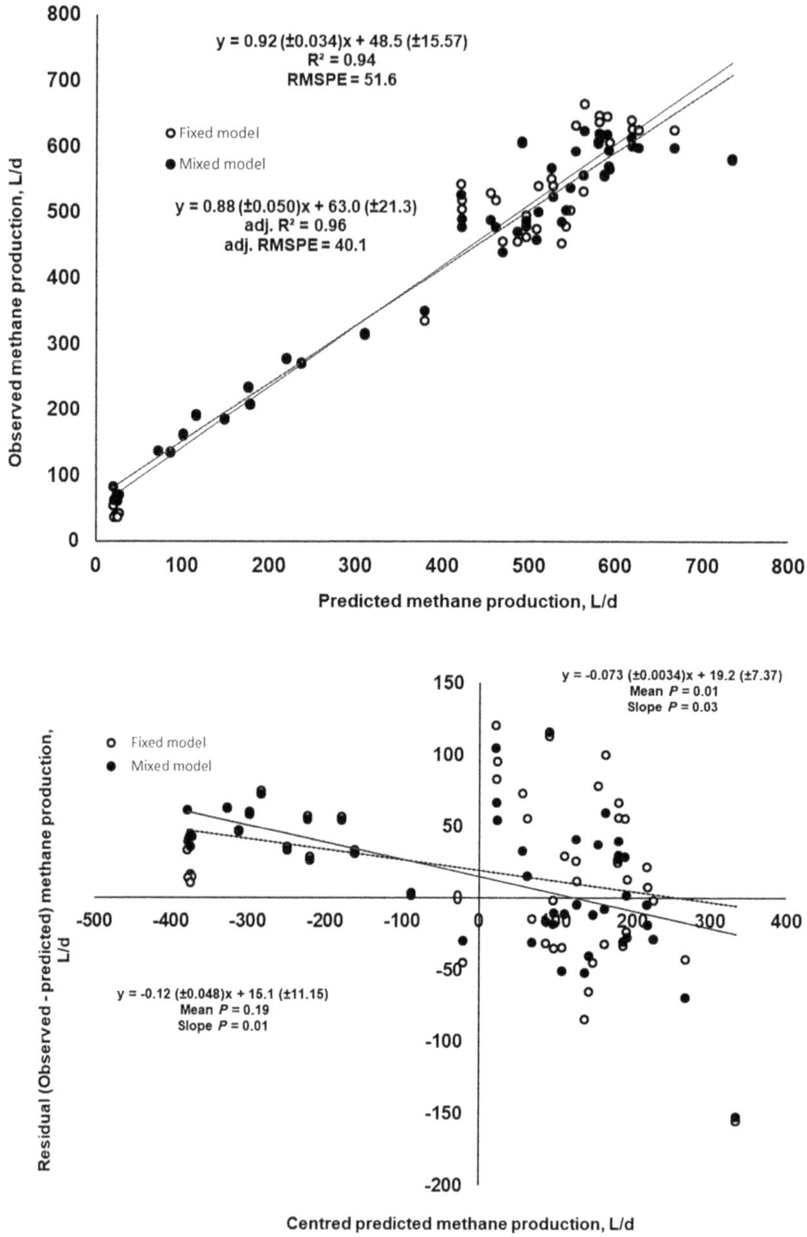

Figure 5 The above panel represents the relationship between predicted methane (CH_4) emission *in vitro*, derived from 48-hour incubation, and observed CH_4 emission *in vivo* (L/d; *n* = 49), presented with both fixed and mixed model regression analyses. The below panel represents centered predicted values and residuals (observed – predicted) of CH_4 emission (L/d; *n* = 49). adj = adjusted, and RMSPE stands for root mean square prediction error. Source: Danielsson et al., 2017b.

8 Conclusion and future trends

Methane emissions from dairy cows result from enteric fermentation, a necessary process for rumen digestion. Ruminants are and will continue producing CH_4. Methanogens, the primary organisms responsible for managing excess hydrogen in the rumen, use hydrogen and CO_2 to produce CH_4, which is primarily expelled from the animal through belching, with smaller amounts released via the rectum. This process represents a loss of energy that could otherwise be utilized by the animal for productivity purposes. Therefore, numerous strategies have been evaluated to reduce rumen methanogenesis, with feed additives emerging as one of the most promising approaches. Based on current research, 3-NOP appears to be the most effective among feed additives. 3-nitrooxypropanol generally reduce CH_4 production (g/d) and CH_4 intensity by about 30%. Complete inhibition is neither possible nor preferable as it will interfere with production performance. When considering the use of additives, factors such as cost, efficacy, and long-term effects must be carefully evaluated. It is essential to determine whether additives can sustainably maintain decreased CH_4 emissions over time, as the rumen may adapt to additives, potentially causing CH_4 emissions to revert to normal or lead to the release of H_2 into the atmosphere. Research should focus on decreasing CH_4 production per unit of feed intake (yield) and productivity (intensity), focusing on redirecting energy toward more efficient use by animals rather than being lost as CH_4 emissions. Today, there is a growing trend among animal scientists to focus on improved feed efficiency, as more efficient cows are expected to consume less feed and produce less CH_4 per unit of output. Recent studies indicate that cows with low residual CO_2 exhibit lower feed intake, reduced CH_4 intensity, and greater efficiency in using metabolizable energy for lactation compared to cows with higher residual CO_2. This highlights the potential of breeding programs that increase feed efficiency to significantly reduce greenhouse gas emissions and lower long-term costs for farmers. Predicting enteric CH_4 emissions from the rumen is a powerful evaluation tool that enables targeted, cost-effective strategies for reducing environmental impact. Prediction of CH_4 emissions can be achieved through empirical or mechanistic models, with varying degrees of robustness and accuracy depending on the dataset from which the models are developed. Finally, accurate measurement is essential for any reduction strategy, as it provides the necessary data to assess, monitor, and refine interventions aimed at effectively lowering enteric CH_4 emissions. The GreenFeed system is currently the most practical method for measuring live gas exchange in dairy cows, enabling comprehensive analysis of various gases to rank cows and identify those with greater efficiency.

9 Where to look for further information

Key publications/conferences:

- Comparison paper by Kass et al. (2022) regarding Karoline and Molly mechanistic models.
- Feed additives mitigating methane emission by Hristov et al. (2025).
- Meta-analysis on the efficacy of 3-NOP by Kebreab et al. (2023).
- *In vitro* method description for predicting *in vivo* CH_4 using kinetic data from the *in vitro* gas production system by Ramin and Huhtanen (2012).
- Evaluation of the GreenFeed method for measuring dairy cow gas emissions on farms by Huhtanen et al. (2015b).
- Comparison study of *in vivo* and *in vitro* CH_4 emissions: an experiment utilizing GreenFeed and *in vitro* gas production by Fant and Ramin (2024).
- American Dairy Science Association conference.
- European Federation of Animal Science conference.

Major international research projects:

- Sustainable food production in Swedish dairy farming. The project aims to identify the most cost-effective sustainable approach for reducing gas emissions from dairy cows.
- https://www.slu.se/en/faculties/vh/research/forskningsprojekt/not/sustainable-food-production-in-swedish-dairy-farming/.

10 References

Alstrup, L., et al. (2015). Effect of fat supplementation and stage of lactation on methane production in dairy cows. *Anim. Feed Sci. Technol.* 207: 10–19.

Arndt, C., et al. (2022). Full adoption of strategies to mitigate enteric methane emissions by ruminants and how they can help to meet the 1.5°C climate target by 2030 but not 2050. *Proc. Natl. Acad. Sci. USA.* 119: e2111294119.

Axelsson, J. (1949). The amount of produced methane energy in the European metabolic experiments with adult cattle. *Ann. R. Agric. Coll. Swed.* 16: 404–418.

Baker, S.K. (2002). Patterns of methane production and feed intake in ruminants. *Anim. Prod. Aust.* 25: 213.

Bayat, A.R., et al. (2018). Plant oil supplements reduce methane emissions and improve milk fatty acid composition in dairy cows fed grass silage-based diets without affecting milk yield. *J. Dairy Sci.* 101: 1136–1151.

Bayat, A.R., et al. (2022). Evaluating the effects of high-oil rapeseed cake or natural additives on methane emissions and performance of dairy cows. *J. Dairy Sci.* 105: 1211–1224.

Beauchemin, K.A., et al. (2020). Fifty years of research on rumen methanogenesis: lessons learned and future challenges for mitigation. *Anim.* 14: s2–s16.

Beauchemin, K.A., et al. (2012). Can the sulfur hexafluoride tracer gas technique be used to accurately measure enteric methane production from ruminally cannulated cattle? *J. Anim. Sci.* 90: 2727–2732.

Benchaar, C. (2020). Feeding oregano oil and its main component carvacrol does not affect ruminal fermentation, nutrient utilization, methane emissions, milk production, or milk fatty acid composition of dairy cows. *J. Dairy Sci.* 103: 1516–1527.

Benchaar, C. (2016). Diet supplementation with cinnamon oil, cinnamaldehyde, or monensin does not reduce enteric methane production of dairy cows. *Animal.* 10: 418–425.

Brask, M., et al. (2013). Methane production and digestion of different physical forms of rapeseed as fat supplements in dairy cows. *J. Dairy Sci.* 96: 2356–2365.

Bruning-Fann, C.S. and Kaneene, J.B. (1993). The effects of nitrate, nitrite, and n-nitroso compounds on animal health. *Vet. Hum. Toxicol.* 35: 237–253.

Bryant, M., et al. (1967). Methanobacillus omelianskii, a symbiotic association of two species of bacteria. *Arch. Mikrobiol.* 59: 20–31.

Calsamiglia, S., et al. (2007). Invited review: essential oils as modifiers of rumen microbial fermentation. *J. Dairy Sci.* 90: 2580–2595.

Caro, D., et al. (2014). Global and regional trends in greenhouse gas emissions from livestock. *Clim. Change.* 126: 203–216.

Cobellis, G., et al. (2016). Critical evaluation of essential oils as rumen modifiers in ruminant nutrition: a review. *Sci. Total Environ.* 545: 556–568.

Crutzen, P.J. (1995). Ruminant physiology: digestion, metabolism, growth and reproduction. In: *The Role of Methane in Atmospheric Chemistry and Climate.* Engelhardt, W.V. et al. (eds). Ventura, Germany. pp. 291–313

Czerkawski, J.W. (1986). *An Introduction to Rumen Studies.* Robert Maxwell, M.C., Oxford, UK.

Czerkawski, J.W. and Breckenridge, G. (1977). Design and development of a long-term rumen simulation technique (Rusitec). *Br. J. Nutr.* 38: 371–384.

Danfær, A., et al. (2006). The Nordic dairy cow model, Karoline-evaluation. In: *Nutrient Digestion and Utilization in Farm Animals Modeling Approaches.* Kebreab, E., Dijkstra, J., Bannink, A., Gerrits, W. J. J. & France, J. (eds). CABI Publishing, USA. pp. 407–415.

Danielsson, R., et al. (2017a). Methane production in dairy cows correlates with rumen methanogenic and bacterial community structure. *Front. microbiol.* 8: 226.

Danielsson, R., et al. (2017b). Evaluation of a gas in vitro system for predicting methane production in vivo. *J. Dairy Sci.* 100: 8881–8894.

Demeyer, D.I. (1991). Quantitative aspects of microbial metabolism in the rumen and hindgut. In: Rumen Microbial Metabolism and Ruminant Digestion. Jouany, J.P (ed). Institut National de la Recerche Agronomique, cop, Paris. pp. 217–237.

Ellis, J.L., et al. (2007). Prediction of methane production from dairy and beef cattle. *J. Dairy. Sci.* 90: 3456–3466.

Fant, P. and Ramin, M. (2024). Relationship between predicted in vivo and observed in vivo methane production from dairy cows fed a grass-silage based diet with barley, oats, or dehulled oats as a concentrate supplement. *Anim. Feed Sci. Technol.* 311: 115955.

Ferris, C.P., et al. (1999). The effect of genetic merit and concentrate proportion in the diet on nutrient utilization by lactating dairy cows. *J. Agric. Sci.* 132: 483–490.

Fouts, J.Q., et al. (2022). Enteric methane mitigation interventions. *Transl. anim. sci.* 6: txac041.

Fregulia, P., et al. (2024). Composition of the rumen microbiome and its association with methane yield in dairy cattle raised in tropical conditions. *Mol. Biol. Rep.* 51: 447.

Fregulia, P., et al. (2022). Taxonomic and predicted functional signatures reveal linkages between the rumen microbiota and feed efficiency in dairy cattle raised in tropical areas. *Front. Microbiol.* 13: 1025173.

Grainger, C., et al. (2007). Methane emissions from dairy cows measured using the sulfur hexafluoride (SF_6) tracer and chamber techniques. *J. Dairy Sci.* 90(6): 2755–2766.

Hanigan, M.D., et al. (2013). Revised digestive parameter estimates for the Molly cow model. *J. Dairy Sci.* 96: 3867–3885.

Hegarty, R.S., et al. (2021). An evaluation of emerging feed additives to reduce methane emissions from livestock. Edition 1. A report coordinated by Climate Change, Agriculture and Food Security (CCAFS) and the New Zealand Agricultural Greenhouse Gas Research Centre (NZAGRC) initiative of the Global Research Alliance (GRA).

Hegarty, R.S. (1999a). Mechanisms for competitively reducing ruminal methanogenesis. *Aust. J. Agric. Res.* 50: 1299–1305.

Hegarty, R.S. and Gerdes, R. (1999b). Hydrogen production and transfer in the rumen. *Rec. Adv. Anim. Nutr. Aust.* 12: 37–44.

Hellwing, A.L.F., et al. (2012). Test of a low-cost and animal-friendly system for measuring methane emissions from dairy cows. *J. Dairy Sci.* 95: 6077–6085.

Henderson, G., et al. (2015). Rumen microbial community composition varies with diet and host, but a core microbiome is found across a wide geographical range. *Sci. rep.* 5: 14567.

Honan, M., et al. (2021). Feed additives as a strategic approach to reduce enteric methane production in cattle: Modes of action, effectiveness and safety. *Anim. Prod. Sci.* 62(14), 1303–1317

Hristov, A.N. (2024). Invited review: Advances in nutrition and feed additives to mitigate enteric methane emissions. *J. Dairy Sci.* 107: 4129–4146.

Hristov, A.N., et al. (2013a). Mitigation of methane and nitrous oxide emissions from animal operations: I. A review of enteric methane mitigation options. *J. Anim. Sci.* 91: 5045–5069.

Hristov, A.N., et al. (2013b). Effect of Origanum vulgare L. leaves on rumen fermentation, production, and milk fatty acid composition in lactating dairy cows. *J. Dairy Sci.* 96: 1189–1202.

Hristov, A. N., et al. (2015a). An inhibitor persistently decreased enteric methane emission from dairy cows with no negative effect on milk production. *Proc. Natl. Acad. Sci.* 112: 10663–10668.

Hristov, A. N., et al. (2015b). The Use of an automated system (GreenFeed) to monitor enteric methane and carbon dioxide emissions from ruminant animals. *JoVE.* 103: 1–8.

Hristov, A. N., et al. (2025). Feed additives for methane mitigation: Recommendations for testing enteric methane-mitigating feed additives in ruminant studies. Journal of Dairy Science, 108(1), 322–355

Huhtanen, P., et al. (2015a). Nordic dairy cow model Karoline in predicting methane emissions: 1. Model description and sensitivity analysis. *Livest. Sci.* 178: 71–80.

Huhtanen, P., et al. (2015b). Comparison of methods to determine methane emissions from dairy cows in farm conditions. *J. Dairy Sci.* 98: 3394–3409.

Huhtanen, P., et al. (2021). Residual carbon dioxide as an index of feed efficiency in lactating dairy cows. *J. Dairy. Sci.* 104: 5332–5344.

Huws, S.A., et al. (2018). Addressing global ruminant agricultural challenges through understanding the rumen microbiome: past, present, and future. *Front. microbiol.* 9: 2161.

Immig, I. (1996). The rumen and hindgut as source of ruminant methanogenesis. *Environ. Monit. Assess.* 42: 57–72.

Janssen, P.H. (2010). Influence of hydrogen on rumen methane formation and fermentation balances through microbial growth kinetics and fermentation thermodynamics. *Anim Feed Sci Technol.* 160: 1–22.

Jentsch, W., et al. (2007). Methane production in cattle calculated by the nutrient composition of the diet. *Arch. Anim. Nutr.* 61: 10–19.

Joblin, K.N. (1999). Ruminal acetogens and their potential to lower ruminant methane emissions. *Aust. J. Agric. Res.* 50: 1307–1313.

Johnson, K.A. and Johnson, D.E. (1995). Methane emissions from cattle. *J. Anim. Scie.* 73: 2483–2492.

Johnson, K., et al. (1994). Measurement of methane emissions from ruminant livestock using a sulfur hexafluoride tracer technique. *Environ. Sci. Technol.* 28(2): 359–362.

Kass, M., et al. (2022). Comparison of Molly and Karoline models to predict methane production in growing and dairy cattle. *J. Dairy. Sci.* 105: 3049–3063.

Kebreab, E., et al. (2023). A meta-analysis of effects of 3-nitrooxypropanol on methane production, yield, and intensity in dairy cattle. *J. Dairy Sci.* 106: 927–936.

Kelly, W.J., et al. (2022). Hydrogen and formate production and utilisation in the rumen and the human colon. *Anim. microbiome.* 4:22.

Kennady, V., et al. (2023). Sustainable livestock production: a systematic review of different constraints associated with sustainable livestock production. *European J. Agric. Food Sci.* 5: 1–11.

Khairunisa, B.H., et al. (2023). Evolving understanding of rumen methanogen ecophysiology. *Front. Microbiol.* 14:1296008.

Kholif, A.E. and Olafadehan, O.A. (2021). Essential oils and phytogenic feed additives in ruminant diet: chemistry, ruminal microbiota and fermentation, feed utilization and productive performance. *Phytochem Rev.* 20:1087–1108.

Khurana, R., et al. (2023). Effect of a garlic and citrus extract supplement on performance, rumen fermentation, methane production, and rumen microbiome of dairy cows. *J. Dairy Sci.* 106: 4608–4621.

Kiehl, J.T. and Trenberth, K.E. (1997). Earth's annual global mean energy budget. *Bull. Amer. Meteor. Soc.* 78: 197–208.

Liu, Y. and Whitman, W.B. (2008). Metabolic, phylogenetic, and ecological diversity of the methanogenic archaea. *Ann. N. Y. Acad. Sci.* 1125: 171–189.

Lovett, D., et al. (2003). Effects of forage/concentrate ratio and dietary coconut oil level on methane output and performance of finishing beef heifers. *Livest. Prod. Sci.* 84: 135–146.

Mackie, R.I., et al. (2024). Hydrogen production and hydrogen utilization in the rumen: key to mitigating enteric methane production. *Anim. Biosci.* 37:323–336.

Maigaard, M., et al. (2024a). The acute effects of rumen pulse-dosing of hydrogen acceptors during methane inhibition with nitrate or 3-nitrooxypropanol in dairy cows. *J. Dairy Sci.* 107: 5681–5698.

Maigaard, M., et al. (2024b). Effects of dietary fat, nitrate, and 3-nitrooxypropanol and their combinations on methane emission, feed intake, and milk production in dairy cows. *J. Dairy Sci.* 107: 220–241.

Makkar, H.P.S. (2018), Review: Feed demand landscape and implications of food-not feed strategy for food security and climate change, *Animal*. 12: 1744–1754.
Martins, L.F., et al. (2024). Lactational performance, enteric methane emission, and nutrient utilization of dairy cows supplemented with botanicals. *J. Dairy Sci.* 107: 242–257.
McAllister, T.A. and Newbold, C.J. (2008). Redirecting rumen fermentation to reduce methanogenesis. *Aust. J. Exp. Agric.* 48: 7–13.
McAllister, T.A., et al. (1996). Dietary, environmental and microbiological aspects of methane production in ruminants. *Can. J. Anim. Sci.* 76: 231–243.
McGinn, S.M., et al. (2006). Assessment of the sulfur hexafluoride (SF_6) tracer technique for measuring enteric methane emissions from cattle. *J. Environ. Qual.* 35(5): 1686–1691.
McGurrin, A., et al. (2023). Anti-methanogenic potential of seaweeds and seaweed-derived compounds in ruminant feed: current perspectives, risks and future prospects. *J Anim Sci Biotechnol.* 14: 145.
Melgar, A., et al. (2020a). Effects of 3-nitrooxypropanol on rumen fermentation, lactational performance, and resumption of ovarian cyclicity in dairy cows. *J. Dairy Sci.* 103: 410–432.
Melgar, A., et al. (2020b). Dose-response effect of 3-nitrooxypropanol on enteric methane emissions in dairy cows. *J. Dairy Sci.* 103: 6145–6156.
Mills, J.A.N., et al. (2003). Alternative approaches to predicting methane emissions from dairy cows. *J. Anim. Sci.* 81: 3141–3150.
Mitsumori, M. and Sun, W. (2008). Control of rumen microbial fermentation for mitigating methane emissions from the rumen. *Asian-Australas J Anim Sci.* 21: 144–154.
Mizrahi, I., et al. (2021). The rumen microbiome: balancing food security and environmental impacts. *Nat. Rev. Microbiol.* 19: 553–566.
Moate, P.J., et al. (2011). Influence of cold-pressed canola, brewers grain and hominy meal as dietary supplements suitable for reducing enteric methane emissions from lactating dairy cows. *Anim. Feed Sci. Technol.* 166–167: 254–264.
Moss, A. R., et al. (2000). Methane production by ruminants: its contribution to global warming. Ann. Zootech. 49: 231–253.
Morgavi, D.P., et al. (2023). Review: Reducing enteric methane emissions improves energy metabolism in livestock: is the tenet right? *Anim.* 17: 100830.
Muizelaar, W., et al. (2021). Safety and transfer study: Transfer of bromoform present in Asparagopsis taxiformis to milk and urine of lactating dairy cows. *Foods*. 10: 584.
Muizelaar, W., et al. (2023). Evaluation of 3 northwest European seaweed species on enteric methane production and lactational performance of Holstein-Friesian dairy cows. *J. Dairy Sci.* 106: 4622–4633.
Murray, R.M., et al. (1976). Rates of production of methane in the rumen and large intestine of sheep. *Br. J. Nutr.* 36: 1–14.
Nielsen, N.I., et al. (2013). A prediction equation for enteric methane emission from dairy cows for use in NorFor. *Acta Agr. Scand. A Anim. Sci.* 16: 126–130.
Niu, M., et al. (2018). Prediction of enteric methane production, yield, and intensity in dairy cattle using an intercontinental database. *Glob Change Biol.* 24: 3368–3389.
NTP, National Toxicology Program. (1989). NTP toxicology and carcinogenesis studies of tribromomethane (Bromoform) (CAS No. 75-25-2) in F344/N Rats and B6C3F1 Mice (Gavage Studies). *Natl. Toxicol. Program. Tech. Rep. Ser.* 350: 1–194.
Olijhoek, D.W., et al. (2016). Effect of dietary nitrate level on enteric methane production, hydrogen emission, rumen fermentation, and nutrient digestibility in dairy cows. *J. Dairy Sci.* 99: 6191–6205.

Olijhoek, D.W., et al. (2018). Methane production, rumen fermentation, and diet digestibility of Holstein and Jersey dairy cows being divergent in residual feed intake and fed at 2 forage-to-concentrate ratios. *J. Dairy Sci.* 101: 9926–9940.
Olijhoek, D.W., et al. (2020). Phenotypic modeling of residual feed intake using physical activity and methane production as energy sinks. *J. Dairy Sci.* 103: 6967–6981.
Panyakaew, P., et al. (2013). Medium-chain fatty acids from coconut or krabok oil inhibit in vitro rumen methanogenesis and conversion of non-conjugated diesc biohydrogenation intermediates. *Anim. Feed Sci. Technol.* 180: 18–25.
Patra, A.K. and Puchala, R. (2023). Methane mitigation in ruminants with structural analogues and other chemical compounds targeting archaeal methanogenesis pathways. *Biotechnol. Adv.* 69: 108268.
Pitta, D.W., et al. (2022). The effect of 3-nitrooxypropanol, a potent methane inhibitor, on ruminal microbial gene expression profiles in dairy cows. *Microbiome.* 10: 146.
Ramin, M. and Huhtanen, P. (2012). Development of an in vitro method for determination of methane production kinetics using a fully automated in vitro gas system–A modelling approach. *Anim. Feed Sci. Technol.* 174: 190–200.
Ramin, M. and Huhtanen, P. (2013). Development of equations for predicting methane emissions from ruminants. *J. Dairy Sci.* 96: 2476–2493.
Ramin, M. and Huhtanen, P. (2015). Nordic dairy cow model Karoline in predicting methane emissions: 2. Model evaluation. *Livest. Sci.* 178: 81–93.
Roehe, R., et al. (2016). Bovine host genetic variation influences rumen microbial methane production with best selection criterion for low methane emitting and efficiently feed converting hosts based on metagenomic gene abundance. *PLoS Genet.* 12: e1005846.
Roque, B.M., et al. (2021). Red seaweed (Asparagopsis taxiformis) supplementation reduces enteric methane by over 80 percent in beef steers. *PLoS One.* 16: e0247820.
Roque, B.M., et al. (2019). Inclusion of Asparagopsis armata in lactating dairy cows' diet reduces enteric methane emission by over 50 percent. *J. Cleaner. Prod.* 234: 132–138.
Sanjorjo, R.A., et al. (2023). In Pursuit of Understanding the Rumen Microbiome. *Fermentation.* 9: 114.
Schils, R.L.M., et al. (2013). Mitigation of greenhouse gases from agriculture: Role of models. *Acta Agr. Scand. A Anim.* Sci. 62(4): 212–224.
Schönhusen, U., et al. (2003). Effects of protozoa on methane production in rumen and hindgut of calves around time of weaning. *Arch. Anim. Nutr.* 57: 279–295.
Shabat, S.K.B., et al. (2016). Specific microbiome-dependent mechanisms underlie the energy harvest efficiency of ruminants. *ISME J.* 10: 2958–2972.
Shelton, A.N., et al. (2019). Uneven distribution of cobamide biosynthesis and dependence in bacteria predicted by comparative genomics. *ISME J.* 13: 789–804.
Shi, W., et al. (2014). Methane yield phenotypes linked to differential gene expression in the sheep rumen microbiome. *Genome Res.* 24: 1517–1525.
Silvestre, T., et al. (2023). Lactational performance, rumen fermentation, nutrient use efficiency, enteric methane emissions, and manure greenhouse gas-emitting potential in dairy cows fed a blend of essential oils. *J. Dairy Sci.* 106: 7661–7674.
Smith, P.H. and Hungate, R. (1958). Isolation and characterization of Methanobacterium ruminantium n. sp. *J. Bacteriol.* 75: 713–718.

Söllinger, A., et al. (2018). Holistic assessment of rumen microbiome dynamics through quantitative metatranscriptomics reveals multifunctional redundancy during key steps of anaerobic feed degradation. *MSystems*. 3: 10–1128.

Stefenoni, H.A., et al. (2021). Effects of the macroalga Asparagopsis taxiformis and oregano leaves on methane emission, rumen fermentation, and lactational performance of dairy cows. *J. Dairy Sci.* 104: 4157–4173.

Thorsteinsson, M., et al. (2024). Potential of 2 Northern European brown seaweeds (Fucus serratus and Fucus vesiculosus) as enteric methane inhibitors in dairy cows. *J. Dairy Sci.* In press.

Thorsteinsson, M., et al. (2023). Effects of dietary inclusion of 3 Nordic brown macroalgae on enteric methane emission and productivity of dairy cows. *J. Dairy Sci.* 106: 6921–6937.

Ungerfeld, E.M. and Pitta, D. (2024). Review: Biological consequences of the inhibition of rumen methanogenesis. *Anim*: 10: 101170.

Ungerfeld, E.M. (2020). Metabolic hydrogen flows in rumen fermentation: principles and possibilities of interventions. *Front. microbiol.* 11: 589.

Ungerfeld, E.M. (2022). Opportunities and Hurdles to the Adoption and Enhanced Efficacy of Feed Additives towards Pronounced Mitigation of Enteric Methane Emissions from Ruminant Livestock. *Methane*. 1: 262–285.

van Gastelen, S., et al. (2024). Long-term effects of 3-nitrooxypropanol on methane emission and milk production characteristics in Holstein Friesian dairy cows. *J. Dairy Sci.* In press.

Van Soest, P.J. (1994). *Nutritional Ecology of the Ruminant*. 2nd ed. Ithaca, NY: Cornell University Press.

van Zijderveld, S.M., et al. (2011a). Dietary inclusion of diallyl disulfide, yucca powder, calcium fumarate, an extruded linseed product, or medium-chain fatty acids does not affect methane production in lactating dairy cows. *J. Dairy Sci.* 94: 3094–3104.

van Zijderveld, S.M., et al. (2011b). Persistency of methane mitigation by dietary nitrate supplementation in dairy cows. *J. Dairy Sci.* 94: 4028–4038.

Wallace, R.J., et al. (2015). The rumen microbial metagenome associated with high methane production in cattle. *BMC Genet*. 16: 1–14.

Wang, K., et al. (2023). Could propionate formation be used to reduce enteric methane emission in ruminants? *Sci. Total Environ*. 855: 158867.

Wang, L.Z., et al. (2016). The Effect of Dietary Replacement of Ordinary Rice with Red Yeast Rice on Nutrient Utilization, Enteric Methane Emission and Rumen Archaeal Diversity in Goats. *PLoS One*. 11: e0160198.

Yan, T., et al. (2000). Prediction of methane energy output in dairy and beef cattle offered grass silage-based diets. *Livest. Prod. Sci.* 64: 253–263.

Zabranska, J. and Pokorna, D. (2018). Bioconversion of carbon dioxide to methane using hydrogen and hydrogenotrophic methanogens. *Biotechnol. Adv.* 36: 707–720.

Zhang, B., et al. (2023). Methane prediction equations including genera of rumen bacteria as predictor variables improve prediction accuracy. *Sci. Rep.* 13: 21305.

Zhou, M., et al. (2009). Assessment of the microbial ecology of ruminal methanogens in cattle with different feed efficiencies. *Appl. Environ. Microbiol.* 75: 6524–6533.

Zimmerman, P. (2011). U.S. Patent 7966971, "Method and system for monitoring and reducing ruminant methane production", U.S. Patent and Trademark Office, June 28, 2011.

Chapter 3

Assessing greenhouse gas emissions from dairy farms

C. Alan Rotz and Curtis Dell, USDA-ARS, USA

1 Introduction

Milk production on dairy farms is a contributor to greenhouse gas (GHG) emissions both nationally and globally. Within the U.S., dairy cattle and their manure are estimated to produce 90.5 million metric tons (MMT) of carbon dioxide equivalent (CO_2e)/year or 1.4% of the total GHG emissions from the country (EPA, 2023). For a broader analysis of milk production in the U.S., Rotz et al. (2021) performed a cradle to farm gate life cycle assessment (LCA) that included anthropogenic CO_2 emissions, emissions in the upstream production of resources used on dairy farms, and allocation of cull animals to beef production. They found total GHG emissions representing all dairy farms to be 99 MMT CO_2e, which was 1.5% of the U.S. total GHG emissions in 2019. Globally, dairy cattle and the manure they produce are estimated to emit 713 MMT CO_2e/year, which is approximately 1.3% of all global GHG emissions (FAO, 2023).

Mitigation or reduction of GHG emissions has become a priority in many countries. A driver is the Paris Agreement, a legally binding international treaty on climate change that took effect in November of 2016 (UNEP, 2022). Through this agreement, goals were set to reduce emissions and sequester carbon

http://dx.doi.org/10.19103/AS.2025.0157.05
Published by Burleigh Dodds Science Publishing Limited, 2026.

(C) to limit global warming to 1.5°C. To date, global GHG emissions continue to rise, falling short of the trajectory needed to meet goals (UNEP, 2022). In response, dairy industries in many countries are setting goals to work toward net-zero production, i.e. no long-term effect on global warming. As an example, the U.S. dairy industry has set a goal of achieving GHG neutrality by 2050 while optimizing water use and improving water quality (DMI, 2023). Many strategies are available and being developed to help meet these goals, but finding economical solutions for producers is challenging (Hristov et al., 2022; Hristov et al., 2013; Montes et al., 2013).

Emission amounts vary among dairy farms due to differences in climate, soil characteristics, and management. Differences are best evaluated by comparing intensities where intensity is usually defined as the emission per unit of fat and protein corrected milk (FPCM) produced. Because milk solids content varies among farms or production systems, a recommended procedure is to convert raw milk to 4% fat and 3.3% protein equivalent (IDF, 2015). A comprehensive intensity including all important GHG emission sources can be referred to as the C footprint of milk production.

Quantifying GHG emissions at the whole farm level is challenging. There are many sources including cattle, manure storage and handling, combustion of fossil fuels, and decomposition of soil amendments. The primary GHG compounds emitted from dairy farms are methane (CH_4), nitrous oxide (N_2O), and carbon dioxide (CO_2). Ammonia (NH_3) is also an important gas emission from dairy farms. It is a pollutant with potential impacts on human health and over-fertilization of sensitive ecosystems. Ammonia also has implications for GHG emission. It can be transported through the atmosphere and redeposited on land in locations where conditions are favorable for transformation to N_2O, further contributing to global warming (IPCC, 2006b). In a national assessment, Rotz et al. (2021) found that dairy farms in the U.S. may emit over 20% of the total NH_3 emission inventory for the country.

Accurately measuring the GHG and NH_3 emissions from all sources on a dairy farm simultaneously through time is essentially impossible. Therefore, modeling of some form is required to integrate all emissions for a farm or production system (Rotz, 2018). Measurement is important though for quantifying emissions from specific sources and to provide data and information for model development and verification. For a comprehensive analysis, LCA is often used to incorporate all important emission sources. Most LCA studies use a partial cradle to farmgate analysis that includes all sources associated with the production of resources used on the farm and emissions from the farm. A full LCA is also used, which additionally includes emissions associated with transportation, processing, retail, consumption, and waste (Thoma et al., 2013a, 2013b).

The objective of this chapter is to discuss the various sources of GHG emissions from dairy farms and models used to represent those sources alone and integrated in whole farm models. Assessments of representative dairy farms provide distributions of GHG emissions among sources for different types of farm management from extensive grazing to intensive confinement systems. Whole farm assessments illustrate the potential reductions of GHG emissions that can be achieved using various mitigation strategies applied to different types of dairy farm management.

2 Emission sources and models

Important GHG emission sources on dairy farms are the animals, manure they produce, and land used to produce feed crops. Animal emissions are predominantly CH_4; whereas emissions from cropland are N_2O. Manure emissions normally include both CH_4 and N_2O, where the relative amounts of each depend upon the type of handling and storage used. Anthropogenic CO_2 comes from the conversion and emission of fossil C through fuel combustion and the decomposition of lime and inorganic fertilizer such as urea. Biogenic C flows (primarily photosynthesis and respiration) are normally ignored in GHG assessments since this C is just being recycled from the atmosphere back to the atmosphere with no long-term effect. A consideration for this assumption, though, is that large amounts of C are exported from dairy farms in milk, animals, and feeds sold. Because that C is assumed to be respired back to the atmosphere downstream from the farm, no credit or offset can be claimed by the producer. This places a burden on the farm when the emission is actually occurring elsewhere.

Models of different levels of complexity are used to predict or estimate GHG emissions. Across this continuum of complexity, models can be categorized as (1) emission factors, (2) process-driven emission factors, (3) empirical or statistical relationships, and (4) mechanistic process simulation (Rotz, 2018). The simplest emission factors use a value to represent an emission source. Process-driven emission factors use factors or relatively simple relationships controlled by influences, such as ambient temperature, to represent a source. These two groups are defined by the Intergovernmental Panel on Climate Change (IPCC) as tier 1 and tier 2 factors (Gavrilova et. al, 2019). Empirical or statistical models use relationships created by statistical regression to represent a source. The most complex is the mechanistic process simulation where dynamic models describe relationships and interactions as they occur through time. Models of each type can be combined to study the net GHG emissions of dairy production at farm, region, or national scales.

The compounds of CH_4, N_2O, and CO_2 are normally modeled using separate factors or relationships. Values determined for each are then combined as CO_2e units considering global warming potential (GWP). Recommended GWP values have varied over the years as more is learned about warming potential.

Most studies conducted in the past ten years have considered a 100-year time horizon with factors of 28 kg CO_2e/kg CH_4 and 265 kg CO_2e/kg N_2O (Myhre et al., 2013). A few have considered the possibility of climate-C feedback where the changing climate may alter the amount of C absorbed or released from sinks, which may further exacerbate climate change. To consider this effect, factors of 34 and 298 were used for CH_4 and N_2O, respectively (Myhre et al., 2013). Considering typical ratios of CH_4 and N_2O emissions from dairy farms, the latter GWP values increase the CO_2e emission about 15% compared to the use of the lower values without feedback. Studies prior to 2013 often used GWP values of 25 and 298 for CH_4 and N_2O (IPCC, 2007), which gave totals for dairy systems about 3% less compared to the use of 28 and 265. The most recent recommended values are 27 for CH_4 and 273 for N_2O (Forster et al., 2021). For most dairy systems, use of these values gives a total CO_2e about 1% less than that using GWPs of 28 and 265. When comparing total GHG emission values of various studies, the GWP values used should be considered.

Another consideration for the global warming effect of cattle is that CH_4 is a short-lived gas in the atmosphere. Methane oxidizes, returning that C back to CO_2 with a half-life in the atmosphere of about eight years. In comparison, the warming effect of CO_2 produced from fossil sources remains in the atmosphere much longer. The use of 100-year GWP values does not accurately represent this difference in time in the atmosphere. Another model referred to as GWP* has been introduced to better represent the effect of short-lived gases such as CH_4 (Smith et al., 2021). Use of GWP* can show a much lower effect of dairy production systems on global temperature rise (Rotz and Mitlhoner, 2025; Pressman et al., 2023). To apply this model, the change in CH_4 emission rate over time is used. This requires emission measures at different points in time, which complicates the modeling procedure compared to the use of a constant factor.

2.1 Animal emissions

Animals are normally the primary GHG emission source on dairy farms. Enteric fermentation, a natural function of digestion in ruminant animals, produces large quantities of CH_4. Enteric CH_4 is normally 40–50% of the total GHG emission from dairy farms (Rotz et al., 2021). In general, the amount emitted is a function of the gross energy intake of feed as influenced by fiber content and digestibility of the diet. Nitrous oxide can also be produced in the rumen (Mitloehner et al., 2009; Parker et al., 2018). The amount produced is normally small compared to other sources, and thus ignored in most farm assessments.

Many models have been developed to predict emissions as a function of the diet consumed (Benaouda et al., 2019). Most of these are statistical models developed from datasets specific to one or more experiments. Recent work has combined international datasets to develop relationships applicable over a broader range of conditions (Appuhamy et al., 2016). Enteric models have covered the full range in complexity from simple emission factors to dynamic simulation of the rumen (Rotz, 2018). If enteric N_2O is considered, a simple emission factor expressed as the portion of nitrogen (N) consumed in the diet can be used (Rotz, 2018). When known, nitrate content in the diet may be a better predicter (Parker et al., 2018; Rotz, 2018).

2.2 Manure emissions

Manure emissions are normally second in the percentage of total GHG emitted by a dairy farm (Rotz et al., 2021). The amount and type emitted vary greatly depending upon the type of manure handling and storage used. When long-term manure storage is not used, GHG emissions from dairy manure are small. With anaerobic lagoon storage in a warm climate, manure CH_4 emissions may be greater than those of enteric emissions of the animals producing the manure (Owen and Silver, 2014).

Manure emissions can include both CH_4 and N_2O depending upon the manure system used. Methane is normally the dominant source of GHG from manure, averaging three times the CO_2e of that from N_2O (Rotz et al., 2021). Emissions occur during animal housing, but long-term manure storage is the primary source. With tie-stall and free-stall type housing, manure does not remain in the facility long enough to produce much CH_4 and N_2O. With open lots, manure is removed less frequently, and greater emissions occur (Waldrip et al., 2016). Use of housing with deep bedding essentially combines housing and long-term storage, which can produce substantially greater emissions, up to 35% of that from the total farm (Rotz, 2018). With anerobic liquid storage, CH_4 emissions can be very large while N_2O emissions may only occur where solids float to the surface, providing some crusting (Owen and Silver, 2014). With solid manure storage and composting systems, the amounts of each emitted may be more balanced.

Models for manure storage can again be found covering the full range of complexity (Rotz, 2018). The most used models are those published by the IPCC (2006a; Gavrilova, 2019). Their tier 1 approach uses an emission factor that is a function of animal type, volatile solids excretion rate, and manure storage type to predict annual CH_4 emissions. Tier 2, process-driven emission factors add more detail in describing the volatile solids excretion rate as a function of the energy intake and digestibility of the diet. Nitrous oxide emissions are estimated as a function of the N excretion rate of the cattle and manure storage type where tier 2 considers diet characteristics to predict excretion rate. Dynamic simulation models have also been developed for liquid, static stack, and active composting manure systems (Li et al., 2012; Bonifacio et al., 2017). These more complex models simulate the biogeochemical processes in the manure through time as influenced by ambient and manure conditions.

Manure is also a major source of NH_3 emissions (Rotz et al., 2014). Emissions occur in the housing facility, during storage, following field application, and from urine and fecal deposits on pasture (Rotz, 2004). Again, models across the full spectrum of complexity are used to predict NH_3 emissions. The IPCC provides emission factors to estimate N loss due to volatilization of NH_3 and N oxides combined (Gavrilova, 2019). Rotz et al. (2014) provide a dynamic simulation model that tracks NH_3 formation, speciation, aqueous-gas partitioning, and mass transfer processes to predict daily emissions as influenced by the type of facility, manure characteristics, and environmental conditions.

2.3 Pasture and cropland emissions

Production of feed crops is another major source of GHG emissions for dairy farms. This source is mostly N_2O, but minor amounts of CH_4 may be released during the application of manure to cropland. Methane emission can be modeled as the amount remaining in the manure at the time of application (Rotz et al., 2022). Soil can also provide a sink where CH_4 is absorbed and oxidized, but the amount decomposed is normally very small compared to that emitted throughout the production system (Chianese et al., 2009). This small amount is normally ignored in the modeling of dairy systems. For manure deposited by grazing animals, the IPCC recommends an emission factor related to the volatile solids excreted (Hergoualc'h, 2019).

Nitrogen transformation processes in the soil include nitrification and denitrification, both of which can create N_2O. The N_2O produced by these processes is an important contributor to total farm GHG emissions. Nitrification is an aerobic process that oxidizes NH_4 to NO_3, with the production of NO and N_2O as intermediates. Denitrification is a facultative anaerobic microbial process that reduces nitrate to N_2 when soil oxygen is depleted. Depending upon the soil conditions, this process can be incomplete, also producing NO and N_2O. These processes are sometimes combined in an emission factor. The IPCC provides a general N_2O-N emission factor of 1% of the inorganic and organic N incorporated into soil, but this factor varies from 1.6% in wet climates to 0.5% in dry climates (Hergoualc'h, 2019). An additional emission of 0.2–0.6% of applied N is recommended for fecal and urine N deposited by grazing cattle.

Mechanistic soil models are also used to predict N_2O emissions from crop and pasture lands. Emissions are predicted through time as influenced by temperature, precipitation, soil N level, and other soil conditions (Del Grosso et al., 2000; Li et al., 2012). These dynamic simulation models provide a better

Published by Burleigh Dodds Science Publishing Limited, 2026.

prediction of N_2O emissions through time, but a high level of uncertainty remains in predicting this emission source (Hergoualc'h et al., 2019).

2.4 Anthropogenic carbon dioxide emissions

Anthropogenic CO_2 is newly formed CO_2 through human activities that is emitted to the atmosphere. This CO_2 primarily comes from the combustion of fossil fuels. Other important sources from dairy farms are the decomposition of lime and urea-based fertilizer applied to cropland producing feed (Hergoualc'h, 2019). This decomposition converts C originally stored in the earth to CO_2 in the atmosphere. This added CO_2 can remain in the atmosphere for 1000 years or more with a long-term effect on global warming (Forster et al., 2021). Modeling emissions from these sources is done with an emission factor (Hergoualc'h, 2019) where the factor is the amount of C in the fuel, lime, or urea converted to CO_2 (multiplied by 44/12). Thus, the amount of CO_2 emitted is dependent upon the type and concentration of C in the fuel, fertilizer or lime used.

2.5 Resource production

Another important source of emissions that should be considered is the upstream emissions that occur during the production and transport of resources used on the dairy farm. These should include electricity, fuel, fertilizer, machinery, pesticides, seed, purchased feed and replacement animals, and any other important material used to produce milk. These are normally modeled using an emission factor of CO_2e per unit of product used. This emission factor may be developed from measured data or modeling of upstream production systems. For example, emissions related to purchased feed and animals can be developed through the dynamic simulation of crop or livestock farms (Rotz et al., 2021). Typical values for important resources used on dairy farms have been reported (Rotz et al., 2021). Values can vary regionally and globally though, dependent upon the processes and resources used to produce the products used on farms and the transportation required. They may also vary through time as production processes become more efficient or as fossil energy sources are replaced with renewable energy.

3 Whole farm models

Emission models are most useful when component sources are combined to estimate the net GHG emission for the farm or dairy production system. They can also be applied at regional, national, and global scales to determine the GHG emissions associated with dairy production. These integrated models can use component models of any type across the spectrum of complexity from simple emission factors to dynamic process simulation. Often integrated models use a combination of emission factors for minor sources and more mechanistic approaches for the major sources (Hutchings et al., 2018). The level of detail and accuracy needed for individual sources on the farm depends upon the magnitude of their contribution to the total system emission.

The IPCC recommended tier 2 approach is widely used in farm-scale GHG assessment models. Models have been developed and applied in many regions of the world. Examples of models currently available for broad use include the Cool Farm Tool (2023), Farm-AC (2023), COMET-Farm (2023), HOLOS (2023), and DairyGEM (2023). DairyGEM is designed specifically for dairy farms, while the other tools have a more general design that can also represent other livestock and cropping systems. The Cool Farm Tool, Farm-AC, and COMET-Farm were developed for use online, whereas HOLOS, and DairyGEM are tools that are downloaded and installed on a computer. Although these models use the IPCC (2006a, 2006b, 2007) relationships for many components, they often incorporate empirical and mechanistic relationships for some GHG sources to be more specific to the region and conditions where they are employed.

The Global Livestock Environmental Assessment Model (GLEAM, 2023) is another online tool that relies heavily on the IPCC tier 2 methodology for estimating GHG emissions. GLEAM is a GIS framework that represents the biophysical processes and activities along livestock supply chains. This tool was developed to estimate emissions on regional, national, and global scales. Emissions can be evaluated for all major livestock species, which includes dairy cattle systems.

A few models applied to dairy farms are primarily mechanistic in representing farm processes. These include Manure-DNDC (Li et al., 2012) and the Integrated Farm System Model (IFSM; Rotz et al., 2021). These models simulate the important component processes through time on a daily and sometimes hourly time-step to follow the conditions influencing emissions. Manure-DNDC was developed with emphasis on the biogeochemical processes in soil and manure that produce emissions while IFSM emphasizes the crop, animal, feeding, and manure handling processes and their influence on emissions.

Life cycle assessment provides an important tool when evaluating dairy production at the farm and larger scales. Life cycle assessment is an accounting procedure that attempts to consider all important impacts of a product or service. Environmental impacts are often emphasized, but other considerations can include social and economic impacts (Chen and Holden, 2017). Within environmental impacts, global warming is one of many metrics that can be considered. Life cycle assessments applied to dairy have included and often emphasized global warming, but many other metrics should be considered for a more complete assessment of environmental sustainability (Kim et al., 2019; Henderson et al., 2023).

An LCA of GWP applied to dairy must include all important direct GHG emission sources as well as indirect sources (LEAP, 2016). Proper allocation of emissions among coproducts must also be considered. For dairy farms, an important co-product is cattle exported to beef production. Different procedures for allocation can be used, but the International Dairy Federation (IDF, 2015) has recommended a biophysical model for allocating between milk and meat production in dairy systems. Most LCAs applied to dairy have used a partial LCA ending at the farmgate, i.e. with the milk leaving the farm. Full LCAs have also been done which consider all emissions associated with milk from the production of resources used through consumption and the waste created (Thoma et al., 2013a, 2013b; Henderson et al., 2023).

4 Farm system comparisons

An important use of farm-scale models is to compare the production efficiency and resulting environmental impacts of dairy production systems. When the same modeling approach is used within a study to compare systems, valuable information can be gained to direct the adoption or modification of management strategies and technologies for use in dairy production. Many differences can occur between modeling approaches, which prevents a direct comparison of production systems assessed in different studies. Emission intensities found in studies using different models and assumptions can only be compared for verification purposes. Better standardization of modeling procedures could reduce this variability among studies; however, there will always be variation in modeling approach and assumptions creating uncertainty in comparing across studies. The most valuable use of farm models is to assess and compare production systems using the same model and baseline assumptions.

Many modeling studies have been conducted to evaluate the GHG emissions of dairy production systems. This type of evaluation normally includes LCA for a comprehensive inclusion of all important GHG sources and sinks. As examples, several studies will be discussed to illustrate the application of farm models in comparing systems.

Dairy farms in many regions of the world have increased in size, i.e. the number of cows maintained and the land used to produce feed. Farm size alone does not necessarily affect GHG emission intensity when animal numbers, land area, and related management change in proportion. With the increase in farm size, though, improvements in production efficiency may be obtained. Based upon 536 farm surveys throughout five regions of the U.S., Thoma et al. (2013a) found farmgate GHG emission intensities varying from 0.6 to 3.0 kg CO_2e/

kg FPCM, with a mean of 1.23 kg CO_2e/kg FPCM. They found that large dairy operations tended to have lower emission intensities through more efficient feed conversion. This was partially offset, though, because manure system emissions were greater on larger farms.

Various studies have compared pasture-based dairy production to confinement feeding of conserved feeds. Conclusions from these assessments have been variable, often influenced by the climate of the region of the assessment. In cool moist climates conducive to grass production, pasture-based systems may provide a lower GHG emission intensity than confinement feeding strategies. O'Brien et al. (2012) compared a seasonal pasture-based dairy farm and a confinement dairy farm in Ireland. In this climate, the GHG emission intensity of the confinement farm was 18% greater than that of the pasture-based dairy. When high-performance grass-based systems in Ireland were compared to confinement systems in the U.S. and U.K., similar GHG emission intensities were found (O'Brien et al., 2013). With potential C sequestration included though, Irish grass-based systems showed a small advantage. A study of New Zealand dairy production found that confinement systems produced an 11% greater GHG emission intensity than pastoral systems (Ilyas et al., 2019).

When climate and soil conditions are more favorable for corn and small grain feed crops, confinement feeding systems can outperform pasture systems. Rotz et al. (2020) did an assessment of the environmental impacts of grass-based dairy systems in the northeastern U.S. Compared to all dairy farms in the region, the GHG emission intensities of the grass-based dairies were 20–50% greater. In an international comparison, use of confinement production systems normally increased the GHG emission intensity in New Zealand compared to pasture-based systems, but often decreased the intensity of production systems in the northeastern U.S. and the Netherlands (McDowell et al., 2022).

For a given production strategy, management intensity can improve performance and reduce GWP. Herron et al. (2021) compared an average pasture-based dairy system in Ireland to a more intensive system with greater stocking rate, herbage production, milk production and other industry targets. With the increased intensity, GHG emission intensity was reduced by 30%. For grass-based systems in New Zealand, Ledgard et al. (2016) found that intensifying production had only a small effect on GHG emission intensity. There was a small increase with the more intensive approach due to the import of more supplemental feed, where the production and transport of this feed had a relatively high C footprint. Among grass-based farms in the northeastern U.S., Rotz et al. (2020) found that the emission intensity of farms that fed only grass was 30% greater than those that fed some concentrate to increase energy intake. This improvement was due to the greater milk production obtained from cows supplemented with concentrate feeds. Lizarralde et al. (2014) studied the emission intensity of 24 grazing-based dairy farms in Uruguay. They found that improved pasture and feeding management reduced the GHG emission intensity of milk produced in this region. In their assessment, increased feeding of concentrate was not effective in reducing global warming unless forage consumption was unconstrained. Increased intensification in dairy production was also studied for dairy farms in Mexico (Rendón-Huerta et al., 2018). Based on a survey of 26 farms, an increase in intensification reduced GHG emission intensity by up to 29%. This reduction in global warming was primarily obtained through increased feed efficiency and milk production.

Global warming impacts of organic dairy production systems have also been assessed. Through an assessment of 14 farms across the U.S., emission intensities of 0.76–1.08 kg CO_2e/kg FPCM were found (Aguirre-Villegas et al., 2022). Comparing alternative management practices on these farms, they found GHG reductions of 15% and 30% for individual and combined practices, respectively, where manure storage and renewable energy production had the greatest benefits. In a comparison to conventional milk production in the Netherlands, GWP per kilogram of organic milk was a little (7%) greater than that of conventional milk (Thomassen et al., 2008). An important consideration in their comparison was the C footprint of the supplemental feeds fed.

In a comprehensive LCA of U.S. dairy production, Rotz et al. (2021) modeled 120 representative farms across six regions. These farms represented a wide range in size and management approach. The farmgate GWP of the milk produced on these farms varied from 0.69 kg CO2e/kg to 1.45 kg CO_2e/kg FPCM with a national mean around 1.0 kg CO_2e/kg FPCM. Dairy production in the U.S. made substantial improvements in efficiency and milk production per cow over the ten years between the survey-based study of Thoma et al. (2013b) and the modeling assessment of Rotz et al. (2021) which contributed to the 19% reduction in the mean intensities found in the 2 studies. Differences in approach also contributed to differences in the mean and ranges in intensities found between the studies.

Many factors influence the emission intensity or C footprint of milk production including the climate, soil characteristics, and animal and manure management. Rotz et al. (2021) found that no production system or approach provided consistently lower C footprints, but some trends were found in the data. Small farms with 100 or fewer cows had an average footprint 4% greater than the mean of all farms and those with 1000 or more cows averaged 4% less than the mean. Farms where cows received a substantial portion of their forage from grazing pasture had a milk footprint 5% less than the mean of all farms. This benefit primarily came through reduction or elimination of long-term manure storage. Farms that used little or no long-term manure storage had a mean footprint 4% less than the mean of all farms. Farms with long-term liquid or slurry storage had a footprint close to the mean of all farms and those storing manure as a solid or compost averaged 9% greater than the mean. Milk production had a small influence where herds with an annual production of more than 10 000 kg FPCM/cow had an average footprint 3% less than those with lower production.

5 Mitigation strategies

Perhaps the most useful purpose for whole farm GHG models is the evaluation of mitigation strategies. With goals being set in the dairy industry for C neutrality, tools are needed to quantify life cycle GHG emissions of alternative production strategies and mitigation technologies. Many strategies have the potential to mitigate GHG emissions from dairy farms. Potential strategies have been reviewed to mitigate emissions from enteric (Hristov et al., 2013a), manure management (Montes et al., 2013), and animal management (Hristov et al., 2013b) sources. In these studies, a wide range of potential strategies were qualitatively assessed for their potential effectiveness and feasibility for implementation. These studies concluded that many interactions occur among sources of CH_4, N_2O, and NH_3, so an assessment of mitigation practices requires evaluation at the whole farm level.

Through a farm survey, actual and potential implementation of GHG mitigation strategies was evaluated in Europe (Beldman et al., 2021). The wide range of potential strategies included those discussed by Hristov et al. (2013a, 2013b) and Montes et al. (2013). They found that the diversity in production approaches prevented prioritizing strategies for industry-wide adoption or recommendation. A number of strategies must be considered to select those most appropriate for individual farms. They concluded that GHG reduction potential was greatest for farming systems with low milk production and the reduction potential was lower for high milk-producing, input-intensive farming systems.

Models are being used to determine the potential benefits of alternative approaches to reduce energy and other resource inputs as well as reducing emissions from various farm sources. This is a relatively new area of research, but a few studies have been published. Feasible and effective practices often depend upon the type of production system used. For example, mitigation approaches are different between pasture-based systems and confinement feeding systems. Some may be more practical for smaller herd sizes while others may only be applicable to large herds.

O'Brien et al. (2016) used an LCA approach to estimate the C footprint of milk produced on 62 grass-based dairy farms. They found that the C footprint was correlated with numerous farm attributes including N fertilizer use, the portion of the diet coming from grazed grass, and milk solids production. They concluded that grass-based dairies could reduce GHG emissions by increasing herd genetic merit, extending the grazing season, and optimizing N fertilizer use.

Five possible mitigation actions were evaluated for Italian dairy farms (Rencricca et al., 2023). A LCA of 63 farms was used to assess management and distribution of livestock manure and fertilizers, anaerobic manure treatment, optimization of the herd composition, feed quality, and heat recovery. The most effective mitigation strategies were anaerobic manure treatment and the management and distribution of livestock manure and fertilizers providing potential reductions in total farm GHG emissions of 6–9%.

Greenhouse gas mitigation strategies were evaluated along with their effects on N and P losses for two representative dairy farms in the Great Lakes region of the U.S. (Veltman et al., 2018). These included five dietary, four manure management, and six field interventions. Implementation of individual strategies provided GHG reductions of 0–20%, and combining the most promising strategies for each farm provided a 36% reduction in the C footprint of the milk produced. Field interventions such as the use of cover crops and subsurface manure injection had little benefit in reducing GHG emissions but provided substantial reductions in N and P losses from the farms. This study was extended to measure potential benefits under projected future climate (Veltman et al., 2021). With the future climate, GHG emissions were predicted to increase, but by implementing the combined mitigation strategies, the increased temperature and precipitation of future climate had little effect on GHG emissions. Strategies with the greatest potential to reduce GHG emissions under current and future climate were covered manure storage with a flare to combust the CH_4 produced or the use of an anaerobic digester, with either providing a 20% reduction (Rotz et al., 2016).

Rotz et al. (2025) evaluated 26 animal feeding, manure handling, cropping and energy saving strategies alone and in combination to quantify their potential for reducing the farmgate life cycle C footprint of milk production on US dairy farms. Strategies with the most benefit included anaerobic digestion of manure, a covered manure storage using a flare for methane destruction, and the use of a feed additive for mitigating enteric methane. A combination of strategies provided an average reduction in the C footprint over all farms of 44%. Offsets that might be achieved on a few farms through soil carbon sequestration or use of food waste in anaerobic digestion provided further potential reductions of up to 7. They concluded that mitigation strategies have the potential for providing large reductions in GHG emissions but achieving GHG neutrality over all US dairy farms appeared improbable.

6 Dairy production system assessments

From the study of Rotz et al. (2021), four dairy production systems were selected to further illustrate the effects of production strategy on GHG emissions. The four systems represented a range in size and management approach of farms found in different climatic regions of the U.S. The farms were (1) a 50-cow organic, grass-based farm in the humid climate of Pennsylvania, (2) a 500-cow free-stall confinement operation in Wisconsin, (3) a 3000-cow free-stall and open lot confinement operation in the dry climate of California, and (4) a 7000-cow open lot confinement operation in the dry climate of Texas. These systems varied in the amount of grazing used, milk production per cow, manure handling and storage practices, and feeds produced on the farm, as well as climate and soils of the location (Table 1). Diversity in farm size and management was found in all regions (Rotz et al., 2021). These farms represented well-managed production systems using different management approaches found in the regions but not all farms in these regions.

6.1 Carbon footprints

Each farm was simulated over 25 years of historical weather for its location using the IFSM (Rotz et al., 2021). A breakdown of the C footprints of the milk produced on each farm is shown in Fig. 1. Total GHG emission intensities varied from 0.87 kg CO_2e/kg FPCM for the small organic operation in Pennsylvania to 1.05 kg CO_2e/kg FPCM for the free-stall and open lot confinement operation in California. The emission intensity of the large open lot dairy in Texas was 0.94 kg CO_2e/kg FPCM and that for the dairy in Wisconsin was 0.97 kg CO_2e/kg FPCM.

Although there were not large differences in the total emissions associated with each, the breakdown among sources varied among the four production systems (Fig. 1). The major differences among systems were emissions from enteric and manure sources. Enteric emissions were 10–20% less for the larger western dairies because less forage and more by-product feeds were fed on these farms. Enteric emissions are sensitive to the fiber content and digestibility of cattle diets with less CH_4 production as less forage and more concentrate feed is fed (Gavrilova et al., 2019). The high forage diet and lower milk production of grazing cows increased their enteric emissions on the smallest farm.

There were major differences in the manure handling methods across the four farms. For the 50-cow grazing farm with daily hauling of the cow manure, the only long-term manure storage was that in a bedded pack heifer barn. This led to low emissions from manure on this farm. On the 500-cow free-stall farm, all manure was stored as slurry. The cow manure was stored in an open tank while the heifer manure was in a pit below a slatted floor under the barn. These gave greater emissions from the heifer housing and moderate emissions from

Table 1 Brief description of four dairy farms in the U.S. used to compare the global warming potential and mitigation practices of different production systems

Farm parameter	50-cow farm	500-cow farm	3000-cow farm	7000-cow farm
Location	Pennsylvania	Wisconsin	California	Texas
Type	Grazing	Confinement	Confinement	Confinement
Number of cows	50	500	3000	7000
Number of heifers	31	410	2350	4150
Breed	Jersey	Holstein	Holstein	Holstein
Diet protein content, %	17.5	17	17	16
Milk production, kg FPCM/cow	7638	10669	11000	12284
Replacement rate, %	28	38	38	36
Total land, ha	50	430	470	1100
Pastureland, ha	35	50	0	0
Primary crops	Grass	Alfalfa, grass, corn	Corn rye silage	Corn silage
Cow housing facility	Tie stall barn	Free-stall barn	Free-stall barn	Open lot
Heifer housing facility	Bedded pack barn	Slatted floor	Open lot	Open lot
Primary manure storage	None	Slurry tank	Liquid basin	Compost stack
Soil type	Silt loam	Silt loam	Clay loam	Loam

Published by Burleigh Dodds Science Publishing Limited, 2026.

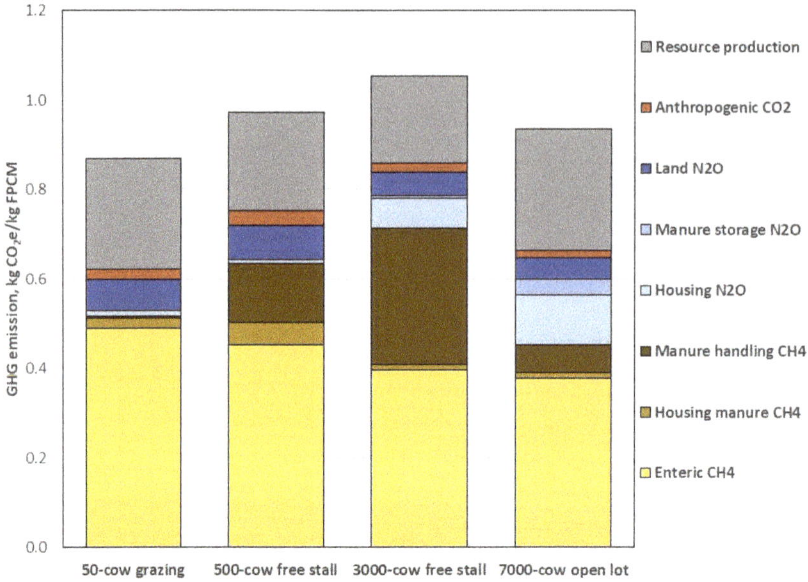

Figure 1 Greenhouse gas (GHG) emission intensities for various sizes and types of dairy farms in the United States. The farms are (1) 50-cow organic with summer grazing and tie stall and bedded pack housing in Pennsylvania, (2) 500 cows with free-stall housing and grazing of heifers in Wisconsin, (3) 3000 cows with free-stall and open lot housing in California, and (4) 7000 cows with open lot housing and composted manure in Texas (Table 1).

the slurry storage tank compared to the other farms. The 3000-cow farm used a combination of stored liquid manure from free-stall barns and stacked dry manure from open lots. The open lots contributed to greater N_2O emissions while the storage of the liquid manure produced greater CH_4 emissions. Emissions were also greater due to the warmer climate of this region. For the 7000-cow open lot dairy in Texas, active composting of manure led to lower CH_4 emissions but greater N_2O emissions. Manure emissions were relatively low from this dairy where most came as N_2O from the open lot.

Emissions associated with the production of resources used on the farms also contributed a major portion of the total (Fig. 1). Major contributors to this source were purchased feeds, electricity, fuel, and fertilizers. Differences among farms for this source were small, but the 7000-cow dairy purchased more feed per animal, which increased this source compared to the other farms.

Anthropogenic CO_2 emissions were relatively small for all dairies, with greater emissions from the smaller farms in the eastern regions. This source was greater on these farms because soils were more acidic, requiring the use of lime, particularly for alfalfa production. The decomposition of the lime increased CO_2 emissions from this source.

6.2 Mitigation strategies

Following the procedure of Rotz et al. (2025), mitigation strategies were evaluated for the four diverse farms of Table 1 to illustrate the potential for reducing GHG emissions. Each farm was simulated with IFSM using up to 12 possible mitigation strategies to measure the potential reduction in the C footprint of the milk produced. Each strategy was simulated independently and then combined with the others where the model considered the interactions among strategies to predict the overall effect on the C footprint of milk produced. The farmgate life cycle GHG emission with each strategy or group of strategies was compared to the initial footprint to determine the reduction achieved (Table 2).

Strategies considered included (1) use of an enteric CH_4 inhibitor, (2) more efficient feeding of protein, (3) improved feed efficiency, (4) an enclosed manure storage with a flare to combust any CH_4 produced, (5) anaerobic digestion of manure to produce electricity or renewable natural gas, (6) more efficient electricity use, (7) photovoltaic solar panels to produce electricity, (8) nitrification inhibitor on cropland, (9) reduced cow replacement rate, (10) reduced animal mortality, (11) subsurface manure injection, and (12) manure processing to extract nutrients and eliminate storage. In addition, different levels of soil C sequestration in the farmland were considered. Not all strategies could be used on all four farms. For the small grazing farm, strategies related to manure handling and processing did not apply since long-term storage was not used and much of the manure was deposited on pasture. Anaerobic digestion was also not compatible with the manure handling methods of the small farm or large open lot dairy.

To model the potential benefit of using an enteric CH_4 inhibitor, a 30% reduction in enteric CH_4 was assumed for cattle fed in confinement. With grazing systems, we assumed the inhibitor was only fed during winter months when the animals were fed in confinement. Use of the inhibitor reduced the life cycle GHG emissions associated with the small grazing farm by about 9%. For the other farms with confinement feeding, the reduction was 11–13% (Table 2). A treatment of this type was not yet approved for use on U.S. dairy farms, and it was unknown whether this treatment could be used in organic production. Nevertheless, these simulations illustrate the potential benefit of their use.

The benefit of more efficient feeding of protein was modeled by decreasing the diet protein content to 15.5%. This was done by altering the diet and using amino acid supplements. The benefit achieved varied depending upon the initial level of protein in the diet and the feeds available. The total farm GHG reduction varied from 8.6% on the small grazing farm to 1.7% on the largest farm (Table 2). The initial diet protein content on the large farm was relatively low (16%), so the GHG reduction achieved by dropping to 15.5% was small.

Table 2 Effect of mitigation strategies on reducing the farmgate life cycle carbon (C) footprint of milk for four different types of dairy farms in the U.S.[1]

	50-cow farm		500-cow farm		3000-cow farm		7000-cow farm	
	CO2e/FPCM2	%	CO2e/FPCM	%	CO2e/FPCM	%	CO2e/FPCM	%
Individual strategies								
Enteric methane inhibitor	0.078	9.0	0.129	13.3	0.119	11.3	0.113	12.1
More efficient protein feeding	0.075	8.6	0.039	4.0	0.044	4.2	0.016	1.7
Improved feed efficiency	0.052	6.0	0.053	5.5	0.079	7.5	0.063	6.7
Covered basin and flare			0.130	13.4	0.258	24.5		
Anaerobic digester			0.143	14.7	0.190	18.0		
Electrical use efficiency improved	0.006	0.7	0.006	0.6	0.006	0.6	0.006	0.6
Solar panels to produce electricity	0.044	5.1	0.030	3.1	0.037	3.5	0.026	2.8
Nitrification inhibitor	0.016	1.8	0.016	1.6	0.004	0.4	0.003	0.3
Reduced replacement rate	0.029	3.3	0.005	0.5	0.039	3.7	0.011	1.2
Animal mortality reduced	0.005	0.6	0.007	0.7	0.006	0.6	0.005	0.5
Subsurface manure injection			0.006	0.6	0.027	2.6		
Manure nutrient extraction					0.310	29.4		
Combined Strategy[3]	0.280	32.2	0.448	46.1	0.559	53.1	0.243	26.0
Potential C sequestration[4]								
0.5 t C/ha/year	0.216	24.9	0.125	12.9	0.021	2.0	0.020	2.1
1.0 t C/ha/year	0.433	49.8	0.251	25.8	0.042	4.0	0.041	4.4

[1]Farms are a 50-cow organic, grass-based farm in Pennsylvania, a 500-cow free-stall confinement operation in Wisconsin, a 3000-cow free-stall and open lot confinement operation in California, and a 7000-cow open lot confinement operation in Texas (see Table 1).
[2]Reduction in greenhouse gas emissions expressed in kg CO_2e per kg of FPCM.
[3]Total reduction combining individual strategies considering the many interacting effects among strategies and farm components.
[4]Additional reduction attained considering the potential short-term sequestration rate.

Improved feed efficiency was modeled by increasing the net energy obtained from feed by 10%, which reduced feed intake without reducing milk production. This might be achieved through genetic selection of cattle. For all farms, this change decreased the C footprint by about 6–7% (Table 2). Much of this benefit came through less feed purchased or more excess feed available to sell off the farm due to the savings in feed intake.

Covering the manure storage and using a flare to combust the CH_4 produced was only used with slurry or liquid manure storages. For the slurry storage on the 500-cow farm, the footprint reduction was about 13% (Table 2). For the liquid system in the warmer climate of the 3000-cow farm, the reduction was 25%. Due to the warmer climate, the CH_4 emission from the open storage was much greater, so the cover was able to provide a greater reduction. Use of an anaerobic digester was also possible on these two farms. For the 500-cow farm, the gas produced was used to generate electricity consumed on the farm. The footprint reduction obtained with this strategy was similar to that from flaring the gas. The benefit through reduced electricity purchase was partially offset by increased CH_4 leakage from the digester and CH_4 loss during storage of the digestate. For the 3000-cow farm, the gas produced was exported for use as renewable natural gas. Use of the digester provided a greater GHG reduction on this farm because the emission from storage on the base farm was much greater in the warmer climate of California (Table 2).

Improvement in electrical use efficiency was modeled by reducing electricity use on the farm by 30%. Because the footprint of producing the electricity was relatively small compared to other GHG sources, this improvement had a relatively small impact. The benefit was similar across all farms with less than 1% reduction in the farmgate GHG emission (Table 2).

Solar photovoltaic generation of electricity was modeled assuming the use of 2 industrial panels (4 m^2) per cow. This size could be designed for installation on a barn roof housing the animals or on shade structures over open lots. Net metering of electricity use was assumed. The electricity produced was primarily used on the farm but could also be exported to the grid depending upon the amount produced versus that needed. The C footprint reduction varied from 2.8% to 5.1% among the farms (Table 2).

A nitrification inhibitor was applied to the cropland and pasture to reduce the direct N_2O emissions from farmland. This reduced the nitrification and denitrification processes resulting from the inorganic fertilizer and manure N applied. The assumption was that direct N_2O emissions from the farmland were reduced by 30%. The benefit received varied with the amount of land on the farm producing feed. For the small grazing farm, a 1.8% reduction in the C footprint was determined. This decreased with confinement operations due to a decrease in the land use per animal. Land area decreased because greater forage yields were obtained on farms where corn silage was produced, particularly when corn was double cropped with rye silage (3000-cow farm). For purchased feeds produced outside the dairy farms, the use of the inhibitor was not considered. Thus, the benefit of using the inhibitor on the large open lot dairy was minimal because most of the feed was imported from other farms (Table 2).

Reduced cow replacement rate and animal mortality were both related to animal health and well-being. Annual cow replacement rate was reduced from 28% to 20% on the grazing farm and from 36% or 38% to 30% on the other farms with the number of replacement heifers reduced proportionately. The resulting benefit in GHG reduction varied from 0.5% to 3.7% among the four farms (Table 2). Reducing animal mortality rates by 20% had a relatively small effect, reducing the C footprint by less than 1%.

Subsurface injection of manure could only be used on farms using slurry or liquid manure. Injecting the manure into the soil reduced NH_3 emissions (indirect GHG source) but increased N_2O emissions. Greater N_2O emissions occurred because the N saved through NH_3 reduction increased the N available in the soil for nitrification and denitrification processes. The concentrated N and moist soil conditions around the injected manure may also contribute to greater N_2O emissions (Dell et al., 2011). This process also required more power for manure application, creating a small increase in fuel consumption. The net effect was a small reduction in C footprint (Table 2).

Technologies are being explored for processing manure or the digestate from anaerobic digesters to remove nutrients and produce stable products. This can potentially eliminate the need for long-term manure storage and the resulting GHG emissions. The nutrients can then be applied to crop and pasture fields, providing better timing and more precise amounts to meet crop needs. This was only modeled for the 3000-cow farm with a liquid manure system. For this farm, 70% of the manure was handled as a liquid, with the remainder being dry manure from open lots. The system was modeled by exporting 5% of the manure dry matter (carbon) and 70% of the manure N, P, and K from the farm. Fertilizers were then applied at rates and timing to best meet the nutrient needs of the crops. With this approach, the total farm GHG emission was reduced by 29%.

Farm-scale modeling and evaluation is particularly useful for studying the combined effect of multiple mitigation strategies. With all the applicable mitigation strategies combined, the C footprint of the milk produced on the small grazing farm was reduced by 0.28 kg CO_2e/kg FPCM or 32% (Table 2). Inclusion of anaerobic digestion and covered digestate storage reduced the C footprint of the 500-cow farm by 46% for a final footprint of 0.529 kg CO_2e/kg FPCM. The 3000-cow farm achieved the largest combined reduction in GHG emissions of 53%. This lowered the C footprint of the milk produced on this farm to 0.49 kg CO_2e/kg FPCM. The added benefit primarily came through the use of manure processing with nutrient extraction and stable storage of the products produced. The least benefit of combined strategies was found for the 7000-cow open lot dairy. With the handling of solid manure, options for reducing GHG emissions were more limited. Considering that the initial footprint for this production system (0.936 kg CO_2e/kg FPCM) was less than that of the other confinement farms, the final value of 0.693 kg CO_2e/kg FPCM was 18–43% greater than the final values of the other production systems.

6.3 Carbon sequestration

Sequestering C on the farm is another option for reducing the C footprint of milk production. By accumulating organic matter in pasture and cropland soils, CO_2 drawn from the atmosphere can be stored long-term in the soil. The potential to sequester soil C on dairy farms is variable and uncertain. Soils seek a long-term C balance where the C lost through respiration equals that incorporated into the soil. Therefore, a change in management is needed to generate sequestration. Possible changes include the transition from annual crops such as corn to perennial grasses and the adoption of no-till or cover cropping management practices. During the transition period following a change in management, substantial amounts of C may be sequestered and stored in the soil.

Perennial pastures are recognized as an important sink for C storage. When soils depleted in organic matter through production of annual crops are converted to perennial grassland, the soil may sequester an average of 0.9 t C/ha per year for many years (Paustian et al., 2019; Franzluebbers, 2010). Eventually the soil will come into equilibrium with the soil C content similar to native, undisturbed soil. Implementing improvements in grazing practices, such as adjusting stocking rates and changing forage species, may also enhance sequestration. For annual cropping systems, there is the potential to sequester 0.1–0.4 t C/ha per year for about 15 years by introducing no-till or reduced tillage in crop establishment (Paustian et al., 2019; Franzluebbers, 2010). The amount achieved is related to soil texture and climate, where lower rates occur in sandy soils and drier conditions. More diverse crop rotations and cover cropping may also increase sequestration up to 0.3 t C/ha per year depending upon soil type and tillage management (West and Post, 2002; Thapa et al., 2023). These sequestration rates give some guidance on the maximum that may occur on dairy farms.

Potential C sequestration rates of 0.5 and 1.0 t C/ha per year were considered for the four diverse dairy farms simulated with IFSM. These rates represented the potential short-term benefit for a change in management, realizing that this potential sequestration rate could not continue indefinitely. The C sequestered provided an offset of the C footprint of the milk produced.

The benefit varied with the potential sequestration rate and the ratio of farmland available per cow or unit of milk produced. The small organic grazing farm showed the greatest potential for offsetting GHG emissions. Assuming a recent change from annual cropping practices, this management approach, which relied upon perennial pasture with little tillage and pasture establishment, would also have the greatest potential to sequester soil C. With an annual sequestration rate of 1 t C/ha combined with the possible mitigation strategies, this farm approached C neutrality (C footprint of 0.15 kg CO_2e/kg FPCM). Considering available data on C sequestration in pastures (Paustian et al., 2019; Franzluebbers, 2010), this rate may be achieved for many years through the regeneration of pasture soils following degraded conditions.

The three larger farms relied heavily on annual cropping systems. With these systems, soil C sequestration would be more difficult to achieve, but there is potential through management changes such as conversion to reduced or no-till systems and/or the use of cover crops. Due to the lower land base used to produce feed on these farms, the offset obtained was relatively small. With an annual sequestration rate of 0.5 t C/ha, the offset was 2–13% of the C footprint of the milk produced (Table 2). Even if a greater sequestration rate could be maintained, the offset remains relatively small, not allowing the farms to approach C neutrality (Table 2).

7 Conclusion and future trends

To counter rising GHG concentrations in the atmosphere, the dairy industry along with many others is working to mitigate emissions. The ultimate goal is C neutrality or net-zero GHG emissions. Only through major reductions in emissions can the current trend of increasing GHG in the atmosphere and the resulting effects on climate be stopped or reversed.

There are many sources of GHG associated with dairy production. Quantifying and integrating all sources from a farm or production system requires a modeling approach. Models across a wide spectrum of detail from simple emission factors to complex process-level simulation are being used to represent and study GHG emissions at the farm scale and beyond. Life cycle assessment is often incorporated to combine all important sources. Most LCA studies use cradle to farmgate assessments which include emissions associated with the production of resources used on the farm as well as the direct sources from the soil, animals, manure, and equipment on the farm.

Farm models are often used to compare production systems. These vary from low-input extensive grazing to high-input intensive confinement systems. Modeling studies have demonstrated that the most appropriate system for reducing GHG emissions is dependent upon the climate and available resources at a location.

Farm models provide valuable tools for determining a breakdown or distribution of emission sources within production systems. This distribution can be used for prioritizing mitigation approaches. Strategies can be aimed toward farm components that provide the greatest emissions.

Farm-scale modeling also provides a valuable tool for assessing mitigation strategies. The effects of different strategies can interact, so the model must account for these interactions to provide net effects. By modeling various strategies alone and in combination, potential reductions in the C footprint of the milk produced can be assessed. Combinations of technologies and strategies are required to move toward net-zero dairy production systems.

Models will continue to develop and improve, providing better assessments at the whole farm, regional, national, and global scales. A need in this development process is more experimental measurement of emission sources and the interacting factors that influence these emissions. As we learn more about individual emission sources within the farm, more accurate component models will develop at all levels of detail. Integration of these improved components into whole farm models will provide more useful tools for assessing mitigation strategies.

Net-zero dairy production is a lofty but important goal. As new strategies and technologies are developed and implemented on farms, models will continue to provide important tools for quantifying emissions and assessing our progress toward net-zero production.

8 Where to look for further information

Further reading

- More discussion on GHG mitigation strategies in dairy and livestock production can be found in 'Special topics – Mitigation of methane and nitrous oxide emissions from animal operations: I. A review of enteric methane mitigation options' (Hristov et al., 2013a); 'Mitigation of methane and nitrous oxide emissions from animal operations: II. A review of manure management mitigation options' (Montes et al., 2013); and 'Mitigation of methane and nitrous oxide emissions from animal operations: III. A review of animal management mitigation options' (Hristov et al., 2013b).
- More detail in modeling GHG emissions can be found in 'Modeling greenhouse gas emissions from dairy farms' (Rotz, 2018).
- More discussion on LCA and the C footprint of dairy can be found in 'Greenhouse gas emissions from milk production and consumption in the

United States: a cradle-to-grave life cycle assessment circa 2008' (Thoma et al. (2013a), 'Assessing the carbon footprint of dairy production systems' (Rotz and Thoma, 2017), and 'Strategies for mitigating greenhouse gas emissions from US dairy farms toward a net zero goal' (Rotz et al. 2025).

Major international projects

- COMET-farm: Whole farm and ranch carbon and GHG accounting system, USDA Natural Resource Conservation Service, Colorado State University, Fort Collins, Colorado, https://comet-farm.com/.
- Cool Farm Tool, Cool Farm Alliance, Westgate, Grantham, Lincolnshire, NG31 6LE England, https://coolfarm.org/.
- HOLOS farm model, Lethbridge Research and Development Centre, Agriculture and Agri-Food Canada, Lethbridge, Alberta, Canada, https://agriculture.canada.ca/en/agricultural-production/holos.
- Integrated Farm System Model, Agricultural Research Service, U.S. Dept. of Agriculture, University Park, Pennsylvania. https://www.ars.usda.gov/northeast-area/up-pa/pswmru/docs/integrated-farm-system-model/.

9 References

Aguirre-Villegas, H.A., et al. (2022), 'Farm level environmental assessment of organic dairy systems in the U.S.' *J. Cleaner Prod.* 363:132390. https://doi.org/10.1016/j.jclepro.2022.132390.

Appuhamy, J.A.D.R.N., et al. (2016), 'Models for predicting enteric methane emissions from dairy cows in North America, Europe, and Australia and New Zealand'. *Global Change Bio.* 22:3039–3056. https://doi.org/10.1111/gcb.13339.

Beldman, A., et al. (2021), '*Mitigation options to reduce GHG emissions at dairy and beef farms; Results from a literature review and survey on mitigation options currently being used within the network of the Sustainable Agriculture Initiative Platform (SAI Platform)*', Wageningen Economic Research, Report 2021-099. 78. https://doi.org/10.18174/554786.

Benaouda, M., et al. (2019), 'Evaluation of the performance of existing mathematical models predicting enteric methane emissions from ruminants: animal categories and dietary mitigation strategies', *Anml. Feed Sci. Tech.* 255(114207):1–20. https://doi.org/10.1016/j.anifeedsci.2019.114207.

Bonifacio, H.F., et al. (2017), 'A process-based model for cattle manure compost windrows: part 1 – model description'. *Trans. ASABE* 60(3):877–892. https://doi.org/10.13031/trans.12057.

Chen, W. and Holden, N.M. (2017), 'Social life cycle assessment of average Irish dairy farm'. *Int. J. Life Cycle Assess.* 22:1459–1472. https://doi.org/10.1007/s11367-016-1250-2.

Chianese, D.S., et al. (2009), 'Simulation of methane emissions from dairy farms to assess greenhouse gas reduction strategies'. *Trans. ASABE* 52(4):1313–1323.

Published by Burleigh Dodds Science Publishing Limited, 2026.

COMET-Farm (2023), *COMET-Farm, Whole farm and ranch carbon and Greenhouse gas accounting system*. Natural Resources Conservation Service, USDA. https://comet-farm.com/ (accessed 16 January 2024).

Cool Farm (2023). Cool Farm Tool, Cool Farm Alliance. https://coolfarm.org/the-tool/ (accessed 16 January 2024).

DairyGEM (2023). Dairy Gas Emission Model. Agricultural Research Service, USDA. https://www.ars.usda.gov/northeast-area/up-pa/pswmru/docs/dairy-gas-emissions-model/ (accessed 16 January 2024).

Del Grosso, S.J., et al. (2000), 'General model for N2O and N2 gas emissions from soils due to denitrification'. *Glob. Biogeochem. Cycles* 14(4):1045–1060.

Dell, C.J., et al. (2011), 'Subsurface application of manures slurries for conservation tillage and pasture soils and their impact on the nitrogen balance'. *J. Environ. Qual.* 40:352–361. https://doi.org/10.2134/jeq2010.0069.

DMI (2023). U.S. Dairy Net Zero Initiative. Dairy Management, Inc. https://www.usdairy.com/sustainability/environmental-sustainability/net-zero-initiative (accessed 16 January 2024).

EPA (2023), *Inventory of U.S. greenhouse gas emissions and sinks: 1990–2021*. U.S. Environmental Protection Agency, EPA 430-R-23-002. https://www.epa.gov/ghgemissions/inventory-us-greenhouse-gas-emissions-and-sinks-1990-2021 (accessed 16 January 2024).

FAO (2023), *FAOSTAT*. Food and Agriculture Organization of the United Nations, Rome, Italy. https://www.fao.org/faostat/en/#data (accessed 16 January 2024).

FarmAC (2023), *The FarmAC model user guide*. Dept of Agroecology, Aarhus University. https://web14.agro.au.dk/projectnet/farmac/Pages/FarmAC%20users%20guide%20v3.htm (accessed 16 January 2024).

Forster, P., et al. (2021), 'The earth's energy budget, climate feedbacks, and climate sensitivity'. In Masson-Delmotte, V., et al. (eds.), *Climate change 2021: the physical science basis. contribution of working group I to the sixth assessment report of the intergovernmental panel on climate change*. Cambridge University Press, pp. 923–1054. https://doi.org/10.1017/9781009157896.009.

Franzluebbers, A.J. (2010), 'Achieving soil organic carbon sequestration with conservation agricultural systems in the Southeastern United States'. *Soil Sci. Soc. Am. J.* 74:347–357. https://doi.org/10.2136/sssaj2009.0079.

Gavrilova, O., et al. (2019), '2019 refinement to the 2006 IPCC guidelines for national greenhouse gas inventories', Volume 4: agriculture, forestry and other land use, chapter 10: emissions from livestock and manure management. *Intergovernmental Panel on Climate Change*. https://www.ipcc-nggip.iges.or.jp/public/2019rf/vol4.html (accessed 16 January 2024).

GLEAM (2023), Global Livestock Environmental Assessment Model, Food and Agriculture Organization of the United Nations. https://www.fao.org/gleam/en/ (accessed 16 January 2024).

Hergoualc'h, K., et al. (2019), 2019 refinement to the 2006 IPCC guidelines for national greenhouse gas inventories, Volume 4: agriculture, forestry and other land use, chapter 11: N_2O emissions from managed soils, and CO_2 emissions from lime and urea application. *Intergovernmental Panel on Climate Change*. https://www.ipcc-nggip.iges.or.jp/public/2019rf/vol4.html (accessed 16 January 2024).

HOLOS (2023), *HOLOS farm model, agricultural production tools and data*. Agricultural and Agri-Food Canada. https://agriculture.canada.ca/en/agricultural-production/holos (accessed 16 January 2024).

Henderson, A.D., et al. (2023), 'Spatialized life cycle assessment of fluid milk production and consumption in the United States'. *Sustainability* 15:1890. https://doi.org/10.3390/su15031890.

Herron, J., et al. (2021), 'Life cycle assessment of pasture-based dairy production systems: current and future performance'. *J.Dairy Sci.* 105:5849–5869. https://doi.org/10.3168/jds.2021-21499.

Hristov A.N., et al. (2013a), 'Special topics--mitigation of methane and nitrous oxide emissions from animal operations: I. A review of enteric methane mitigation options'. *J. Anim. Sci.* 91(11):5045–5069. https://doi.org/10.2527/jas.2013-6583.

Hristov, A.N., et al. (2013b), 'Mitigation of methane and nitrous oxide emissions from animal operations: III. A review of animal management mitigation options'. *J. Anim. Sci*. 91:5095–5113. https://doi.org/10.2527/jas2013-6585.

Hristov, A.N., et al. (2022), 'Effective nutritional strategies to mitigate enteric methane in dairy cattle'. *J. Dairy Sci.* 105:8543–8557. https://doi.org/10.3168/jds.2021-21398.

Hutchings, N.J. et al. (2018), How do farm models compare when estimating greenhouse gas emissions from dairy cattle production? Animal (12:10):2171–2180 https://doi:org/10.1017/S175173111700338X.

IDF (2015), *A common carbon footprint approach for dairy*. Bulletin 479/2015, International Dairy Federation. https://www.fil-idf.org/wp-content/uploads/2016/09/Bulletin479-2015_A-common-carbon-footprint-approach-for-the-dairy-sector.CAT.pdf (accessed 16 January 2024).

Ilyas, H.M.A., et al. (2019), 'The carbon footprint of energy consumption in pastoral and barn dairy farming systems: A case study from Canterbury, New Zealand. *Sustainability* 11:4809. dx.doi.org/10.3390/su11174809.

IPCC (2006a), Guidelines for national greenhouse gas inventories'. Volume 4, agriculture, forestry and other land use. Chapter 10, emissions from livestock and manure management. *Intergovernmental Panel on Climate Change*. http://www.ipcc-nggip.iges.or.jp/public/2006gl/pdf/4_Volume4/V4_10_Ch10_Livestock.pdf (accessed 16 January 2024).

IPCC (2006b), Guidelines for national greenhouse gas inventories. Volume 4, Agriculture, forestry and other land use. Chapter 11, N2O emissions from managed soils, and CO2 emissions from lime and urea application. *Intergovernmental Panel on Climate Change*. http://www.ipcc-nggip.iges.or.jp/public/2006gl/pdf/4_Volume4/V4_11_Ch11_ N2O&CO2.pdf (accessed 16 January 2024).

IPCC (2007), *Climate change 2007: the physical science basis. contribution of working group I to the fourth assessment report of the intergovernmental panel on climate change*. Solomon, S., eds.) Cambridge University Press, 996 pp. https://archive.ipcc.ch/pdf/assessment-report/ar4/wg3/ar4_wg3_full_report.pdf (accessed 16 January 2024).

Kim, D., et al. (2019), 'Analysis of beneficial management practices to mitigate environmental impacts in dairy production systems around the Great Lakes'. *Agric. Syst.* 176:1–12. https://doi.org/10.1016/j.agsy.2019.102660.

LEAP (2016), *Environmental performance of large ruminant supply chains: guidelines for assessment*. Food and Agriculture Organization. http://www.fao.org/3/a-i6494e.pdf (accessed 16 January 2024).

Ledgard, S.F., et al. (2016), 'Life cycle assessment of dairy production systems in New Zealand.' In Currie, L.D. and Singh, R. (eds), *Integrated nutrient and water management for sustainable farming*. Occasional Report No. 29. Fertilizer and Lime Research Centre, Massey University, 8 pages. http://flrc.massey.ac.nz/publications.html (accessed 16 January 2024).

Li, C., et al. (2012), 'Manure-DNDC: a biogeochemical process model for quantifying greenhouse gas and ammonia emissions from livestock manure systems'. *Nutr. Cycl. Agroecosys*. 93:163–200. https://doi.org/10.1007/s10705-012-9507-z.

Lizarralde, C., et al. (2014), 'Practices to reduce milk carbon footprint on grazing dairy farms in southern Uruguay: case studies'. *Sustain. Agric. Res.* 3(2):1–15. https://doi.org/10.5539/sar.v3n2p1.

McDowell, R.W., et al. (2022), 'Limiting grazing periods combined with proper housing can reduce nutrient losses from dairy systems'. *Nature Food*. 3:656–664. https://doi.org/10.1038/s43016-022-00644-2.

Mitloehner, F.M., et al. (2009), 'Direct measurements improve estimates of dairy greenhouse-gas emissions'. *Calif. Agric*. 63:79–83. https://doi.org/10.3733/ca.v063n02p79.

Montes, F., et al. (2013), 'Mitigation of methane and nitrous oxide emissions from animal operations: II. A review of manure management mitigation options'. *J. Anim. Sci.* 91:5070–5094. https://doi.org/10.2527/jas.2013-6584.

Myhre, G., et al. (2013), 'Anthropogenic and natural radiative forcing'. In: Stocker, T.F., et al. (eds.), *Climate change 2013: the physical science basis. contribution of working group I to the fifth assessment report of the intergovernmental panel on climate change*. Vol. 23. Cambridge University Press. https://doi.org/10.1021/ie801542g.

O'Brien, D., et al. (2012), 'A life cycle assessment of seasonal grass-based and confinement dairy farms'. *Agric. Syst.* 107:33–46. https://doi.org/10.1016/j.agsy.2011.11.004.

O'Brien, D., et al. (2013), 'A case study of the carbon footprint of milk from high-performing confinement and grass-based dairy farms'. *J. Dairy Sci.* 97:1835–1851. https://doi.org/10.3168/jds.2013-7174.

O'Brien, D., et al. (2016), 'How can grass-based dairy farmers reduce the carbon footprint of milk?' *Anim. Prod. Sci.* 56:495–500. https://doi.org/10.1071/AN15490.

Owen, J.J. and Silver, W.L. (2014), 'Greenhouse gas emissions from dairy manure management: a review of field-based studies'. *Global Change Bio*. 20(2):425–438. https://doi.org/10.1111/gcb.12687.

Parker, D.B., et al. (2018), 'Enteric nitrous oxide emissions from beef cattle'. *Prof. Anml. Sci.* 34:594–607. https://doi.org/10.15232/pas.2018-01769.

Paustian, K., et al. (2019), 'Soil C sequestration as a biological negative emission strategy'. *Front. Clim.* 1:8. https://doi.org/10.3389/fclim.2019.00008.

Pressman, E.M., et al. (2023), 'Methane emissions from California dairies estimated using novel climate metric global warming potential star show improved agreement with modeled warming dynamics. *Front. Sustain. Food Syst*. 6:1072805. https://doi.org/10.3389/fsufs.2022.1072805.

Rencricca, G., et al. (2023), 'Mitigation actions scenarios applied to the dairy farm management systems. *Foods* 12:1860. https://doi.org/10.3390/foods12091860.

Rendón-Huerta, J.A., et al. (2018), 'Comparison of greenhouse gas emissions from Mexican intensive dairy farms'. *S. Afr. J. Anim. Sci.*, 48(1):48–55. https://doi.org/10.4314/sajas.v48i1.6.

Rotz, C.A. (2004), 'Management to reduce nitrogen losses in animal production'. *J. Anim. Sci.* 82(E. Suppl.):E119–E137.

Rotz, C.A. (2018), 'Modeling greenhouse gas emissions from dairy farms'. *J. Dairy Sci.* 101:6675–6690. https://doi.org/10.3168/jds.2017-13272.

Rotz, C.A. and Thoma, G. (2017), 'Assessing the carbon footprint of dairy production systems'. In: D.K. Beede (ed.), *Large dairy herd management*, 3rd ed., American Dairy Science Association, pp. 19–31.

Rotz, C.A., et al. (2014), 'Ammonia emission model for whole farm evaluation of dairy production systems. *J. Environ. Quality* 43:1143–1158. https://doi.org/10.2134/jeq2013.04.0121.

Rotz, C.A., et al. (2016), 'Evaluating greenhouse gas mitigation and climate change adaptation in dairy production using farm simulation'. *Trans. ASABE* 59(6):1771–1781. https://doi.org/10.13031/trans.59.11594.

Rotz, C.A., et al. (2020), 'An environmental assessment of grass-based dairy production in the northeastern United States'. *Agric. Syst.* 184. https://doi.org/10.1016/j.agsy.2020.102887.

Rotz, C.A., et al. (2021), 'Environmental assessment of United States dairy farms'. *J. Cleaner Prod.* 315:128153. https://doi.org/10.1016/j.jclepro.2021.128153.

Rotz, C.A., et al. (2022), *Integrated farm system model: reference manual*, version 4.7. USDA Agricultural Research Service. https://www.ars.usda.gov/ARSUserFiles/80700500/Reference%20Manual.pdf (accessed 16 January 2024).

Rotz, C.A. and Mitloehner, F. M. (2025), United States dairy farms and global warming. J. Dairy Sci. 108(3):2610-2619. https://doi.org/10.3168/jds.2024-25360.

Rotz, C.A., et al. (2025), Strategies for mitigating greenhouse gas emissions from US dairy farms toward a net zero goal. J. Dairy Sci. https://doi.org/10.3168/jds.2025-26392.

Smith, M.A., et al. (2021), 'Further improvement of warming-equivalent emissions calculation'. *Clim. Atmos. Sci.* 4:19. https://doi.org/10.1038/s41612-021-00169-8.

Thapa, V.R., et al. (2023), 'Soil organic carbon sequestration potential of conservation agriculture in arid and semi-arid regions: a review'. *J. Arid Environ.* 217:105028. https://doi.org/10.1016/j.jaridenv.2023.105028.

Thoma, G., et al. (2013a), 'Greenhouse gas emissions from milk production and consumption in the United States: a cradle-to-grave life cycle assessment circa 2008'. *Int. Dairy J.* 31:S3–S14. https://doi.org/10.1016/j.idairyj.2012.09.010.

Thoma, G., et al. (2013b), 'Regional analysis of greenhouse gas emissions from USA dairy farms: a cradle to farm-gate assessment of the American dairy industry circa 2008'. *Int. Dairy J.* 31:S29–S40. https://doi.org/10.1016/j.idairyj.2012.09.010.

Thomassen, M.A., et al. (2008), 'Life cycle assessment of conventional and organic milk production in the Netherlands'. *Agric. Syst.* 96:95–107. https://doi.org/10.1016/j.agsy.2007.06.001.

United Nations Environment Programme (2022), *Emissions Gap Report 2022: The Closing Window – Climate crisis calls for rapid transformation of societies*. Nairobi. https://www.unep.org/emissions-gap-report-2022 (accessed 16 January 2024).

Veltman, K., et al. (2018), 'A quantitative assessment of beneficial management practices to reduce carbon and reactive nitrogen footprints and phosphorus losses of dairy farms in the Great Lakes region of the United States'. *Agric. Syst.* 166:10–25. https://doi.org/10.1016/j.agsy.2018.07.005.

Veltman, K., et al. (2021), 'Assessing and reducing the environmental impact of dairy production systems in the northern US in a changing climate'. *Agric. Syst.* 192:1–14. https://doi.org/10.1016/j.agsy.2021.103170.

Waldrip, H.M., et al. (2016), 'Nitrous oxide emissions from open-lot cattle feedyards: a review'. *J. Environ. Qual.* 45:1797–1811. https://doi.org/10.2134/jeq2016.04.0140.

West, T.O. and Post, W.M. (2002), 'Soil organic carbon sequestration rates by tillage and crop rotation: a global data analysis'. *Soil Sci. Soc. Am. J.* 66:1930–1946.

Chapter 4

Advances in measuring the water footprint of dairy farming

António Cardoso Monteiro, Polytechnic Institute of Viseu, Portugal; and Özdal Gökdal, Aydın Adnan Menderes University, Turkey

1 Introduction

Water is crucial on a dairy farm. It is essential for crop production (where crops are grown, e.g. for feed), animal consumption, cow comfort through water-based cooling, milk cooling, sanitation operations (such as maintaining animal hygiene and facility cleaning) and waste collection and transport (Robinson et al., 2016; Naranjo et al., 2020; Palhares et al., 2020). This requirement for water in milk production can lead to increased pressure to regulate and monitor water use, which creates significant challenges – especially financial – for farmers (Palhares et al., 2020).

There are currently studies that estimate water consumption in ruminant production systems; however, most are based solely on direct water use (Murphy et al., 2017; Palhares et al., 2020; Al-Bahouh et al., 2021). However, water consumption in livestock production should not be considered in isolation in calculating a water footprint (WF); it must include water consumption by others, e.g. industry, services and people, taking into account the available

http://dx.doi.org/10.19103/AS.2025.0157.07

water resources and their needs (Hoekstra & Chapagain, 2007; Hoekstra, 2017; Mekonnen & Hoekstra, 2010; Zhang et al., 2013).

Given increasing demand from population growth, agriculture is responsible for approximately 90% of freshwater consumption, of which 30% is used for livestock production (Mekonnen & Hoekstra, 2012; Zhang et al., 2013). Climate change impacts, including temperature increases, extreme meteorological phenomena and disturbances in ecosystems, are exerting more pressure on freshwater resources (Drastig et al., 2010).

Adaptation strategies for water consumption in agriculture seek to increase water use efficiency (WUE), e.g. through changes in agronomic practices, crop selection and optimization of dairy breeds. These strategies may include advances in feed production techniques, adoption of modern irrigation methods and enhancement of livestock system productivity. Changes in crop management practices – such as adjusting sowing dates and conserving soil moisture – can significantly increase water efficiency (Drastig et al., 2010; Monteiro et al., 2023).

This chapter aims to evaluate the findings from recent research and case studies. It focuses on the latest developments in measuring the WF of dairy farming, highlighting some strategies for optimizing water use and addressing sustainability challenges in the industry.

2 The water footprint concept

The WF concept, which was introduced by Hoekstra (Hoekstra, 2003) and later developed by Hoekstra and Chapagain (Hoekstra & Chapagain, 2008), links human consumption to global water resources and offers a comprehensive assessment of water consumption and its environmental impact (Drastig et al., 2010; Romaguera et al., 2010).

The WF concept offers a differentiated approach to understanding water use, addressing both water consumption and its impact on water quantity and quality. There are two globally accepted WF concepts (Palhares et al., 2020). These are the Water Footprint Standard (Hoekstra et al., 2011; Hoekstra, 2017) and life cycle assessment (LCA) (Drastig et al., 2010; Palhares et al., 2020). The concept of the Water Footprint Standard was used to evaluate the WF of dairy milk in a cradle-to-farm gate perspective which categorizes water use into (Hoekstra et al., 2011):

- Green water: Rainwater used for irrigation or other purposes.
- Blue water: Groundwater or surface water extraction.
- Grey water: Wastewater from household activities (suitable for irrigation).

The WF is a comprehensive metric for assessing the use of freshwater resources and goes beyond the conventional metrics that focus exclusively on water

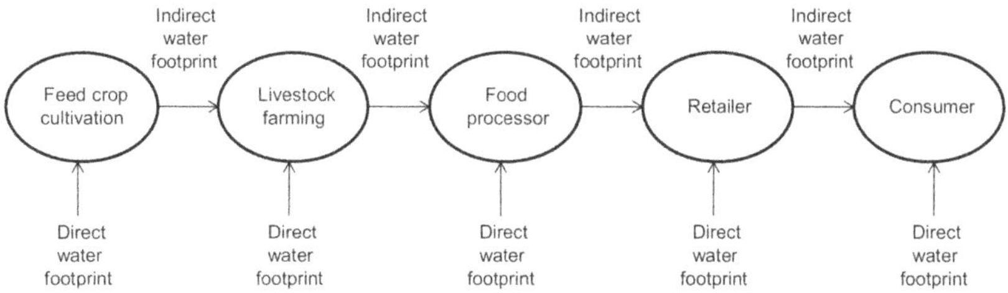

Figure 1 The direct and indirect water footprint in each stage of the supply chain of an animal product product. Source: Adapted from: Hoekstra et al. (2011).

abstraction. The WF is defined as the total volume of freshwater used in the production of a given product. This measurement encompasses calculations derived from primary data on direct and indirect water consumption, the nitrogen load of effluents and processes related to feed and milk production (Fig. 1). A necessary distinction exists between direct and indirect WF in dairy systems, where direct WF includes on-farm water usage, and indirect WF captures off-farm water usage for feed production (Bronts et al., 2023). Consequently, the WF of a dairy cow comprises the indirect WF of feed production and the direct WF of drinking and service water components (Chapagain & Hoekstra, 2003, 2004; Mekonnen & Hoekstra, 2012).

The concept of WF is closely related to the concept of virtual water, which refers to the volume of water required to produce a good or service (Hoekstra & Chapagain, 2007). This emphasizes the flow of 'virtual water' through international trade, which can be calculated for various entities, including goods, services, activities, companies, organizations, individuals and communities (Chapagain & Hoekstra, 2003, 2004; Hoekstra, 2003; Hoekstra & Chapagain, 2007, 2008; Mekonnen & Hoekstra, 2012). Agricultural commodities incorporate significant amounts of green, blue and grey water. Understanding these flows allows countries and industries to assess their indirect water use and manage resources more effectively (Drastig et al., 2010; Hoekstra et al., 2011).

LCA integrates environmental impacts throughout the production chain, ranging anywhere from on-farm processes to retail. Although LCA has been used to assess greenhouse gas emissions and other impacts, its application to water use in dairy farming remains limited. Studies suggest that most environmental impacts occur during the farming phase, including crop production, effluent management and the use of drinking water (Drastig et al., 2010; Høgaas Eide, 2002).

3 Water footprint assessment

Dairy farming has come under public scrutiny due to concerns about both animal welfare and environmental impact. In response, the dairy industry is attempting to cope with climate change by enhancing animal adaptability, improving feeding efficiency, adopting less water-intensive cultivation practices (for feed crops) and increasing water reuse.

The WF of a product consists of the total amount of water consumed (which cannot then be re-used without further treatment) by that product. For instance, the WF of 1 L of milk has been estimated at approximately 1020 L of water, which considers water used for feed production, drinking water and servicing (Mekonnen & Hoekstra, 2012). Water use can be classified into two categories: consumptive and non-consumptive use. Consumptive use involves removing water from its natural source, namely transforming it from liquid to vapour, thus reducing its availability in terms of place and time. Non-consumptive use (Fig. 2) returns almost all the water to its source, although there may be some changes in the timing of its availability (Monteiro et al., 2023; Perry et al., 2023).

It is necessary to quantify water consumption on farms in order to manage water use and reduce consumption efficiently. This quantification can be done directly through monitors, such as water meters, or through estimates based on models or reference values (Cullens, 2011; Monteiro et al., 2023).

Early assessments of the WF of dairy production primarily focused on volumetric WFs, such as litres of water per litre of milk. However, these evaluations often lacked comprehensiveness. They did not account for the types of water used (blue, green, grey) or the specific water scarcity conditions in different regions, necessitating more nuanced approaches to fully capture environmental impacts. This highlights the need for a more nuanced approach that recognizes the environmental impacts of water consumption (Ridoutt et al., 2010). Research indicates that farms with high WUE require fewer resources (water and land) to produce the same amount of milk. Moreover, effective farm

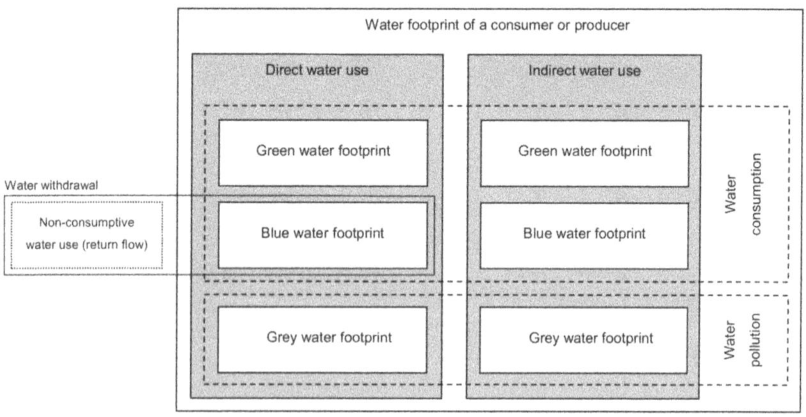

Figure 2 Representation of the components of a water footprint (Hoekstra et al., 2011).

management practices play a more significant role in WUE than the type of production system (Nagypál et al., 2020).

The WF of an animal can be calculated using the formula of Mekonnen & Hoekstra's (2012) (Equation 1):

$$WF_{(a,c,s)} = WF_{feed(a,c,s)} + WF_{drink(a,c,s)} + WF_{serv\,(a,c,s)} \tag{1}$$

where WF(a,c,s) represents the WF of an animal by category (a), in country (c) and in production system (s) considering feed, drinking water and service-water consumption.

Researchers use the total WF value per kilogram of fat- and protein-corrected milk (FPCM) to calculate the WF in a dairy farm (Grossi et al., 2022) (Fig. 3).

The FPCM, expressed in kilograms per year, can be determined using the International Dairy Federation (IDF, 2010) equation, which accounts for milk yield, fat and protein content (Equation 2) (Velarde-Guillén et al., 2023):

$$\begin{aligned} FPCM\left(\frac{kg}{year}\right) &= Milk\,yield\left(\frac{kg}{year}\right) \\ &\times \left(0.1226 \times Milk\,fat(\%) + 0.0776 \times Milk\,protein(\%) + 0.2534\right) \end{aligned} \tag{2}$$

Alternatively, Hodúr et al. (2022) used the FAO (2021) formula (Equation 3) to determine energy-corrected milk (FPCM):

$$\text{FPCM} = \text{Milk amount} \times \left(0.337 + 0.116 \times \text{fat\%} + 0.06 \times \text{protein\%}\right) \tag{3}$$

The WF of milk production can be estimated using the methodology set out by Mekonnen and Hoekstra (2012). According to this approach, the WF of milk production (M) is calculated by dividing the sum of WFs for feed (F), consumption (C) and services (S) by the annual FPCM, as shown in Equation 4:

Figure 3 Overall milk water footprint. Source: Adapted from: Grossi et al. (2022). FPCM = Fat and Protein Corrected Milk.

$$WF_M\left(\frac{m^3}{year}\right)=\frac{WF_F+WF_C+WF_S}{FPCM\left(\frac{\frac{kg}{farm}}{year}\right)} \tag{4}$$

The green, blue and grey WFs have been estimated by Mekonnen and Hoekstra (2012) for different countries, taking into account the type of production system (grazing, mixed and industrial) and differences in feed conversion efficiency, local climate, soil conditions and irrigation data. These authors estimated a total WF value of 1299 litres of water (Lw) per kg of FPCM, of which 91% was attributed to green water use, 5% to grey water use and 4% to blue water use (Shine et al., 2020a).

Green water was responsible for around 90% of the total WF. Feed production was the main driver of blue water use, accounting for up to 86% of total blue water use (Mekonnen & Hoekstra, 2010; Mekonnen & Hoekstra, 2012; Sultana et al., 2014; Shine et al., 2020a).

The WF of an animal related to the feed it consumes consists of two parts: the WF of the various feed ingredients and the water that is used to mix the feed. Feed footprint calculations include many variables and can be calculated by considering feed ingredients, volume and composition of feed, feed conversion efficiencies and estimation of feed composition (Mekonnen & Hoekstra, 2012).

Drinking water consumption (DWC) per lactating cow (litres/day) can be calculated using by equation set out by Meyer et al. (2004) (Equation 5):

$$DWC=-26.12+1.516\left(t_{average}\right)+1.29\left(m_{milk\ prod}\right)+0.058\left(m_{bodyweight}\right)+0.46\ m_{Na} \tag{5}$$

where $t_{average}$ is the average temperature value, $m_{milk\ prod}$ is the daily milk yield (kg/day), $m_{bodyweight}$ is the bodyweight of animals (kg) and m_{Na} is the quantity of daily sodium intake (g/day).

The Grey Water Footprint (WF_{grey}) was estimated by Hodúr et al. (2022) based on Equation 6 as the water volume required to dilute pollutants in wastewater to acceptable levels:

$$WF_{process,gray}=\frac{L}{\left(c_{\max-c_{nat}}\right)}=\frac{\left(Effl\times c_{effl}-Abstr\times c_{act}\right)}{\left(c_{\max-c_{nat}}\right)} \tag{6}$$

where $WF_{process,grey}$ is the grey WF of a process, it shows the pollutant load (mass/time), c_{max} is the maximum acceptable concentration of pollutant (mass/volume), c_{nat} is the natural concentration in the receiving water body (mass/

volume), Effl is the effluent volume/time, c_{effl} is the concentration of the pollutant in the effluent (mass/volume), Abstr is the water volume of the abstraction (volume/time), and c_{act} is the actual concentration of the intake water (mass/volume).

As mentioned in Section 2, assessment methods are approached through two different frameworks: the LCA and the Water Footprint Network (WFN) (Pfister et al., 2017; Bai et al., 2018).

The LCA protocol can be applied to analyse water consumption in raw milk production from cradle to farm gate, while also assessing the stress-weighted water scarcity footprint (WSF) and water scarcity productivity (WSP) (Liao & Su, 2019). The WSF is used to assess the potential environmental impacts of water scarcity resulting from an enterprise's activities within a specific area, such as a watershed or basin, without accounting for water quality (Bai et al., 2018) (Equation 7):

$$WSF = \sum_{i=0}^{n} = (\alpha SF, i \times Vi) \tag{7}$$

The WSF (m^3 H_2O-eq); Vi (m^3) represents freshwater consumption per unit time in position i; αSF,i is for characteristic factors corresponding to Vi; i is for different geographical location, e.g. watershed, region, nation, etc.

In their study, Liao and Su (2019) combined the WSI concept with WSP (Fig. 4).

Water productivity (WP) was defined as the physical quantity or economic value produced by a unit of water (Cai and Rosegrant, 2003). WP with the WSI as the equation can be combined by the formula used by (Liao and Su, 2019) (Equation 8):

$$\mathrm{WSP}\left(\mathrm{kg}/m^3\right) = P_i(\mathrm{kg}) / /\mathrm{WC}_i\left(\mathrm{m}^3\right)\mathrm{WSI}_i \tag{8}$$

WSP (kg/m^3) represents water scarcity productivity; P_i (kg) is for FPCM production per unit time in position; WC_i (m^3) is for water consumption per

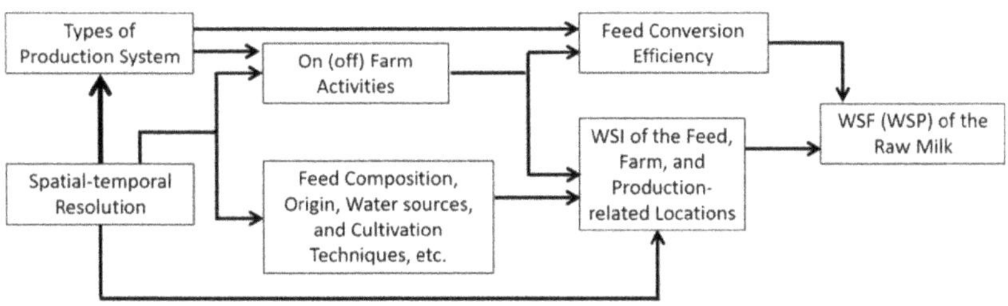

Figure 4 Factors that might affect the water scarcity footprint (WSF) or water scarcity productivity (WSP) of raw milk. Source: Adapted from: Liao and Su (2019).

unit time in position; WSI_i is for characteristic factors corresponding to P_i; *i* is for different geographical location (e.g. watershed, region, nation etc.). WSP can be increased by either increasing FPCM yield or reducing WC and maintaining the yield level or changing the production location to a low WSI region.

Some studies (Owusu-Sekyere et al., 2017; Ibidhi & Ben Salem, 2020) have associated the WF indicator with the economic water productivity (EWP) of livestock products, which is used to calculate the income generated per cubic meter of water used.

Studies have also used two other WF methods (Boulay et al., 2018; Higham et al., 2024):

- the WFN-based Blue Water Footprint Impact Index (WFIIblue) and
- the Available WAter REmaining (AWARE) water scarcity footprint (WFAWARE).

The Blue Water Footprint Impact Index (WFIIblue) is a method for assessing the water scarcity impact of blue water consumption during the production of a product. This method, based on the WFN approach, first calculates the consumptive blue water footprint (WFblue), which is the volume of blue water used per unit of product (m^3/kg of FPCM). In this method, WSblue becomes 1 when all the available blue water in the region has been consumed, significantly impacting the environmental water requirements. When WFIIblue is calculated in this way, it helps differentiate the water scarcity impacts of blue water consumption across regions with varying water resources and highlights areas where water scarcity might be more pronounced due to increased consumption (Higham et al., 2024).

The AWARE method is a recommended approach for assessing the impacts of water scarcity. It quantifies the relatively available water remaining per unit area after meeting the demands of humans and aquatic ecosystems. This method addresses the question: 'What is the potential to deprive another user (human or ecosystem) when consuming water in this area?'. The resulting characterization factor (CF) ranges from 0.1 to 100 and is used to calculate water scarcity footprints, as defined by the ISO standard. AWARE is based on the available water remaining in each watershed (measured relative to the global average) after accounting for human and ecosystem water requirements (Higham et al., 2024).

Waltner et al. (2023) quantified the total water footprint (WFanimal) of dairy cattle on a per-animal basis, focusing on climatic factors such as Temperature-Humidity Indices (THIs). The WFanimal was calculated as the sum of indirect water consumption (WFfeed) and direct water use for drinking (WFdrink) and servicing (WFserv). For crops such as maize, alfalfa hay and barley, green WFs were estimated using evapotranspiration (ET) and yield data.

Water consumption by dairy cows can be influenced by various factors, such as dry matter intake, milk production, feed composition (dry matter, protein content) and climate. Water temperature also affects water consumption with higher intakes resulting from temperatures that are closer to body temperature (Meyer et al., 2004; Erina et al., 2013; Sultana et al., 2014; Golher et al., 2021; Salcedo et al., 2022).

Dry matter intake has an impact on the frequency and duration of water intake, with evidence that an increase in dry matter intake also increases water consumption. Production systems also have an impact on water use, with small-scale farms and those with lower production showing higher water consumption per unit of milk (Sultana et al., 2014).

Dairy cattle require approximately 80 litres of water/cow/day (ranging from 14 L/day to 171 L/day) for animal consumption, and if the volume of water for cleaning facilities, equipment and (in some cases) cooling the environmental temperature is added, the result could be a consumption of water use between 130 L/day/cow and 195 L/day/cow (Meyer et al., 2004; Nagypál et al., 2020; Naranjo et al., 2020; Grossi et al., 2022).

Modelling water use on dairy farms provides the advantage of being less expensive than installing metering equipment. Models can be developed to predict animal water intake, parlor water use and green and blue water requirements for milk production.

Models obtained by multiple linear regression use several variables, such as those related to farm area, milk production, herd size, concentrates, grass growth, imported feed and fresh water consumption measured on the farm (direct water) to predict green and blue water consumption (Higham et al., 2017; Shine et al., 2020a). The variables collected from the farm must be easily obtainable and collected on a large scale without the use of specialized equipment.

The use of machine learning or deep learning algorithms could contribute to the development of water consumption prediction on dairy farms, especially if there is a large amount of data collected in a standardized way and at an international level, establishing a global model for predicting water consumption on dairy farms (Higham et al., 2017; Shine et al., 2018, 2020a; Osaki et al., 2024).

4 Water management technologies

As can be seen, several factors can influence WF, and it is possible to control these factors to reduce the WF either directly or indirectly. Reducing the WF should be included in the initial planning of a farm to optimise the efficient use of water through the installation of equipment and the construction of facilities that consider the use of water, waste water treatment and irrigation systems.

To minimise water losses from drinkers installed in facilities, it is essential to carefully select the type, number and location of these devices. In cattle housing, the use of constant-level drinkers is recommended to reduce water wastage over the use of valve drinkers. In addition, any drinker, whether a trough or a bowl, must allow each animal to consume water at a minimum rate of 10 L/min (Flaba et al., 2014).

Hydraulic waste dragging systems streamline floor cleaning operations but significantly increase effluent production. Water consumption in cleaning dairy cow facilities can be minimized by optimizing floor and equipment washing procedures. For instance, effective strategies to clean efficiently and quickly include using hoses with flow-controlled nozzles or employing high-pressure systems (USDA, 2012). Additionally, incorporating V-shaped floors with a 3% slope and central drains or perforated rough flooring with a 0.5% perforation for urine removal or separation further reduces water usage by enabling the use of mechanical dragging systems (Monteiro et al., 2023).

Pre-cooling milk with water using a plate cooler can reduce the energy required to cool milk by up to 50%, but it is important to consider the financial and environmental impact of the water use (Shine et al., 2020b).

Many dairy farmers recycle the water used to pre-cool the milk for farm washing or animal drinking, provided it is potable. Some producers pre-cool their milk without recycling the water, while others don't use a milk pre-cooling strategy at all. Consequently, the milk pre-cooling method chosen will affect the total water and energy consumption associated with cooling the milk, depending on factors such as the water/milk ratio of the plate cooler, the type of wash hose used, the size of the parlour, the size of the plate cooler and the volume of the wash water tank (Shine et al., 2020a).

Technologies such as rainwater harvesting and wastewater treatment have been shown to reduce reliance on groundwater and minimise environmental impact (Willers et al., 1999; Nagypál et al., 2020).

Effluent management technologies include physical-chemical separation of solids, anaerobic or aerobic treatment, and vegetative treatment. Solids can be separated using press or centrifuge equipment, with or without chemical agents. Anaerobic treatment can be done using digesters or anaerobic lagoons, while aerobic treatment can be performed in lagoons or tanks with various aeration methods. These treatment systems for dairy wastewater offer dual benefits of water reuse and pollution mitigation. For example, aerobic fermentation has been used to convert liquid waste into organic soil conditioners, reducing nutrient run-off and pollution risks (Harlia et al., 2024). Likewise, woodchip filters have been effective in removing organic matter and nutrients, making effluent suitable for reuse in farm operations (Ruane et al., 2011). Reverse osmosis technology in dairy plants has similarly demonstrated substantial reductions in wastewater volume and enhanced water reuse efficiency (Vourch et al., 2008). Vegetative treatment involves using rooted plants in constructed wetlands or floating aquatic plants in lagoons (Burton & Turner, 2003; Sommer et al., 2013; USDA, 2012).

Advances in irrigation technology contributed to more sustainable irrigation practices, such as automated precision irrigation and sensor-based monitoring, which have significantly reduced water usage on dairy farms by optimizing water use in dairy pastures, enhancing productivity while reducing water waste (Jordan et al., 2021; Monteiro et al., 2021; Vallejo-Gómez et al., 2023). For example, sensor-based systems and crop models optimize irrigation efficiency by addressing spatial and temporal water demand variability (Hills et al., 2016).

Remote sensing provides significant advantages in efficient irrigation management. Techniques such as using satellite imagery (e.g. Landsat) to estimate ET and map water depth enable the development of intelligent systems that generate prescription maps for variable rate irrigation, thereby optimizing water application (Farg et al., 2017; Mendes et al., 2019). Thermal and multispectral images obtained by unmanned aerial vehicles or drones also help identify different irrigation needs that allow improved irrigation management (Cozzolino, 2017).

Wireless sensor technologies have advanced, providing low-cost, energy-efficient smart sensors to monitor soil moisture and incorporate meteorological data. Wireless sensor networks (WSNs) consist of spatially distributed devices connected wirelessly, often using technologies such as ZigBee, LoRa and GPRS for irrigation control due to their range, cost, energy efficiency and reliability. These sensors measure humidity, temperature and soil moisture, automate irrigation, enable real-time control and reduce water consumption to increase WUE (Aqeel-Ur-Rehman et al., 2014; Jawad et al., 2017; Ojha et al., 2015).

The integration of technological innovations can lead to significant improvements in both environmental sustainability and farm productivity. While some of these technologies resulted in increased water and energy consumption, the overall benefits in terms of milk yield, quality and animal health can be outweighed by additional resource usage. There is potential for precision livestock farming (PLF) and big data and artificial intelligence (AI) technologies to reduce the WF of dairy farming in the future (Lovarelli et al., 2024; Neethirajan, 2024). For example, the integration of PLF systems, such as sprinklers that activate automatically based on cow presence, demonstrated significant decrease in water consumption with the potential to reduce consumption from water sprinklers up to 25% (Lovarelli et al., 2024).

5 Case studies

Water is a critical resource in dairy farming that is utilized at every stage of production from feed cultivation to animal drinking and facility cleaning. As freshwater availability becomes increasingly constrained due to climate change and population growth, managing the WF in dairy systems has become crucial. This section presents several case studies from diverse geographies, methodologies and innovative solutions that optimize water use and improve sustainability in dairy farming.

One of the studies on WF in dairy farming was conducted by Palhares and Pezzopane (2015), who evaluated the WF of conventional and organic dairy production systems in Brazil. Their study revealed that green water was the largest contributor to the total WF for both systems, and accounted for 39–57% in conventional systems and 32–59% in organic systems. Irrigation water dominated blue water use, with 95% in conventional and 96% in organic systems. Grey WF, calculated using phosphorus as an element, were 1.5 and 1.9 times higher for conventional and organic, respectively. The blue water scarcity index was slightly higher for organic systems (0.13) than for conventional (0.11). The study concluded that WF values alone do not fully capture the environmental impacts, as these depend on regional water availability and production factors. The study emphasized the importance of improving irrigation practices and optimizing green and blue water use to enhance water efficiency. Despite being in high rainfall regions, both systems faced green water scarcity with an index of 1.1. The study identified key management strategies to enhance WUE such as improving irrigation practices, implementing best management practices to optimize green and blue water use and reducing grey water emissions through better nutrient and wastewater management.

Nutritional interventions also play a crucial role in reducing WF. Palhares et al. (2019) evaluated dietary strategies in Brazilian dairy systems and found that adjusting protein levels in feed from 23% to 14.5% reduced the total WF by 10%, with green water consumption accounting for most of the savings. This study highlights the potential of feed optimization in improving WUE.

Liao & Su (2019) explored the environmental impacts of milk production in Taiwan through a LCA approach. Their analysis incorporated the WSF and WSP across five farms. Results revealed significant variability in water use, with WSF ranging from 2.2 H_2Oeq/kg to 44.8 H_2Oeq/kg FPCM, depending on regional water stress levels. Farms with lower stress levels achieved a higher WSP of 0.749 kg FPCM/m^3 water, underscoring the importance of targeting water scarcity hotspots in dairy farming. The study applied the methodology developed by Ridoutt et al. (2010), which focuses on assessing water scarcity by prioritizing blue water consumption and its availability for human and environmental needs. This method is distinguished between blue, green and grey water, highlighting that only blue water directly contributes to scarcity, while green water is linked to land use impacts and grey water reflects water quality degradation.

In arid regions like Kuwait, Al-Bahouh et al. (2020) assessed the blue and grey WFs of confined dairy farms, finding significant seasonal variations. The blue WF was higher during the summer (54.5 L/kg FPCM) compared to winter (19.2 L/kg FPCM), while the grey WF, influenced by phosphate concentration in effluent, averaged 23.0 L/kg FPCM. The study emphasized the need for sustainable water management in arid regions, and recommended the adoption of best management practices to enhance WUE.

In Tunisia, Ibidhi and Ben Salem (2020) evaluated the WF and EWP of milk production in eight dairy farms. The study focused on direct (drinking, servicing and cooling water) and indirect (feed production) water consumption. The researchers found that the average WF of 1 kg of FPCM was 1.36 m^3. The green water accounted for 0.93 m^3 and blue water 0.42 m^3. Feed production represented 87% of the total WF, while drinking and servicing water contributed 3.75% and 9%, respectively. The EWP of milk was relatively low, averaging \$0.05/m^3 of water. The authors also analysed scenarios to reduce WF, such as using feed with lower water requirements and improving WUE. This could result in up to 16% reductions in consumptive water use. The authors calculated that efficient use of servicing water could reduce blue WF of milk by up to 4%.

In a study by Al-Bahouh et al. (2021), the blue WF of milk production in Ontario, Canada, was assessed using the WFN methodology. Various water conservation options were estimated by the authors using the AgriSuite software. The study demonstrated that adaptive water conservation practices, such as reusing plate cooler and milk house water, could reduce total water use to 182.7 L/cow/day (a 25.8% reduction) and lower the blue WF to 5.8 L/kg FPCM (a 21.6% reduction). These measures achieved a 77.7% reduction in milk house wash water use from 74.3 L/cow/day to 16.6 L/cow/day.

In the Anand district of India, Saha et al. (2022a) studied effective water use (EWU) while focusing on direct water consumption. They revealed that a single lactating animal consumed an average of 89.02 litres daily, including drinking water, water in feed and cleaning. Unlike traditional approaches that include indirect water consumption such as the water used in producing fodder, this study focused on calculating the moisture content of feed and fodder directly consumed by animals, thereby isolating EWU for milk production. Similarly, the same research group (Saha et al., 2022b) explored the WF of milk production in Banaskantha, Gujarat, focusing on direct water consumption for drinking, feed and cleaning. The study found that a lactating animal consumes an average of 57.34 litres of water per day, including drinking water and water consumed through feed fodder, a lactating animal requires 98.13 litres of water per day in Banaskantha. Daily cleaning water consumption per dairy animal is 12.14 L. The study showed that drinking water has a significant impact on milk production of lactating animals.

In Hungary, Hodúr et al. (2022) investigated, with a focus on milking technologies, the factors contributing to differences in water use and WFs across three large-scale dairy farms. The study revealed that the parallel milking system had the highest blue WF, while the robotic milking system consumed significantly less water. However, the robotic system exhibited the highest grey WF due to increased cleaning requirements. The authors concluded that, from the perspective of minimizing the overall WF, the polygon milking arrangement was the most advantageous compared to both the parallel and robotic milking systems. The findings highlight significant potential for further improvements in milking technology to enhance water efficiency.

Grossi et al. (2022) investigated the impact of cooling management on the WF of milk production in dairy cows. The mean annual WF of the farms was 805 ± 225 litres of water per kg of FPCM, with green water representing the largest share (67.9%) of the total WF. The study found that cooling operations had a minimal impact on the overall milk WF, accounting for only 0.04%.

In Bangladesh, Uddin et al. (2022) applied the methodology developed by the International Farm Comparison Network, which is based on three key pillars: the Typical Farm Approach, the Technology Impact Policy Impact Calculations model and the Dairy Networking Approach (DNA). This methodology was used to evaluate WUE in dairy production systems and to identify the main drivers influencing WUE. The study revealed that approximately 98% of the water required for milk production originated from feed and fodder cultivation, while only 2% was attributed to drinking and service water. Additionally, larger, more productive farms demonstrated higher WUE, achieving values of 0.26 compared to 0.22 observed in smaller farms.

In Hungary, Waltner et al. (2023) studied the interplay between green and blue water use in dairy farming. Feed production accounted for 99.1% of the total WF, with maize silage contributing significantly to green water consumption. Drinking water needs ranged from 74.7 L/cow/day to 101.9 L/cow/day and were influenced by THIs. Servicing water, calculated at 18 L/cow/day, contributed to blue water use. The green water consumption from feed at the studied cattle farm averaged 13 352 lL/cow/day, emphasizing the dominant role of feed production in the overall WF. The authors provided a detailed methodological framework for evaluating both direct and indirect water consumption. The study recommended adaptive strategies, such as adopting thermotolerant breeds and optimizing feeding practices, to mitigate heat stress and improve WUE.

Velarde-Guillén et al. (2023) analysed WF in the arid Peruvian central coast, finding that feed production accounted for 99% of the total WF, with green and blue water contributing 60% and 30%, respectively. The overall WF of dairy production across five dairy farms was calculated to be 0.66 m^3/kg FPCM. The authors stated that imported feed represented a significant portion of the WF, highlighting the need for local, low-water-demand resources to enhance sustainability.

Bronts et al. (2023) analysed the WF of dairy system in the Netherlands and Spain, highlighting the impact of system boundaries and indicator choices on results. The Dutch system demonstrated lower WF values – 0.62 m^3 (green), 0.09 m^3 (blue) and 0.14 m^3 (grey) per kg of milk – due to higher efficiency. In contrast, the Spanish system showed slightly higher figures, with WF values of 0.67 m^3 (green), 0.15 m^3 (blue) and 0.09 m^3 (grey) per kg of milk. Organic systems, characterized by lower efficiency, exhibited larger footprints compared to the Dutch conventional system. Expanding system boundaries to include calves resulted in an 8–15% increase in carbon footprint. The study also noted that green water dominated total WFs, an aspect often excluded in LCA studies. Regarding grey WFs, the authors stated that earlier research included only nitrogen; however, they added that including pesticides in the analysis could yield less favourable outcomes for systems reliant on feed crops rather than grasslands. Additionally, the authors found that water quality standards significantly influenced grey WF calculations.

Expanding the scope of analysis, Rebolledo-Leiva et al. (2024) applied advanced frameworks integrating the water–energy–food (WEF) nexus to assess sustainability in dairy systems. They developed a novel WEF+Circularity Indicator (CEi) that combined water, energy and carbon footprints with food productivity metrics. Applied to 30 dairy farms in Galicia, Spain, the study identified farms with higher circularity percentages as more sustainable. This approach not only incorporated a life cycle perspective but also proposed a framework for guiding a more sustainable and circular food industry. Similarly, Du et al. (2024) studied the eco-efficiency of milk production systems through the water–energy–labour–food nexus in China. The study found that as milk production increased, the resource burdens of large-scale farms were relatively weakened, with the WF dropping by 17% to 8.0 L/kg. The authors' results showed that the resource investment pattern in scale farms shifted from water- and labour-oriented to energy-oriented. Advanced mechanization and on-farm energy production contributed to a 24% improvement in eco-efficiency, underscoring the role of technological innovation in reducing resource burdens.

In Brazil, Palhares et al. (2024) conducted a comprehensive water consumption assessment on 876 dairy farms from January 2021 to December 2022. The study categorized farms into pasture-based, semi-closed and closed systems, analysing water consumption for drinking and cleaning, focusing on two key indicators: litres of water per lactating cow per day (L water/cow/day) and litres of water per kilogram of milk produced (L water kg/milk/day). Closed systems showed the highest water consumption, with an average of 87.5 L/cow/day for drinking and 84.4 L/cow/day for cleaning. In contrast, pasture-based systems showed lower water usage, while semi-confined systems showed the lowest average for drinking water (54.4 L water/cow/day) and pasture systems recorded the lowest cleaning water usage (45.2 L water/cow/day). These findings highlight the need for site-specific water management strategies tailored to farm operations. Methodologically, the researchers implemented continuous monitoring using analogue water meters and adopted some statistical approaches to ensure data reliability.

Similarly, Higham et al. (2024) examined the WF of pastoral dairy farming in New Zealand, analysing 88 farms across three regions. Using the WFN-based blue WF impact index (WFIIblue) and the AWARE water scarcity footprint (WFAWARE), the study identified significant regional disparities. For example, Canterbury's irrigated farms exhibited up to five times higher blue WF compared to non-irrigated farms in Manawatu and Waikato due to lower rainfall and higher irrigation demands. The study emphasized the importance of using local data and catchment-scale analyses to capture spatial and temporal variations in water availability.

In the Mediterranean region, Ruiz-Colmenero et al. (2024) examined cow milk production in Catalonia, emphasizing the critical role of water availability. Their study revealed that feed production was the largest contributor to the farm stage's WF, particularly when irrigating crops like maize. They proposed strategies such as incorporating low-water-scarcity feed ingredients, improving irrigation systems and enhancing supply chain traceability to mitigate water scarcity impacts.

Firdayati et al. (2024) conducted a holistic analysis of water use in Indonesian dairy farms, incorporating direct and indirect water inputs, including rainwater and grey water reuse. Dairy cattle demonstrated higher WP (0.08 kg milk/L) compared to beef cattle (0.0297 kg beef/L), excluding water required for growing feed. The study's integrated approach emphasized the importance of combining multiple water sources and reuse strategies to enhance sustainability.

Mohammadi et al. (2024) evaluated livestock WF in Tehran province of Iran, revealing that milk production had the highest WF, averaging 13 007 m^3/ton. Their recommendations included recycled water use and wastewater treatment development to enhance sustainability. In this study, the authors used the methodology introduced by Mekonnen and Hoekstra (2012) for a comprehensive analysis of the WFs associated with various livestock productions is conducted.

In Brazil, Osaki et al. (2024) presented a model based on artificial neural networks to predict water consumption in dairy farms. The model demonstrated good accuracy in predicting water consumption, with an average absolute error of 28.4%. The study suggested that even with limitations on input variables, it was possible to obtain some prediction, although not totally accurate.

Neethirajan (2024) evaluated the transformative potential of big data and AI in advancing the dairy industry towards net zero emissions, with a focus on the Canadian dairy sector. The study highlighted AI-based predictive analytics for optimizing feed efficiency, real-time monitoring through sensor networks and improved energy utilization as key applications that indirectly influence the WF of dairy farming. The author further emphasized that refining processes such as feed management and manure handling can help reduce overall resource use, including water. This research underscored the broader applicability of digital technologies in agriculture, aligning with global efforts towards environmental sustainability and minimizing water usage in dairy production systems.

Fan et al. (2025) explored the environmental impacts of recombining livestock and crop production through recycling animal manure. The study found that full incorporation of manure significantly reduced the WF by 29% compared to semi-incorporation. The improved full incorporation mode, designed based on the water–energy–food nexus, further reduced the WF by 34%, in addition to reductions in energy use and GHG emissions. The authors employed a methodology for calculating the WF in livestock systems based on the WF Assessment Manual (Hoekstra et al., 2011), while excluding the green WF due to its negligible environmental impact.

In Ontario, in Canada Figliuzzi's (2025) study analysed various factors affecting the milking and washing processes, focusing particularly on the volume of water used and its environmental impacts. The study highlighted that the type of milking system, such as tie-in stalls or robotic systems, significantly affects water usage and therefore the farms' water footprint (MCWW). In particular, tie-in stall systems, which are most common in Ontario, tend to use higher water volumes per cow, resulting in higher water consumption per unit of milk production.

These studies collectively emphasize the complexity of assessing and managing the WF in dairy farming. Regional factors, methodological choices and technological advancements play critical roles in determining WF outcomes. By adopting integrated strategies, optimizing feed production and improving WUE, the dairy industry can address sustainability challenges while minimizing environmental impacts.

6 Lessons from case studies

The WF of dairy farming is a crucial metric for evaluating the sustainability of milk production systems. One of the primary concerns in dairy farming WF determination is changes and differences of water use among green, blue and grey water.

The Water Footprint Standard (Hoekstra et al., 2011) is the method most often chosen by researchers because of its popularity and its ability to distinguish between green, blue and grey water. Due to the current and future impacts of climate change, it is necessary to analyse the water cycle in both the agricultural and housing sectors in order to optimize the process chains in the face of increasing water scarcity. Increasing WP and maximizing WUE in milk production are two strategies for minimizing climate change impacts (Drastig et al., 2010).

In pastoral systems with relatively high rainfall, water scarcity is rarely a concern, but the production of grey water due to pollution remains a significant issue. The presence of excess nitrogen and phosphorus from manure runoff contributes to water contamination, requiring better management practices (Ledgard et al., 2019; Varma et al., 2021). Advanced wastewater treatment technologies can be a potential solution for reducing pollutant loads (Vaishnav et al., 2023)

For intensive dairy units in arid and semi-arid regions, the foremost priority is reducing blue water consumption. As case studies from Kuwait and Tunisia (Al-Bahouh et al., 2020; Ibidhi & Ben Salem, 2020) highlight, drinking and servicing water contribute significantly to the total blue WF in hot climates. Precision irrigation, efficient cooling strategies and feed optimization are essential in these systems to mitigate water stress (Palhares et al. 2019; Ibidhi & Ben Salem, 2020; Grossi et al., 2022; Waltner et al., 2023; Velarde-Guillén et al., 2023). Sensor-based irrigation systems, which adjust water delivery based on real-time soil and crop conditions, have proven effective in minimizing blue water use (Pramanik et al., 2022).

The case studies reviewed emphasize the importance of management interventions in reducing the WF across various dairy production systems. Feed production remains the dominant contributor to overall water consumption. Strategies such as selecting low-water-requirement crops, optimizing feeding efficiencies and utilizing alternative sources can help reduce the indirect WF associated with feed (Palhares et al., 2019; Waltner et al., 2023).

A key takeaway from the studies is the need for integrated assessment methods. Combining WF evaluation with economic and environmental sustainability metrics – such as the WEF or circular economy approaches – can provide a more comprehensive understanding of dairy farm sustainability (Du et al., 2024; Rebolledo-Leiva et al., 2024).

To enhance the clarity and coherence of the analysis, mitigation strategies can be categorized into three key components: feed management, animal management and dairy unit efficiency (Table 1). By implementing these strategies, the dairy industry can improve resource efficiency, minimize environmental impact and contribute to global water sustainability goals.

The water-related processes involved in the life cycle of dairy production are complex and diverse but have not been thoroughly investigated in a

Table 1 Categorized mitigation strategies for reducing water footprint in dairy farming

Categories	Grey WR* strategies	Blue WR strategies
Feed management	Nutrient management, optimized manure application, alternative feed sources	Drought-resistant feed crops, precision irrigation
Animal management	Improved feed conversion efficiency, optimized stocking rates	Selective breeding for water efficiency, behavioural adaptation strategies
Dairy unit efficiency	Wastewater treatment, effluent recycling	Water recycling, energy-efficient cooling

*WR: water reduction.

comprehensive, site-specific manner. While research on water use in dairy farming does exist, it often focuses on general estimates rather than detailed, site-specific data that takes into account variations in climate, soil type and farming practices.

The challenge of reducing the WF of dairy production will require a multidisciplinary approach that integrates technological innovation, best management practices and supportive policies. The application of artificial intelligence and big data in dairy management is promising for predictive optimization of water use. Future research should focus on regionalized strategies adapted to specific climatic and environmental conditions, with systems to monitor on-farm consumption in real time, assess the impact of climate change and develop adaptation strategies to ensure that sustainability measures are in line with the availability of local resources.

7 Conclusion

The WF of dairy farming is a critical metric for assessing the sustainability of milk production systems globally. This chapter has outlined key methodologies and frameworks for evaluating the green, blue and grey WFs, highlighting the importance of integrating these metrics with local environmental and climatic conditions. Feed production remains the dominant contributor to the total WF, underscoring the need for sustainable crop and feed management practices.

Technological advancements, such as sensor-based irrigation, PLF and wastewater recycling, present significant opportunities to reduce water use while improving resource efficiency. Case studies from diverse geographies demonstrate the variability in water consumption patterns and the importance of region-specific strategies to enhance WP.

Future efforts should focus on the adoption of innovative water management technologies, the development of comprehensive lifecycle assessments and the integration of artificial intelligence and big data to predict and optimize water usage in dairy systems. By prioritizing these measures, the dairy industry can achieve greater environmental sustainability and contribute to the global agenda of water conservation.

8 Acknowledgements

The authors thank the Portuguese Foundation for Science and Technology (FCT) for the financial support to the Research Centre for Natural Resources, Environment and Society – CERNAS (UIDB/00681).

9 Where to look for further information

A recommended website is the Water Footprint Network (https://www.waterfootprint.org/).

Key literature includes:

- Hoekstra, A., et al. (2011). *The water footprint assessment manual: Setting the global standard*. Earthscan.
- Mekonnen, M. M. & Hoekstra, A. Y. (2012). A global assessment of the water footprint of farm animal products. *Ecosystems*, 15(3), 401–415. https://doi.org/10.1007/S10021-011-9517-8/.
- Hoekstra, A. Y. (2017). Water footprint assessment: evolvement of a new research field. *Water Resources Management*, 31(10), 3061–3081. https://doi.org/10.1007/S11269-017-1618-5.

10 References

Al-Bahouh, M., et al. (2020). Blue and grey water footprints of dairy farms in Kuwait. *Journal of Water Resource and Protection*, 12(7), 618–635. https://doi.org/10.4236/jwarp.2020.127038.

Al-Bahouh, M., et al. (2021). Blue water footprints of Ontario dairy farms. *Water*, 13(18), 2230. https://doi.org/10.3390/w13162230.

Aqeel-Ur-Rehman, A. Z., et al. (2014). A review of wireless sensors and networks' applications in agriculture. *Computer Standards & Interfaces*, 36(2), 263–270. https://doi.org/10.1016/J.CSI.2011.03.004.

Bai, X., et al. (2018). Comprehensive water footprint assessment of the dairy industry chain based on ISO 14046: a case study in China. *Resources, Conservation and Recycling*, 132, 369–375.

Boulay, A. M., et al. (2018). The WULCA consensus characterization model for water scarcity footprints: assessing impacts of water consumption based on available water remaining (AWARE). *The International Journal of Life Cycle Assessment*, 23, 368–378.

Bronts, S., et al. (2023). The water, land, and carbon footprint of conventional and organic dairy systems in the Netherlands and Spain: a case study into the consequences of ecological indicator selection and methodological choices. *Energy Nexus*, 11, 100217. https://doi.org/10.1016/j.nexus.2023.100217.

Burton, C. H. & Turner, C. (2003). *Manure Management: Treatment Strategies for Sustainable Agriculture*, Silsoe Research Institute, Bedford, p. 451.

Cai, X. and Rosegrant, M.W. (2003). Chapter 10–World Water Productivity: Current Situation and Future Options. In *Water Productivity in Agriculture: Limits and Opportunities for Improvement,* Kijne, J.W., et al., Eds.; CAB International: Wallingford, UK; pp. 163–178.

Chapagain, A. K. & Hoekstra, A. Y. (2003). Virtual water flows between nations in relation to trade in livestock and livestock products value of water. Research Report Series No. 13. https://www.waterfootprint.org/resources/Report13.pdf.

Chapagain, A. K. & Hoekstra, A. Y. (2004). Water footprints of nations value of water. Research Report Series No. 16. https://www.waterfootprint.org/resources/Report16Vol1.pdf.

Cozzolino, D. (2017). The role of near-infrared sensors to measure water relationships in crops and plants. *Applied Spectroscopy Reviews*, 52(10), 837–849. https://doi.org/10.1080/05704928.2017.1331446.

Cullens, F. (2011, October 8). Water use on dairy farms - MSU Extension. Michigan State University Extension. https://www.canr.msu.edu/news/water_use_on_dairy_farms.

Drastig, K., et al. (2010). Water footprint analysis for the assessment of milk production in Brandenburg (Germany). *Advances in Geosciences*, 27, 65–70. https://doi.org/10.5194/ADGEO-27-65-2010.

Du, X., et al. (2024). Assessing the eco-efficiency of milk production systems using water-energy-labor-food nexus. *Science of the Total Environment*, 955, 176812.

Erina, S., et al. (2013). Research on drinking behavior in the feed consumption time in romanian black and white primiparous cows. *Scientific Papers: Animal Science and Biotechnologies*, 46(1), 387–390.

FAO (2021). Fat–Corrected Milk. Available from: https://web.archive.org/web/20221210141330/. https://www.fao.org/dairy-production-products/resources/glossary/en/?index=f (accessed 29/09/2025).

Fan, X., et al. (2025). Water-energy-food nexus in the sustainable management of crop-livestock coupled systems. *Applied Energy*, 378, 124824. https://doi.org/10.1016/j.apenergy.2023.124824.

Farg, E., et al. (2017). Evaluation of water distribution under pivot irrigation systems using remote sensing imagery in eastern Nile delta. *The Egyptian Journal of Remote Sensing and Space Science*, 20, S13–S19. https://doi.org/10.1016/J.EJRS.2016.12.001.

Figliuzzi, T. (2025). *Characterization of Milking Centre Wash Water for Treatment Feasibility Using On-Site Septic Systems* (Master's thesis, University of Guelph, Guelph, Ontario, Canada). University of Guelph, Guelph.

Firdayati, M., et al. (2024). Study on the use of water and feed resources on beef cattle and dairy farms in Bandung Regency, West Java. *E3S Web of Conferences*, 485, 01007. https://doi.org/10.1051/e3sconf/202448501007.

Flaba, J., et al. (2014). The Design of Dairy Cow and Replacement Heifer Housing Report of the CIGR Section II Working Group. In *Report of the CIGR Section II Working Group N° 14 Cattle Housing*. International Commission of Agricultural and Biosystems Engineering. https://doi.org/10.6092/unibo/amsacta/4272.

Golher, D. M., et al. (2021). Factors influencing water intake in dairy cows: a review. *International Journal of Biometeorology*, 65(4), 617–625. https://doi.org/10.1007/s00484-020-02038-0.

Grossi, G., et al. (2022). Impact of summer cooling management on milk water footprint in dairy cows. *Journal of Cleaner Production*, 367, 133062. https://doi.org/10.1016/j.jclepro.2022.133062.

Harlia, E., et al. (2024). Management of water sources and liquid waste in dairy farming environments. *BIO Web of Conferences*, 123, 01047. https://doi.org/10.1051/BIOCONF/202412301047.

Higham, C. D., et al. (2017). Water use on nonirrigated pasture-based dairy farms: combining detailed monitoring and modeling to set benchmarks. *Journal of Dairy Science*, 100(1), 828–840. https://doi.org/10.3168/jds.2016-11822.

Higham, C. D., et al. (2024). The water footprint of pastoral dairy farming: the effect of water footprint methods, data sources, and spatial scale. *Water*, 16(3), 391. https://doi.org/10.3390/w16030391.

Hills, J., et al. (2016). Smarter Irrigation for Profit - Increasing Farm Profit Through Efficient Use of Irrigation Input to Dairy Pastures. *7th Australian Dairy Science Symposium (2016)*. Sydney, Australia 16-8.

Hodúr, C., et al. (2022). Blue and gray water footprint of some Hungarian milking parlors. *Water Practice & Technology*, 17(7), 1378. https://doi.org/10.2166/wpt.2022.073.

Hoekstra, A. Y. (2003). Virtual water trade. Proceedings of the International Expert Meeting on Virtual Water Trade, Value of Water Research Report Series No. 12. https://ihedelftrepository.contentdm.oclc.org/digital/collection/p21063coll3/id/10335/.

Hoekstra, A. Y. (2017). Water footprint assessment: evolvement of a new research field. *Water Resources Management*, 31(10), 3061–3081. https://doi.org/10.1007/S11269-017-1618-5/.

Hoekstra, A. Y. & Chapagain, A. K. (2007). Water footprints of nations: water use by people as a function of their consumption pattern. *Water Resources Management*, 21(1), 35–48. https://doi.org/10.1007/S11269-006-9039-X/METRICS.

Hoekstra, A. Y. & Chapagain, A. K. (2008). *Globalization of Water: Sharing the Planet's Freshwater Resources*, Blackwell Publishing, Oxford, pp. 1–208. https://doi.org/10.1002/9780470696224.

Hoekstra, A., et al. (2011). *The Water Footprint Assessment Manual: Setting the Global Standard*. Earthscan, Oxford.

Høgaas Eide, M. (2002). Life cycle assesswment (LCA) of industrial milk production. *International Journal of Life Cycle Assessment*, 7(2), 115–126. https://doi.org/10.1007/BF02978855/METRICS.

IDF. (2010). A Common Carbon Footprint for Dairy. The IDF Guide to Standard Lifecycle Assessment Methodology for the Dairy Industry. *Bulletin of the International Dairy Federation* 445. International Dairy Federation, Brussels, Belgium.

Ibidhi, R. & Ben Salem, H. (2020). Water footprint and economic water productivity assessment of eight dairy cattle farms based on field measurement. *Animal*, 14(1), 180–189. https://doi.org/10.1017/S1751731119001526.

Jawad, H. M., et al. (2017). Energy-efficient wireless sensor networks for precision agriculture: a review. *Sensors*, 17(8), 1781. https://doi.org/10.3390/S17081781.

Jordan, C., et al. (2021). Measuring the effect of improved irrigation technologies on irrigated agricultur: a study case in Central Chile. *Agricultural Water Management*, 257, 107160. https://doi.org/10.1016/J.AGWAT.2021.107160.

Ledgard, S. F., et al. (2019). Nitrogen and carbon footprints of dairy farm systems in China and New Zealand, as influenced by productivity, feed sources and mitigations.

Agricultural Water Management, 213, 155–163. https://doi.org/10.1016/j.agwat.2018.11.016.

Liao, W.-T. & Su, J.-J. (2019). Evaluation of water scarcity footprint for Taiwanese dairy farming. *Animals*, 9(11), 956. https://doi.org/10.3390/ani9110956.

Lovarelli, D., et al. (2024). Reducing life cycle environmental impacts of milk production through precision livestock farming. *Sustainable Production and Consumption*, 51, 303–314. https://doi.org/10.1016/j.spc.2024.01.013.

Mekonnen, M. & Hoekstra, A. Y. (2010). The green, blue and grey water footprint of animals and animal products, Value of Water Research Report Series No. 48. Unesco-IHE Institute for Water Education. https://research.utwente.nl/en/publications/the-green-blue-and-grey-water-footprint-of-animals-and-animal-pro.

Mekonnen, M. M. & Hoekstra, A. Y. (2012). A global assessment of the water footprint of farm animal products. *Ecosystems*, 15(3), 401–415. https://doi.org/10.1007/S10021-011-9517-8/.

Mendes, W. R., et al. (2019). Fuzzy control system for variable rate irrigation using remote sensing. *Expert Systems with Applications*, 124, 13–24. https://doi.org/10.1016/J.ESWA.2019.01.043.

Meyer, U., et al. (2004). Investigations on the water intake of lactating dairy cows. *Livestock Production Science*, 90(2–3), 117–121. https://doi.org/10.1016/J.LIVPRODSCI.2004.03.005.

Mohammadi, A., et al. (2024). Evaluation of Tehran Province livestock production from water footprint prospective. *Environmental Energy and Economic Research*, 8(2), S082. https://doi.org/10.22097/eeer.2024.437629.1311.

Monteiro, A., et al. (2021). Precision agriculture for crop and livestock farming–brief review. *Animals*, 11(8), 2345. https://doi.org/10.3390/ANI11082345.

Monteiro, A., et al. (2023). Efficient water use in dairy cattle production: a review. *The Open Agriculture Journal*, 17(1), e18743315270668. https://doi.org/10.2174/0118743315270668231127190323.

Murphy, E., et al. (2017). Water footprinting of dairy farming in Ireland. *Journal of Cleaner Production*, 140, 547–555. https://doi.org/10.1016/J.JCLEPRO.2016.07.199.

Nagypál, V., et al. (2020). Sustainable water use considering three Hungarian dairy farms. *Sustainability (Switzerland)*, 12(8), 3145. https://doi.org/10.3390/SU12083145.

Naranjo, A., et al. (2020). Greenhouse gas, water, and land footprint per unit of production of the California dairy industry over 50 years. *Journal of Dairy Science*, 103(4), 3760–3773. https://doi.org/10.3168/JDS.2019-16576/ATTACHMENT/3323AF58-6304-4415-B568-A319FBCE590B/MMC1.PDF.

Neethirajan, S. (2024). Net zero dairy farming–advancing climate goals with big data and artificial intelligence. *Climate*, 12(2), 15. https://doi.org/10.3390/cli12020015.

Ojha, T., et al. (2015). Wireless sensor networks for agriculture: the state-of-the-art in practice and future challenges. *Computers and Electronics in Agriculture*, 118, 66–84. https://doi.org/10.1016/J.COMPAG.2015.08.011.

Osaki, M. R., et al. (2024). Artificial neural network model for water consumption prediction in dairy farms. *Bioscience Journal*, 40, e40256. https://doi.org/10.14393/BJ-V40N0A2024-68845.

Owusu-Sekyere, E., et al. (2017). Economic water productivities along the dairy value chain in South Africa: implications for sustainable and economically efficient water-use Policies in the dairy industry. *Ecological Economics*, 134, 22–28.

Palhares, J. C. P. & Pezzopane, J. R. M. (2015). Water footprint accounting and scarcity indicators of conventional and organic dairy production systems. *Journal of Cleaner Production*, 93, 299–307. https://doi.org/10.1016/j.jclepro.2015.01.016.

Palhares, J. C. P., et al. (2024). Water performance indicators and benchmarks for dairy production systems. *Water*, 16(2), 330. https://doi.org/10.3390/w16020330.

Palhares, J. C. P., et al. (2019). Best practice production to reduce the water footprint of dairy milk. *Ambiente & Água - An Interdisciplinary Journal of Applied Science*, 14(6), 1–9. https://doi.org/10.4136/ambi-agua.2454.

Palhares, J. C. P., et al. (2020). Best practice production to reduce the water footprint of dairy milk. *Revista Ambiente & Água*, 15(1), e2454. https://doi.org/10.4136/AMBI-AGUA.2.

Pfister, S., et al. (2017). Understanding the LCA and ISO water footprint A response to Hoekstra (2016) "A critique on the water-scarcity weighed water footprint in LCA. Ecological Indicators, 72, 352-259. http dx.doi.og 10.1016 j.ecolind.2016.07.051.

Pramanik, M., et al. (2022). Automation of soil moisture sensor-based basin irrigation system. *Smart Agricultural Technology*, 2, 100032. https://doi.org/10.1016/j.ter.2022.100032.

Perry, C., et al. (2023). Water Consumption, Measurements and Sustainable Water User (Technical Report). https://watercommission.org/publication/water-consumption-measurements-and-sustainable-water-use/.

Rebolledo-Leiva, R., et al. (2024). Embedding water-energy-food nexus and circularity assessment for organization benchmarking: a case study for dairy farms. *Waste Management*, 189, 410–420.

Ridoutt, B. G., et al. (2010). Short communication: the water footprint of dairy products: case study involving skim milk powder. *Journal of Dairy Science*, 93(11), 5114–5117. https://doi.org/10.3168/JDS.2010-3546.

Robinson, A. D., et al. (2016). Usage and attitudes of water conservation on Ontario dairy farms. *Professional Animal Scientist*, 32(2), 236–242. https://doi.org/10.15232/pas.2015-01468.

Romaguera, M., et al. (2010). Potential of using remote sensing techniques for global assessment of water footprint of crops. *Remote Sensing*, 2(4), 1177–1196. https://doi.org/10.3390/RS2041177.

Ruane, E. M., et al. (2011). On-farm treatment of dairy soiled water using aerobic woodchip filters. *Water Research*, 45(20), 6668–6676. https://doi.org/10.1016/J.WATRES.2011.09.055.

Ruiz-Colmenero, M., et al. (2024). Challenges when assessing water-related environmental impacts of livestock farming: a case study of a cow milk production system in Catalonia. *Water*, 16, 1299. https://doi.org/10.3390/w16091299.

Saha, U. S., et al. (2022a). Water footprint of milk production in India: a case of Anand district of Gujarat (Working Paper No. 336). Institute of Rural Management Anand. https://www.irma.ac.in.

Saha, U. S., et al. (2022b). Water footprint of milk production in the Banaskantha district of Gujarat (Working Paper 337). Institute of Rural Management Anand. https://www.irma.ac.in.

Salcedo, G., et al. (2022). Water footprint of dairy farms according to typology of feeding. *ITEA Informacion Tecnica Economica Agraria*, 118(4), 547–564. https://doi.org/10.12706/itea.2021.040.

Shine, P., et al. (2018). Multiple linear regression modelling of on-farm direct water and electricity consumption on pasture based dairy farms. *Computers and Electronics in Agriculture*, 148, 337–346. https://doi.org/10.1016/j.compag.2018.02.020.

Shine, P., et al. (2020a). A global review of monitoring, modeling, and analyses of water demand in dairy farming. *Sustainability (Switzerland)*, 12(17), 7012. https://doi.org/10.3390/SU12177201.

Shine, P., et al. (2020b). Energy consumption on dairy farms: a review of monitoring, prediction modelling, and analyses. *Energies*, 13(5), 1288. https://doi.org/10.3390/EN13051288.

Sommer, S. G., et al. (2013). *Animal Manure Recycling: Treatment and Management*, Wageningen Academic Publishers, Wageningen, The Netherlands, pp. 1–364. https://doi.org/10.1002/9781118676677.

Sultana, M. N., et al. (2014). Comparison of water use in global milk production for different typical farms. *Agricultural Systems*, 129, 9–21. https://doi.org/10.1016/j.agsy.2014.05.002.

Uddin, M. M., et al. (2022). Estimation of water requirement and water use efficiency in typical dairy farms in Bangladesh. *Journal of Innovative Agriculture and Social Development*, 1(1), 23–32.

USDA. (2012). Part 651-Agricultural Waste Management Field Handbook. In *Handbook Natural Resources Conservation Service*. United States Department of Agriculture (USDA). https://directives.nrcs.usda.gov/sites/default/files2/1712930985/Part%20651%20-%20Agricultural%20Waste%20Management%20Field%20Handbook.pdf.

Vaishnav, S., et al. (2023). Livestock and poultry farm wastewater treatment and its valorization for generating value-added products: recent updates and way forward. *Bioresource Technology*, 382, 129170. https://doi.org/10.1016/j.biortech.2023.129170.

Vallejo-Gómez, D., et al. (2023). Smart irrigation systems in agriculture: a systematic review. *Agronomy*, 13(2), 342. https://doi.org/10.3390/AGRONOMY13020342.

Varma, V. S., et al. (2021). Dairy and swine manure management – challenges and perspectives for sustainable treatment technology. *Science of the Total Environment*, 778, 146319. https://doi.org/10.1016/j.scitotenv.2021.146319.

Velarde-Guillén, J., et al. (2023). Water footprint of small-scale dairy farms in the central coast of Peru. *Tropical Animal Health and Production*, 55(1), 25. https://doi.org/10.1007/s11250-022-03437-8.

Vourch, M., et al. (2008). Treatment of dairy industry wastewater by reverse osmosis for water reuse. *Desalination*, 219(1–3), 190–202. https://doi.org/10.1016/J.DESAL.2007.05.013.

Waltner, I., et al. (2023). Influence of climatic factors on the water footprint of dairy cattle production in Hungary–a case study. *Water*, 15(18), 4181. https://doi.org/10.3390/w15234181.

Willers, H. C., et al. (1999). Potential of closed water systems on dairy farms. *Water Science and Technology*, 39(5), 113–119. https://doi.org/10.1016/S0273-1223(99)00092-X.

Zhang, G. P., et al. (2013). Water footprint assessment (WFA) for better water governance and sustainable development. *Water Resources and Industry*, 1–2, 1–6. https://doi.org/10.1016/J.WRI.2013.06.004.

Chapter 5

Measuring methane emissions from dairy farms

Luís O. Tedeschi, Texas A&M University, USA; Marcos I. Marcondes, William H. Miner Agricultural Research Institute, USA; Sushil Paudyal, Texas A&M University, USA; André F. Brito, University of New Hampshire, USA; and Hugo F. Monteiro, Cornell University, USA

1 Introduction

Dairy farming is a vital part of the global agricultural economy, providing essential nutrients through milk and dairy products. In addition, world milk production has been steadily increasing, reflecting a growing demand. As shown in Figure 1, the black line represents the annual milk production, showing an overall increasing trend over the years, and the blue line depicts the number of dairy cows, which has fluctuated but shows a general upward trend. This increase in milk production is driven by both a rise in the number of dairy cows and an improvement in milk yield per cow due to better breeding and management practices. Methane (CH_4) is a potent greenhouse gas with a global warming potential approximately 23–28 times greater than that of carbon dioxide (CO_2) over 100 years, depending on the recommendation used by the Intergovernmental Panel on Climate Change (Tedeschi and Beauchemin, 2023).

The Food and Agriculture Organization (FAO, 2006) stated that CH_4 emissions from livestock activities represent about 35% to 40% of total anthropogenic

http://dx.doi.org/10.19103/AS.2025.0157.08

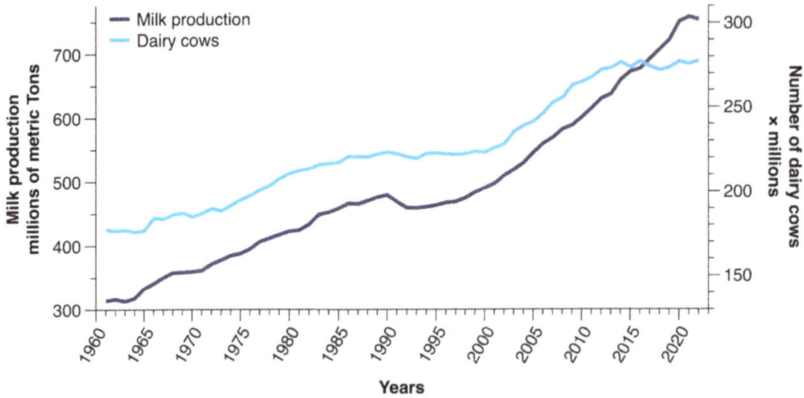

Figure 1 Trends in world milk production and dairy cow population (1961–2022). Based on the data from FAO Stats (https://www.fao.org/faostat/en/#data/QCL).

CH_4 emissions, which together represent about 80% of the agricultural CH_4 output. In contrast, Jackson et al. (2020) and Smith et al. (2021) both agree that agriculture is a significant source of anthropogenic CH_4 emissions, contributing approximately 145 Tg in 2017, which represents about 24% of the total global CH_4 emissions, amounting to roughly 596 Tg per year. However, this is in contrast to the FAO (2006) assessment, which estimated that agriculture, particularly livestock, contributed 35% to 40% of total anthropogenic CH_4 emissions. The discrepancy is believed to be due to differences in methodologies, with more recent reports potentially using advanced data collection and analysis techniques, changes in definitions and scope, and improvements in agricultural practices over the years. These methodological and definitional differences result in variations in how the data is interpreted and presented. In general, the scientific community agrees that enteric fermentation, which occurs when cows produce CH_4 during the digestion process, represents a significant portion of total anthropogenic CH_4 emissions, accounting for approximately 30% to 32%. It is responsible for about 90% of all livestock-derived CH_4 emissions, with cattle being the dominant source, contributing about 77% of the total emissions. Manure management also contributes to CH_4 emissions from storage and handling when it is an anaerobic system. Manure management emissions are predominantly from pigs (approximately 42%) and cattle (approximately 41%), underscoring the substantial role of these animals in overall agricultural CH_4 emissions (Gerber et al., 2013). Similar to other ruminant production systems, unfortunately, a figure for dairy cattle alone is not available, likely due to the lack of specific field methodologies capable of accurately isolating and measuring CH_4 emissions from dairy cattle independently (Tedeschi et al., 2022).

The number of lactating dairy cows and milk yield per cow have been the driving metrics in studying the efficiency of milk production and its environmental impacts. Therefore, if the estimated CH_4 emissions are based on the number of animals (i.e. lactating cows and replacement heifers), then the estimated CH_4 emission is supposed to decrease. On the other hand, if the increase in milk production is due to an increase in feed intake, then it is expected that CH_4 emissions will increase (Ramin and Huhtanen, 2013). However, over the past few decades, there has been a significant increase in the average milk yield per cow in response to advancements in genetics, nutrition, and farm management (Figure 1). Unfortunately, the number of dairy cows has shown different trends in different regions, with some areas experiencing increases due to higher demand while others have stabilized or decreased due to improvements in productivity and changing dietary patterns. Methane emissions per liter of milk produced have generally decreased due to improved efficiency and management practices, even though total emissions may rise with increased total production (Gerber et al., 2013, Hristov et al., 2013b, Capper et al., 2009, Place and Mitloehner, 2010). Regardless, the question remains: what are the specific sources of CH_4 emissions in dairy farming, and what factors influence these emissions?

2 Sources of methane in dairy farming

The CH_4 emissions in dairy farming primarily arise from two sources: enteric fermentation and manure management. Understanding these sources is crucial for developing effective strategies to measure and mitigate emissions.

2.1 Enteric fermentation

Enteric fermentation is a natural digestive process in ruminants, such as cows, where microbes in the stomach (rumen) break down food, particularly fibrous plant material, and produce CH_4 as a byproduct (Goff, 2015, Hofmann, 1988). The animal then expels CH_4 through eructation or belching. The rumen hosts a complex community of microbes that decompose cellulose and other carbohydrates (Nagaraja, 2016, Owens and Basalan, 2016). During this process, hydrogen and CO_2 are produced, with methanogenic microorganisms (i.e. Archaea) converting hydrogen and CO_2 into CH_4 (Tedeschi and Fox, 2020, Van Soest, 1994). Therefore, the composition of a cow's diet significantly affects CH_4 production. High-fiber diets typically result in higher CH_4 emissions compared to more easily digestible feeds (Johnson and Johnson, 1995, Beauchemin et al., 2008, Moss et al., 2000). However, more efficient animals with higher milk production per unit of feed intake tend to have lower CH_4 emissions per liter of milk produced, even though their total CH_4 emissions might be higher. Several strategies can be employed to mitigate CH_4 emissions from enteric fermentation. Dietary adjustments, including the incorporation of more digestible feeds, fats, and certain feed additives like nitrates, tannins, or seaweed, can reduce CH_4 production (Hristov et al., 2013b, Knapp et al., 2014, Kinley et al., 2016, Hristov et al., 2013a, Beauchemin et al., 2008). Selecting and breeding cows with lower CH_4 emissions without compromising productivity is another viable approach (de Haas et al., 2011, Pickering et al., 2015, de Haas et al., 2021). Additionally, feed supplements such as 3-nitrooxypropanol (3-NOP), which specifically targets methanogenesis, have shown promise in reducing CH_4 emissions from enteric fermentation (Kebreab et al., 2023, Uddin et al., 2022, Yu et al., 2021).

2.2 Manure management

Methane emissions from manure management occur when manure is stored or treated in systems that promote anaerobic conditions, such as lagoons, pits, or anaerobic digesters. In anaerobic conditions, methanogenic archaea thrive and produce CH_4. This is more pronounced in liquid manure systems than in solid manure systems (Montes et al., 2013, Amon et al., 2006). The type of feed and the way manure is handled and stored also affect CH_4 emissions, with higher moisture content and specific nutrient composition (e.g. high fiber) increasing CH_4 production. Anaerobic digesters can be used to mitigate CH_4 emissions from manure management by capturing CH_4 for use as a biogas, thereby reducing emissions and providing renewable energy (Holm-Nielsen et al., 2009, Møller et al., 2004). Frequent removal of manure from barns and immediate application to fields can minimize anaerobic conditions, while separating solids from liquids in manure can reduce CH_4 emissions by promoting the aerobic decomposition of solids (Montes et al., 2013). Although CH_4 is a significant concern, it is crucial to consider the overall environmental impact of dairy farming. Manure management also affects emissions of other greenhouse gases, such as nitrous oxide and pollutants like ammonia (Rotz, 2004). Integrated approaches that address multiple environmental impacts can provide more sustainable solutions in the long run. Implementing comprehensive strategies that address feed efficiency, manure management, animal health, and productivity can lead to overall reductions in greenhouse gas emissions. Technological advancements, such as precision feeding, automated manure scraping systems, and real-time CH_4 monitoring, can enhance the effectiveness of mitigation efforts.

Addressing CH_4 emissions in dairy farming requires a multifaceted approach that considers both enteric fermentation and manure management. By understanding these sources and applying integrated, science-based strategies, the dairy industry can significantly reduce its environmental footprint while maintaining productivity and profitability. Another question to answer is:

What are the actual CH_4 emissions from dairy cows under real-world production conditions? To answer this, we need methodologies and techniques that are practical, reliable, and cost-effective.

3 Methods for measuring methane emissions

Measuring CH_4 emissions from dairy farms is essential for understanding the environmental impact and developing effective mitigation strategies. This section aims to critically assess the strengths and limitations of the current measurement methods and discuss their relative merits in specific contexts. We will focus on discussing the most appropriate methodologies depending on the accuracy, practicality, and application needs at animal or group levels, as depicted in Figure 2.

3.1 Animal-level methods

Measurement of CH_4 production at the farm level starts with monitoring the emission at the individual animal level. There is documented evidence that indicates that CH_4 measurement at the individual level corresponds to the total CH_4 production by the given herd (Garnsworthy et al., 2012). At the animal level, direct methods of measuring CH_4 emissions involve capturing and analyzing CH_4 exhaled by individual animals. This can be done using devices like respiration chambers (RC), GreenFeed (GF) systems, or portable accumulation chambers, where CH_4 is directly measured from the animal's breath. Indirect methods at the animal level, on the other hand, estimate

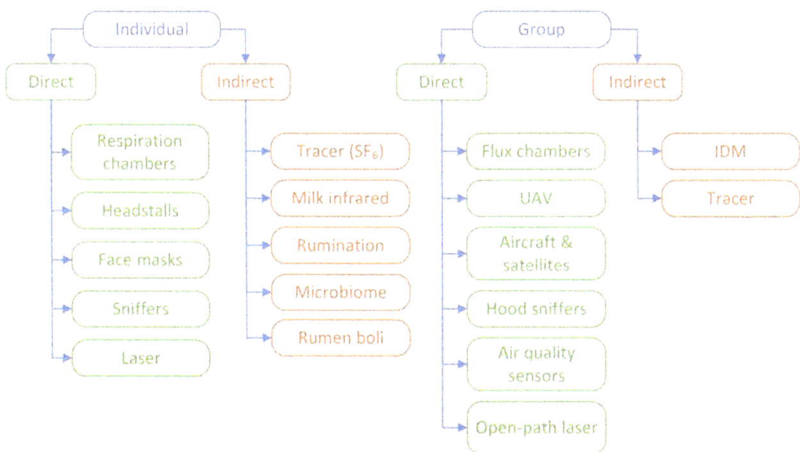

Figure 2 Different methods for determining methane directly and indirectly at the animal or group levels. SF_6 is sulfur hexafluoride, UAV is unmanned (unperson) aerial vehicles, and IDM is inverse dispersion methods (e.g. Eddy covariance).

CH_4 emissions through proxies such as the sulfur hexafluoride (SF_6) tracer technique, where CH_4 concentration is inferred by comparing it to a tracer gas or by analyzing milk fatty acid profiles, which are linked to CH_4 production through rumen fermentation patterns. These methods allow CH_4 emissions to be attributed to individual animals, even when they are housed or fed in groups. For individual methods, some sensors are placed inside the animal (mainly in the rumen), while others analyze animal behavior (e.g. rumination time) or products (e.g. milk) or are externally associated with the animal (e.g. indirect methods).

3.1.1 Direct methods

Direct methods measure the CH_4 emitted by the animals. These typically involve collecting and analyzing gas emissions, allowing for a precise quantification of CH_4 output.

3.1.1.1 Respiration chambers

Respiration chambers are used primarily in research settings due to their high cost and labor-intensive operation. They are ideal for studies that require detailed, individual-level data on CH_4 emissions, energy balance, or metabolism. The RC method, often referred to as open-circuit, indirect calorimetry, involves placing an animal in a sealed, temperature-controlled chamber designed to continuously measure the gases it exhales, primarily CH_4 and CO_2, as well as oxygen (O_2) it inhales (Tedeschi et al., 2022). The air inside the chamber is circulated, and incoming and outgoing air are analyzed to determine gas concentrations. Researchers can calculate the amount of CH_4 emitted by the animal by measuring the differences in gas levels and the airflow rate. The chambers allow the animal to move, eat, and drink normally during the measurement period, typically 24 to 48 hours, providing a detailed and precise assessment of its CH_4 production. While RC are highly accurate, their use is limited to research settings due to their cost and the artificial environment they create, which may not fully replicate real-world feeding conditions (Tedeschi et al., 2022).

3.1.1.2 Headstalls

The GF spot-sampling system, introduced commercially in 2010 by C-Lock Inc. (Rapid City, South Dakota, USA), offers a promising alternative for measuring CH_4 emissions from cattle. Unlike traditional RC, GF allows animals to voluntarily visit an open feeder where they are rewarded with a pelleted concentrates supplement. The system utilizes head position sensors and decision rules to validate the obtained measurements. Typically, CH_4 emissions are measured

Figure 3 A dairy cow using a GreenFeed machine to have her emissions measured while eating. The photo is courtesy of Dr Marcos I. Marcondes.

over short periods (3 to 7 minutes; Figure 3) for several times a day, with data collected over multiple days to extrapolate a 24-hour individual CH_4 emission profile (Hammond et al., 2015).

The study of Della Rosa et al. (2021) indicates that GF measurements are predominantly conducted in confinement settings, with over 77% of studies utilizing such environments. A significant portion of these studies (57%) extended their measurements over 15 days or more, while a smaller percentage conducted measurements over shorter durations. To ensure precise CH_4 estimates, Della Rosa et al. (2021) recommended collecting 20–50 spot samples, which would typically require 7–17 measurement days if animals visit the unit three times daily. Early studies often had fewer measurement days, but recent guidelines have emphasized the importance of adequate sampling to improve accuracy. The GF system, programmed with C-Lock Inc. software, controls the timing of pelleted feed availability to encourage animals to distribute their visits throughout the day, enhancing the accuracy of the CH_4 emission profile. Visitation frequency to GF units is influenced by dietary treatment and possibly the level of feed intake. In the study of Hammond et al. (2015), more visits were observed when heifers grazed on flowers compared to ryegrass and clover, potentially due to the location of the GF unit relative to the paddocks and the dry matter and nutrient content of the swards grazed (Table 1).

Waghorn et al. (2016) used the GF system to examine CH_4 emissions in dairy cows under different stocking rates (SR) in a pasture-based system. Two herds were involved, one with 28 cows (Low SR) and another with 40 cows (High SR). Methane emissions averaged 331 g/cow/day for the Low SR and 305 g/cow/

Table 1 Animal visitation to the GreenFeed (GF) unit across three treatments in rotational strip grazed

Items	Number of measurement days	Total number of GF visits (methane measurements)	Total number of GF visits (methane measurements) per animal per day
Ryegrass		219	1.24
Clover		229	1.30[a]
Flowers		432	2.26[b]
Total	48	880	1.60
SD			1.09
SEM			0.07
P value (forage)			<0.001

Different letters indicate significant differences from ryegrass control according to the Dunnetts test ([a]$P < 0.10$, [b]$P < 0.001$).
Adapted from Hammond et al. (2015).

day for the High SR, with corresponding milk yields of 19.6 kg/day and 16.5 kg/day, respectively. The GF system effectively distinguished the impact of SR on emissions; however, fewer cows from the High SR herd visited the GF units consistently, a behavior not fully explained but possibly influenced by diet. This suggests that less favorable diets may increase GF visitation when a favorable feed reward is provided. This variability in visitation rates and sampling times may impact CH_4 estimates, especially in nutrition experiments exploring diet effects on CH_4 emissions. Unlike RC and SF_6 techniques, the voluntary nature of GF measurements can introduce potential biases, with concordance correlation coefficients showing low agreement (0.1) for GF vs. RC and moderate (0.6) for GF vs. SF_6 (Figure 4). Despite this variability, GF shows a strong overall correlation with RC, both with and without 3-NOP (Ma et al., 2024), as illustrated in Figure 5A, and across multiple studies (Huhtanen et al., 2019), as shown in Figure 5B, highlighting GF's utility as a reliable indicator of CH_4 emissions.

However, in the study by Hristov et al. (2016), the coefficient of variation (CV) for GF and SF_6 were 25.8% and 38.6%, respectively, indicating more significant variability in SF_6 measurements. Similarly, Dorich et al. (2015) compared a portable automated open-circuit gas quantification system (GQS) and the SF_6 technique in 16 mid-lactation Holstein cows. Methane emissions were 472 g/d for ad libitum intake and 458 g/d for restricted intake, with GQS showing lower CV (14.1% to 22.4%) and a moderate relationship ($r^2 = 0.42$) between dry matter intake (DMI) and CH_4 emissions. The SF_6 technique, however, exhibited higher variability (16% to 111%) and a poor relationship ($r^2 = 0.17$) between DMI and CH_4 emissions, likely due to limited barn ventilation and high background gas concentrations. Both studies highlight significant issues with the SF_6 technique,

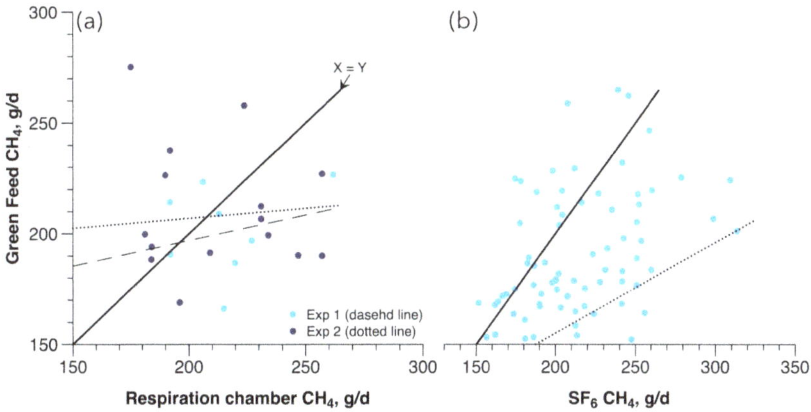

Figure 4 (a) Relationships between measured methane production (g/day) using GreenFeed and respiration chamber techniques of confined dairy heifers in experiments 1 (light blue; n = 8) and 2 (dark blue; n = 16), with a concordance correlation coefficient between techniques of 0.1. (b) Relationships between methane production (g/day), determined using GreenFeed and sulfur hexafluoride tracer (SF_6) techniques, of dairy heifers rotationally strip grazed (n = 136), with a concordance correlation coefficient between techniques of 0.6. Adapted from Hammond et al. (2015).

notably its high variability and sensitivity to environmental conditions such as ventilation and background gas concentrations. Hristov et al. (2016) and Dorich et al. (2015) suggest that the SF_6 technique should be used in well-ventilated environments or outdoors to reduce these effects. They recommend cautious interpretation of SF_6 data, especially in poorly ventilated settings. Conversely,

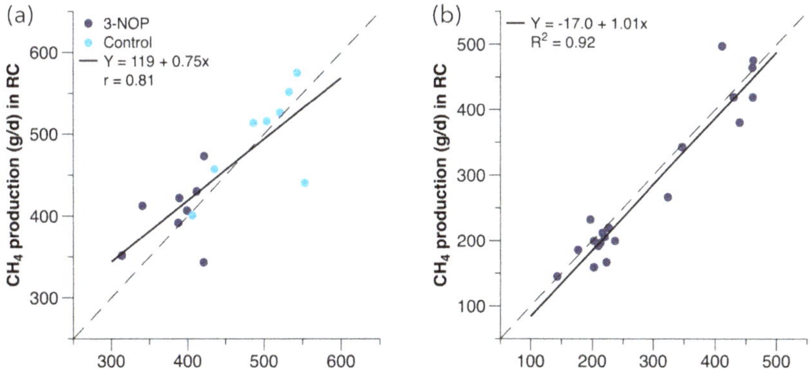

Figure 5 Relationship between estimated daily methane production (g/day) using GreenFeed (GF) and respiration chamber technique (RC) with or without the administration of 3-NOP (A) (Ma et al., 2024) and the average of six studies (B) compiled by Huhtanen et al. (2019).

the GF system (and the similar GQS in Dorich et al.'s (2015) study) provided more consistent and reliable CH_4 emission measurements with lower variability and a better relationship with DMI.

The GF system requires periodic calibration, like RC systems. Calibration frequencies varied from twice daily to weekly, depending on whether manual or auto-calibration methods were used. Proper calibration is essential for accurate CH_4 measurement, involving the use of 'zero gas' and 'span gas' with known gas concentrations (Hammond et al., 2015). Overall, the GF system provides a practical and effective method for estimating CH_4 emissions, with careful attention to calibration and sampling protocols ensuring reliable data. There is no requirement for an analytical laboratory to support the measurements, as is the case for the SF_6 tracer technique, and it is not as expensive as RC. For this reason, GF has been the most used and recommended method for commercial farms looking for tools to measure CH_4 in their system. However, the voluntary nature of GF measurements and the influence of research dietary treatments on visitation frequency must be carefully managed to avoid bias and ensure the validity of CH_4 emission estimates.

Overall, the GF system offers a practical and effective method for estimating CH_4 emissions, with reliable data ensured through meticulous calibration and sampling protocols. Unlike the SF_6 tracer technique, GF does not require an analytical laboratory for measurement support, making it more accessible and cost-effective than RC. Consequently, GF has become the preferred and recommended method for commercial farms and research centers seeking to measure CH_4 emissions. However, the voluntary nature of GF measurements and the influence of dietary treatments on visitation frequency must be carefully managed to avoid bias and ensure the validity of CH_4 emission estimates.

However, despite its practical advantages, it does have notable limitations. As further discussed in Section 5, GF results show a high variability (Figure 4) despite the high correlation with RC (Figure 5), which may cause doubt about its overall adequacy. Therefore, while the GF system can be useful, particularly in practical farm settings, caution should be exercised when interpreting its data for CH_4 emission estimates, especially in research contexts.

3.1.1.3 Face masks

There is a growing need for inexpensive and convenient methods to estimate livestock CH_4 emissions. The RC is currently considered the gold standard for quantifying enteric CH_4 emissions. However, its use is limited in practical research and farm settings due to costly infrastructure, lengthy observation periods, and alterations to animal behavior (Huhtanen et al., 2015). Consequently, researchers have investigated alternative measurement techniques, primarily focusing on calculating daily emissions from spot samples of eructated and respired air.

The face mask (FM) technique has also been cited as an alternative system to measure CH_4 emissions from ruminants (Johnson and Johnson, 1995). The FM technique involves 30-minute measurement periods taken every 2–3 hours throughout the day, often leading to altered animal behavior and inaccurate emission data. However, due to the strong correlation between total daily emissions and spot samples at specific times post-feeding, it has been suggested that the number of samples could be reduced to only once a day at 6 hours after morning feeding for 2–3 days (Oss et al., 2016, Silveira et al., 2019), as shown in Figure 6, since there is evidence that the sampling conducted at that time is strongly correlated with total daily CH_4 emissions (Crompton et al., 2011). Thus, the FM method's accuracy, practicality, reduced sampling frequency, correlation with total emissions, and minimal disruption to animals make it a suitable tool for measuring CH_4 emissions in farm situations.

The FM technique quantifies CH_4 emissions by covering the animal's muzzle and collecting gas samples through a tube connected to a mass flow controller and gas analyzers (Oss et al., 2016). Although this method restricts animals from eating or drinking during short measurement periods (e.g. 30 minutes; Figure 7), it is cheaper, more portable, and simpler than the SF_6 tracer technique (we will discuss that later in this chapter) and RC. Moreover, the FM method can generate CH_4 measurements comparable to those obtained using the RC and SF_6 (Oss et al., 2016, Silveira et al., 2019), as shown in Figure 8.

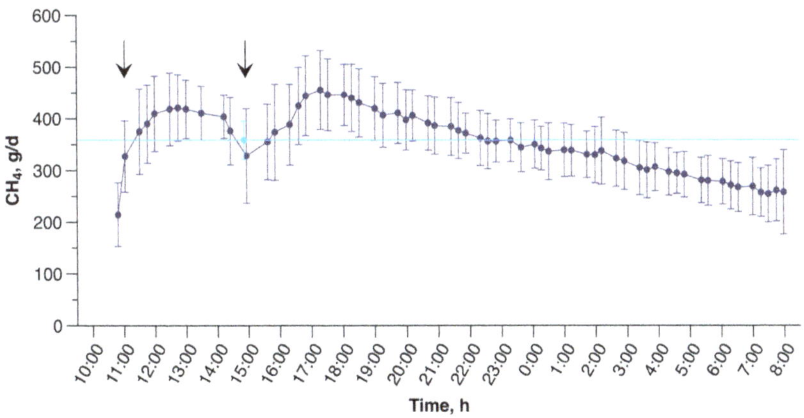

Figure 6 Diurnal pattern of methane (CH_4) production (g/d) as measured by the respiration chambers (dark blue) and by the face mask (light blue) after feeding. The arrows represent the schedule of feeding (1000 and 1500 h). Adapted from Silveira et al. (2019).

Figure 7 Galvanized iron frame for animal restraint and face-mask construction and application to estimate daily methane production in cattle. Heifers are also fitted with a heart rate monitor. The photo is courtesy of Dr Marcos I. Marcondes.

Despite its advantages, the FM method requires animal cooperation and may cause discomfort when animals are confined in a squeeze chute during measurements. Furthermore, variations such as feeding regimes and emission fluctuations may introduce linear bias, limiting the method's accuracy. Therefore, while the face-mask technique offers a more practical and cost-effective approach to measuring CH_4 emissions, it requires careful consideration to ensure accurate and meaningful results.

3.1.1.4 Sniffers

Portable CH_4 detectors offer another approach to direct measurement. These handheld devices, also known as 'sniffers,' can be used to measure CH_4 concentrations in different parts of the farm, including barns and outdoor areas where cows graze. Although portable detectors are versatile and easy to use, their accuracy can be affected by environmental conditions such as wind and temperature. Negussie et al. (2017) compared the estimates from the non-invasive photoacoustic IR spectroscopy technique implemented in a portable gas analysis equipment (PAS-F10) against open-circuit indirect respiration calorimetric chamber measurements. They reported that although DMI, level of milk production, lactation stage, and diurnal variation had significant effects on daily CH_4 output, the repeatability of daily CH_4 output from the PAS-F10 technique ranged from 0.40 to 0.46, indicating that some between-animal variation exists.

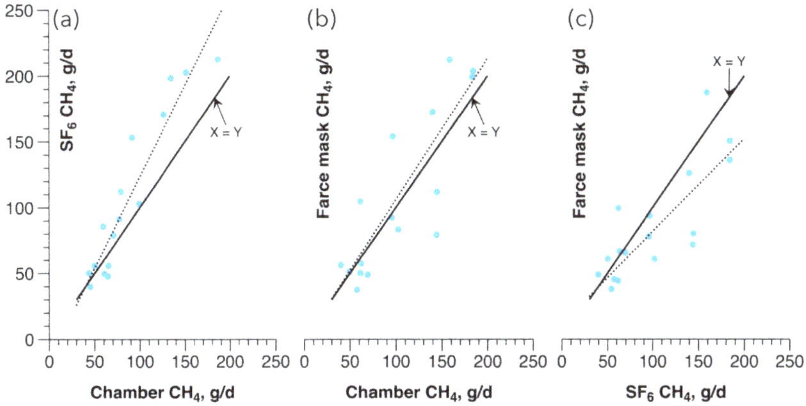

Figure 8 Relationships between CH_4 emission (g/day) determined using SF_6 vs. respiration chamber (RC) techniques (a: y = 1.4016x – 16.573; R^2 = 0.8927; $P < 0.0001$), face mask (FM) vs. RC techniques (b: y = 1.0625x – 3.047; R_2 = 0.7178; $P < 0.0001$) and FM vs. SF_6 techniques (c: y = 0.6779x + 16.942; R^2 = 0.6263; P = 0.0002), of individual yearling bulls fed at different intake levels. The solid line indicates y = x. The correlation coefficients were 0.9448, 0.8472, and 0.7914 for panels a to c, respectively. The Lin's concordance values (CCC) were 0.8232, 0.8244, and 0.7398, respectively. Adapted from Oss et al. (2016).

In this 'sniffer' approach, exhaled air is sampled near the animal's nostrils through a tube fixed in a feed bin and connected directly to a gas analyzer (Lassen et al., 2012, Garnsworthy et al., 2012). The feed bin might be in an automatic milking station, as used by Lassen et al. (2012), or in a concentrate feeding station (Garnsworthy et al., 2012). Different types of gas analyzers (nondispersive infrared, Fourier-transform infrared, or photoacoustic infrared) and different sampling intervals (1, 5, 20, or 90 to 120 s) have been tested. However, there is evidence of interchangeability with either technique in commercial farm conditions (Sypniewski et al., 2019) as they yield similar results. In general, CH_4 concentration during a sampling visit of typically between 3 and 10 min may be specified as the overall mean or the mean of eructation peaks. The CH_4 emission is calculated as a product of peak frequency and mean peak area. As another strategy, Huhtanen et al. (2015) derived a CH_4/CO_2 ratio, which is then multiplied by estimated CO_2 production to predict CH_4 fluxes. Sniffer measurements are more variable than flux methods, with factors like airflow in the barn and head positions increasing measurement error (imprecision).

3.1.1.5 Laser detectors

Emerging technologies are also playing a significant role in improving CH_4 measurement. For example, laser-based CH_4 sensors can provide continuous and real-time measurements of CH_4 concentrations in the air. These sensors can be deployed in barns or outdoor areas to monitor emissions continuously, offering a more dynamic and detailed understanding of CH_4 release patterns. The laser CH_4 detector is a highly responsive, handheld device that is pointed at an animal's nostrils and measures CH_4 column density along the length of the laser beam. Typically, animals are restrained either manually or in head gates at a feed fence for the required length of time. The operator has to stand at the same distance (1 to 3 m) from each animal every time and must be careful to keep the laser pointed at the animal's nostrils throughout the measurement period. Measurements for each cow are taken over periods of 15 to 25 s between eructation events and could detect CH_4 emitted each time the animal breathed out (Chagunda et al., 2009). Responsiveness can be improved with a measurement made once every 0.1 to 0.5 s, allowing for the characterization and separation of eructation and respiration peaks (Roessler et al., 2018). Because laser detectors measure CH_4 in the plume originating from the animal's nostrils, results can be affected by factors such as distance from the animal, pointing angle, animal's head orientation and head movement, air movement and temperature in the barn, adjacent animals, and operator variation (Sorg et al., 2017). The structure of the barn and the resulting ventilation conditions and wind speed at the location of the measurement are also considerable sources of variation in recorded CH_4.

3.1.2 Indirect methods

Indirect methods of measuring CH_4 emissions are also commonly used, especially when direct measurements are impractical or too costly. One such method involves analyzing the composition of milk and feed intake, as indicated below. By monitoring the cows' diet and correlating it with milk production, researchers can estimate CH_4 emissions. This approach relies on the known relationship between feed composition, digestion, and CH_4 production. However, detailed data on feed intake and milk yield is required, which can be challenging to obtain consistently. Another indirect method is using mathematical models and simulation tools to estimate CH_4 emissions based on farm management practices and animal data. These models consider various factors, including the type of feed, the number of cows, and manure management practices, to predict CH_4 emissions (Tedeschi et al., 2022). While these models can provide a comprehensive data set of emissions, their accuracy depends on the quality of the input data and the assumptions made in the model (Tedeschi and Fox, 2020).

3.1.2.1 Tracer techniques

The SF_6 tracer technique is a method for measuring CH_4 emissions from ruminants, initially developed by Zimmerman (1993) and Johnson et al. (1994).

This method involves inserting a known quantity of SF_6, an inert tracer, into the animal's rumen (Figure 9). Unlike the RC technique, which captures all emissions quantitatively, the SF_6 tracer technique is an indirect method that collects a representative sample of the gas produced by the animal. SF_6 is considered ideal for this purpose due to its similar dispersion characteristics to CH_4, its non-toxicity to rumen microbes and the animal, and its detectability at trace levels.

The principle of the SF_6 tracer technique relies on a known release rate of SF_6 from a permeation tube in the animal's rumen. Breath samples are continuously collected via a sampling line from the nostrils to a pre-evacuated canister (Johnson and Johnson, 1995), as shown in Figure 9. The CH_4: SF_6 ratio in the canister, corrected for background concentrations and multiplied by the SF_6 permeation rate, enables the calculation of CH_4 output. Critical aspects of this method include the permeation tubes, sampling system, and background gas records (Lassey and Hegarty, 2020). Subsampling and gas analysis, sampling duration, and quality control significantly impact data integrity.

An advantage of the SF_6 tracer technique is its applicability to grazing or group-housed animals. Approximately 60% of studies using this technique were conducted in confinement, with most studies collecting breath samples over 5–10 days (Della Rosa et al., 2021). Despite its historical variability (Figure 10),

Figure 9 A halter and strap were used to mount the gas collection apparatus. The sample collection point is fixed on the muzzle using the leather halter. The evacuated sample collection canister and orifice plate flow restriction apparatus are securely located on the crossbred (Holstein x Gyr) bull's neck. Adjustable girth and rump straps prevent dislodgement of the mesh strap. The photo is courtesy of Dr Marcos I. Marcondes.

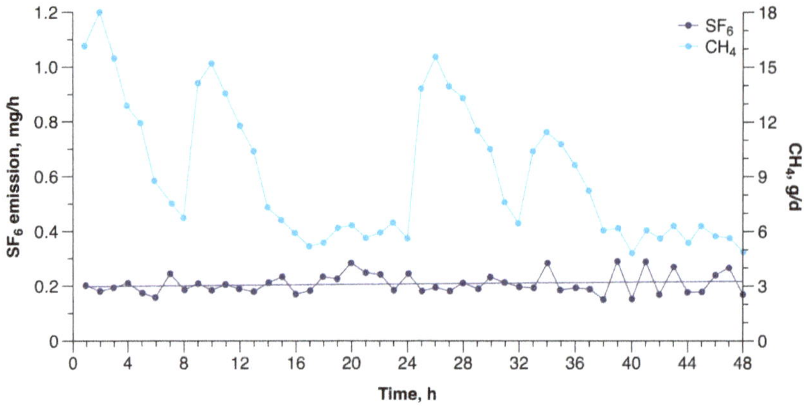

Figure 10 An example of the typical hourly sulfur hexafluoride (dark blue) and CH_4 (light blue) emission patterns from a single Holstein-Friesian cow measured within a respiration chamber over 48 h. Adapted from Deighton et al. (2014).

recent modifications to the SF_6 tracer technique have improved its accuracy and precision, decreased between-animal CV for CH_4 yield (i.e. g/kg of DMI) to levels comparable to CV obtained using chambers (Deighton et al., 2014). These improvements have addressed systemic errors, enhancing the reliability of the tracer technique for quantifying enteric CH_4 emissions.

3.1.2.2 Milk infrared spectra

One of the promising proxies for CH_4 emission is based on the mid-infrared spectra of milk. Some studies suggest the use of milk mid-infrared spectra for widespread screening of dairy herds in order to develop management and genetic selection tools to reduce CH_4 emissions (Vanlierde et al., 2016). Subsequently, several studies have focused on the relationship between CH_4 emissions and the milk FA profiles of dairy cows (Dehareng et al., 2012). The CH_4 mainly originates in the rumen as a result of the fermentation of feed components by rumen microbes (Tedeschi and Fox, 2020). The CH_4 formation in the rumen is related to the formation of specific volatile fatty acids that are transferred to the milk, and that can be used as predictors for CH_4 production at the herd level (Mohamad Salleh et al., 2023). However, this is currently limited to lactating cattle for which the milk recording scheme is in place and calibrated. The predictions are based on the underlying equation that has been continuously refined by extending calibration and reference CH_4 measurement data sets to achieve a higher accuracy (Massaro et al., 2024). The best predictions of CH_4 are obtained from fatty acid spectra when a measure of milk yield is also included in the model to calculate daily CH_4 production and CH_4 intensity (Massaro et al., 2024).

3.1.2.3 Rumination time

With the widespread adoption of precision livestock technologies in dairy farms, animal-level cow monitoring has become feasible (Eckelkamp, 2019). As the large-scale recording of CH_4 emission is expensive to measure and very labor-intensive, the possibility of using metrics that are already recorded in dairy herds as a proxy for CH_4 emission has been extensively explored (Zetouni et al., 2018). Rumination is essential for fiber digestion in ruminants, reduces particle size, increases the surface area available for fermentation by the rumen microbiota, and increases the passage rate, resulting in effective digestion and absorption of nutrients (Beauchemin, 2018, Tedeschi and Fox, 2020). Studies have identified that a longer rumination time is connected with lower daily CH_4 emissions as well as lower CH_4 production per milk unit in high-yielding dairy cows (Mikuła et al., 2022). Recent studies have indicated the potential for these new traits to be used in the genetic selection of cows for lower CH_4 emissions (Weller and Ezra, 2024). Several studies have proposed the use of rumination time as an indicator trait to be used as a proxy for CH_4 production (Byskov et al., 2015, Zetouni et al., 2018), but only a few studies have demonstrated satisfactory evidence. Recent studies have demonstrated a genetic association ($r = -0.53$) between daily rumination time and CH_4 production in dairy cattle, providing a basis for use as an indirect means to measure CH_4 production (Lopes et al., 2024). There are similar attempts to use other sensor-derived variables, including eating time to be used as indicators for CH_4 production (Sepulveda et al., 2022). There are numerous commercial sensors available that work on different mechanisms for rumination data collection. For example, some neck collars use microphone auscultations (Schirmann et al., 2009), ear tag-based accelerometers use head and ear movements (Pereira et al., 2018), and rumen indwelling boli (Capuzzello et al., 2023). The selection of the type of sensor to use on a farm for rumination time measurements has been directed by producer preference and feasibility.

3.1.2.4 Rumen microbiome

Despite the availability of several methods to measure and predict CH_4 emitted from a ruminant, the production of substrates for CH_4 synthesis and CH_4 production itself are actions of the microbes present in the rumen, lower gut, manure, or any CH_4 emitter environment. There is still considerable variability and uncertainty regarding the relationship between the rumen microbiome and methane production. Difford et al. (2018) reported that approximately 6% of bacterial and 12% of archaeal operational taxonomic units in the rumen were significantly heritable, with heritability estimates ranging between 16–44% for bacteria and 18–33% for archaea. This suggests that non-genetic factors drive the majority of variation in rumen microbial abundance. The study also found that host genetics and the rumen microbiome together explained about 34% of the variation in methane emissions in dairy cattle, though these influences were largely independent. These findings confirm the complexity of the interactions between the host, its rumen microbiome, and methane production, with much still unknown about the specific microbial and genetic factors that control methanogenesis.

Therefore, since the advancements of high-throughput sequencing technologies, there is an emerging field of research directly collecting these contents, characterizing the microbial population and abundance in those samples, and using that to predict CH_4 emissions (Ross and Hayes, 2022). The rationale is to tackle the correlation between the abundance, activity, or metabolites of a microbe that produces either substrate for methanogenesis or CH_4 itself with CH_4 production (Ross et al., 2020, Ross et al., 2013, Wallace et al., 2019). Although some predictions have been performed in dairy cattle (Peng et al., 2023, Wallace et al., 2019), most of this work was done in other ruminant species due to the need for a large number of experimental units for the predictions (Ross and Hayes, 2022, Zhang et al., 2023). Across these studies, the rumen microbiome yields a moderate correlation with CH_4 production, intensity, and yield when used as a standalone method, but stronger predictive abilities are reported when integrated with other methods, such as genomics (Ross and Hayes, 2022).

For the predictions of enteric CH_4 emissions, the collection of rumen samples is often performed using the esophageal tubing procedure, given the animal welfare advantages compared to the rumen cannula and, consequently, ease of screening larger cohorts. Although the pH and some metabolites differ between both methods (da Cunha et al., 2023, Duffield et al., 2004, Pathak et al., 2024), several studies have highlighted consistency across microbiome compositions in the two approaches when evaluating the whole rumen contents of dairy cows (da Cunha et al., 2023, Pathak et al., 2024, Paz et al., 2016). These evaluations were performed considering specific DNA sequencing methods, which, together with other specificities of microbiome studies such as temporal variation of the microbiome, DNA extraction procedure, and sequencing approach, should be standardized for better consistency in the predictions (Henderson et al., 2013).

The vast majority of the studies performing microbiome predictions have used amplicon sequencing due to the low level of complexity in the analysis, cost-effectiveness, and for being the first widely adopted next-generation sequencing method to characterize microbiome populations (Peng et al., 2023, Wallace et al., 2019, Zhang et al., 2023). The principle behind amplicon sequencing is to analyze microbial composition and abundance based on a highly conserved gene present across all microbes but with enough nucleotide variability to distinguish microbial groups. By targeting hypervariable regions within these genes using polymerase chain reactions, the bulk of these amplified genes can be differentiated at a level of resolution enough to make specific inferences about microbial groups present in the sample (e.g. genus level for *Ruminococcus* bacteria). The most commonly targeted genes in amplicon studies are those that, once transcribed, become part of the structure of ribosomes present in all living organisms, from prokaryotes to eukaryotes. Targeting simultaneously the genes for 16S rRNA (bacteria and archaea), 18S rRNA (protozoa), and interspaced transcribed spacer (ITS; fungi), these predictions have helped to target specific microbial groups contributing to enteric CH_4 production and also improve genomic predictions for the identification of low-CH_4 emitting animals (Ross and Hayes, 2022, Wallace et al., 2019, Zhang et al., 2023).

Newer sequencing approaches focusing on the entire genetic material of the rumen (e.g. all DNA or RNA) instead of using a single amplified gene for predictions, as in amplicon sequencing, are emerging as promising techniques to accurately predict not only enteric CH_4 emissions in dairy cattle but overall phenotypic traits (Hernández Medina et al., 2022, Ross and Hayes, 2022). Popularly called shotgun sequencing, metagenomics, and metatranscriptomics are currently the 'gold standard' methods in microbiome characterization due to their ability to accurately distinguish microbes and their genes at the species and sometimes strain level. Specifically, characterizing variables (microbial species or genes through shotgun sequencing) that makeup one variable in amplicon sequencing (e.g. a given genus) allows the discovery of better and less noisy microbiome predictors to assess their methanogenic potential and mitigation strategies. There is a scarcity of studies using these approaches for the predictions of enteric CH_4 emissions in dairy cattle, primarily because of the complexity these newer approaches bring to the analysis. Integrating these approaches with current animal and environmental predictors in more robust artificial intelligence approaches such as deep learning should pave a path to unravel microbes' interactions toward more effective methane mitigation efforts.

3.1.2.5 Rumen-based boli

Although rumen-based boli sensors have been widely used for the measurement of rumination, temperature, and pH, they have the potential for CH_4 emissions, as discussed in the next section. New technologies (i.e. https://www.metech.io/) are emerging that are able to measure CH_4 emissions directly from the rumen. Although these technologies have not been validated, they demonstrate promise for use in dairy farms. These sensors (110 mm x 38 mm) can measure CH_4 (0 to 150 000 ppm) in the rumen, providing more accurate measurements from the source.

3.2 Group-level methods

Each of the animal-based methods has its own set of strengths and weaknesses. Direct methods might offer precise measurements but may not capture the variability of emissions across a farm. Indirect methods and models provide broader estimates but rely on accurate, detailed input data. Emerging technologies promise to provide continuous and real-time data but are still being refined and validated for widespread use. Nevertheless, even these new technologies are not immune to failures and limitations.

Whole farm assessment of CH_4 emissions includes measuring CH_4 from critical areas of dairy facilities, including feed production (silage pile, feed mill), farm machinery operations, fertilizer and liquid manure, enteric fermentation, manure management, and exhausts from covered lagoons and settling basins (Borhan and Khanaum, 2022). Measuring CH_4 at the whole farm or barn levels has been attempted by several authors (Borhan et al., 2011a, 2011b). Using multiple flux measurements with a portable gas chromatograph, the group determined CH_4 emissions from ground-level area sources. The study investigated the barn, bedding area, lagoon, settling basin, silage, and walkway as major test sites. The sampling was conducted for five consecutive days and repeated in summer and winter to capture the seasonal dynamics of the CH_4 emissions at the farm scale.

At the group level, direct methods involve measuring CH_4 emissions from the entire group without distinguishing between individual animals. This includes techniques like barn gas emission measurements, where the overall CH_4 concentration in a barn or room is quantified, or open-path laser absorption, where a laser measures CH_4 concentration across a group of animals. Indirect methods at the group level estimate CH_4 production using collective data, such as predicting emissions based on the feed intake and energy balance of the herd or using techniques like Eddy covariance (EC), which measures CH_4 flux over a pasture to estimate emissions from the entire group. These methods are less precise for individual animals but are often more practical and scalable for large populations.

3.2.1 Direct methods

Direct methods at the group level for measuring CH_4 emissions involve capturing and quantifying CH_4 gas released from a group of animals or their environment. These methods measure methane emissions in real-time and directly from the air or environment, typically without attributing emissions to individual animals.

3.2.1.1 Flux chambers

One of the most direct methods for measuring CH_4 emissions is using flux chambers based on indirect calorimetry. These chambers are placed over a specific area, such as a manure storage pit or a grazing field, to capture and measure the amount of CH_4 released. While flux chambers provide precise measurements, they are often limited to small areas and may not represent the entire farm accurately. Norris et al. (2020) successfully utilized this method to quantify gas exchange from soil, demonstrating its applicability in specific agricultural contexts. These systems are considered the 'gold standard' for measuring animal emissions because they measure emissions from the surface with or without manure accumulation (Powers and Capelari, 2016). The basic principles include direct measurement of gases built up in a chamber over time (static) or emitted from a chamber with measured flow rate (passive or flux hood). Field chambers are inexpensive to construct and function with manual or automated systems. Similar to the tracer gas techniques used to determine enteric CH_4 emissions from animals in research settings, tracer gas has also been successfully used to determine CH_4 and other gaseous emissions from animal housing, including buildings and lots (Kaharabata et al., 2000). These methods require that a tracer gas be emitted at a known rate near the CH_4 source and that the concentrations of CH_4 and the tracer gas be measured downwind (Kaharabata et al., 2000). However, the accuracy of the method heavily relies on the assumption that the tracer emissions are representative of the CH_4 emissions and move similarly across the landscape. Chamber methods (static or dynamic) also offer a means of collecting samples for future analyses using gas chromatography. These methods allow users to obtain instantaneous flux estimates over a short period (15 to 30 minutes), and samples are stored in vials and caps that prevent leakage and sample breakdown for laboratory analyses. A suitable number of chambers must be placed to represent the surface area measurement (Rochette, 2011). Consequently, areas that have uniform emission surfaces require fewer chambers placed.

3.2.1.2 Unperson aerial vehicles

Additionally, drones (i.e. unmanned (unperson) aerial vehicles (UAV)) equipped with CH_4 sensors are being developed to provide aerial surveys of CH_4 emissions across large farm areas. Drone-based active AirCore systems associated with a mass balance approach have been used to quantify CH_4 emissions (Andersen et al., 2018). Although challenges exist, these innovative approaches can complement traditional methods and provide more comprehensive data on CH_4 emissions. Vinković et al. (2022) investigated the application of a drone-based active AirCore system to quantify CH_4 emissions from a dairy farm in the Netherlands. The drone collected approximately 350 mL of atmospheric air samples during flights downwind of the farm that were analyzed using cavity ring-down spectroscopy and quantum cascade laser spectrometry. The mass balance approach was employed to calculate emission rates from the measured concentration enhancements above background levels. After subtracting estimated manure emissions, the enteric CH_4 emission factors were calculated to be 0.20–0.51 kg CH_4/animal daily. The authors indicated that the primary sources of uncertainty were variabilities in wind speed and the angle between wind direction and flight transect. Nonsimultaneous sampling in the vertical direction of the plume was identified as a limiting factor for accurate quantification. The study incorporated mobile van measurements using an inverse Gaussian approach for comparison. While the agreement between UAV and van measurements varied (17% to 51%), the combined use of both platforms demonstrated potential for improved emission quantification (Vinković et al., 2022). The potential of drone-based active AirCore systems as a valuable tool for quantifying CH_4 emissions from agricultural sources exists. The method's accuracy improved when combined with precise on-site wind measurements. However, challenges included short flight durations (7–13 minutes), difficulty in optimal sampling location selection due to wind variability, and nonsimultaneous vertical plume sampling.

3.2.1.3 Aircraft and satellite imagery

Each measurement technique brings its strengths and limitations to the quantification of CH_4 emissions. A multi-method approach often provides the most comprehensive assessment of farm-scale CH_4 emissions, including aircraft and satellite imagery (Tedeschi et al., 2022). As these methodologies continue to be refined, they collectively contribute to reducing uncertainty in emission estimates across various agricultural systems. However, adequate evaluation of top–down approaches, commonly used for large-scale assessments, must be systematically compared with individual animal or group-based measurements to increase confidence in the data and identify potential biases. Future research should focus on quantifying the discrepancies between these approaches across different farm sizes and management practices, as well as developing standardized protocols for integrated measurement strategies. This cross-validation is crucial for improving the accuracy of greenhouse gas inventories and informing effective mitigation strategies in animal agriculture.

3.2.1.4 Hood sniffers

As dairy cattle are brought to the milking parlor at least twice a day to harvest milk, opportunities exist for CH_4 measurement when they are restrained during milking. Some studies have used hood sniffers by adjusting for milking parlor use (Calderón-Chagoya et al., 2019). A portable system was developed to be used in milking parlors for group-level emission measurement. Hoods with one open side for head placement are constructed above feed troughs to enclose the cow's head during milking. Each hood was connected to nondispersive infrared CH_4 analyzers, and a polyethylene tubing with an air sampling rate of 2 L/min was used (Calderón-Chagoya et al., 2019). The system should be designed to minimize air currents inside the head boxes. This system, however, helps to obtain CH_4 concentration measurements with minimal disruption to daily milking routines.

3.2.1.5 Air quality sensors

Most recently, group scale emissions measurement in the milking parlor setting was evaluated (Manriquez et al., 2024) using air quality sensor platforms (IEQMAX, CO). The sensors were situated in the middle of the milking parlor and the middle of tunnel ventilation at a height of 10 feet from the ground. The sensors measure emissions at 1-minute intervals. The authors were not able to detect differences in CH_4 levels at the barn and ventilation, indicating either spot would provide accurate measurements.

3.2.2 Indirect methods

Indirect methods at the group level for measuring CH_4 emissions involve estimating methane output using proxy data or models, rather than directly measuring CH_4 gas in the air. These methods rely on indicators like feed intake, animal behavior, or environmental conditions to infer CH_4 emissions from a group of animals.

3.2.2.1 Inverse dispersion methods

Micrometeorological measurements are the strategy to measure CH_4 as measured by a flux tower attached to towers around 8 m above the soil surface (Taylor et al., 2017). The micrometeorological approach uses inverse dispersion methods (IDM) from point sources such as farms. This approach is considered economical, requires modest human resources, uses low-power portable instrumentation, can be used with naturally ventilated barns, which are most common among modern dairy farms, and does not interfere with farm activities (VanderZaag et al., 2014).

Flesch et al. (2004) demonstrated this technique, which uses downwind concentration measurements and atmospheric dispersion models to infer emission rates. While IDM can cover large areas non-intrusively, it requires specific atmospheric conditions and complex modeling. Bühler et al. (2021) evaluated IDM alongside an in-house tracer ratio method for a comparison of CH_4 emissions from a naturally ventilated dairy housing in Switzerland. The IDM yielded average CH_4 emissions of 317 g/animal/d, closely aligned with the in-house tracer method, differing by only 1–8%. The authors reported substantial data loss (57–64%) due to quality filtering and instrument failures, highlighting the method's technical demands. The EC has been used for continuous measurements with high temporal resolution, but it demands homogeneous terrain and comes with the challenges of expensive equipment and complex data processing (Peltola et al., 2014).

3.2.2.2 Tracer methods

Alternative methods, less reliant on complex meteorological data and dispersion modeling, have been employed to quantify CH_4 emissions. Mønster et al. (2014) applied the tracer ratio method to quantify CH_4 emissions from fugitive sources, a technique adaptable to agricultural settings. This method involves releasing a tracer gas at a known rate and measuring concentrations of both tracer and target gases downwind, operating under the assumption of similar dispersion characteristics for both gases. While effective, this assumption may introduce uncertainties in certain atmospheric conditions. The flux chamber method offers precise measurements for small, enclosed areas. Livingston and Hutchinson (1995) described this approach's capacity to isolate specific emission sources but noted significant challenges in scaling up to whole-farm measurements. These methods provide valuable complementary data to broader-scale techniques, though each has its limitations in representing farm-wide emissions.

4 Comparison of methods

4.1 Animal-level methods

As reported above, there are several technologies available to measure enteric CH_4 emissions in ruminants. However, all of them have positive and negative aspects that must be considered before collecting gaseous samples. For instance, RC are considered the 'gold standard' yielding the most accurate and precise measurements of gaseous emissions, including enteric CH_4 production from the rumen and the lower gastrointestinal tract (Tedeschi et al., 2022, Bekele et al., 2022). On the other hand, RC are expensive to construct and maintain, technically challenging, and restrict normal animal behavior and movement, thus not applicable for grazing studies. The automated head chamber GF system allows for multiple spot gaseous measurements from a large number of animals over extended periods under typical management conditions with minimal or no interference with cows' natural behaviors (Dorich et al., 2015, Hegarty, 2013). Despite these positive aspects, the GF apparatus requires the use of feed baits to attract the animals to the unit, does not measure hindgut gas production, visitations may not be uniform across cows, and purchase costs have skyrocketed over the past ten years. According to Garnsworthy et al. (2019), when running method comparison studies, simultaneous repeated measures per cow with two or more techniques are required for assessing differences between methods (averages), as well as random differences (precision) and correlation between methods free of residual error. Nevertheless, not all methods can be used simultaneously, and enteric CH_4 emissions change throughout the day and over lactation. Therefore, crossover or matched-pair repeated measures designs are required in conditions whereby simultaneous measurements are not possible (Garnsworthy et al., 2019).

Table 2 illustrates selected studies in which different methods were used to compare enteric CH_4 production in ruminants, including dairy cows. Specific to lactating dairy cows, Grainger et al. (2007) reported that enteric CH_4 production was similar between RC (mean = 322 g/day) and the SF_6 tracer technique (mean = 331 g/day) when measurements were taken simultaneously with

Table 2 Selected studies in which different methods were used to compare enteric methane emissions in ruminants

Animal used	Diets	Method comparison	Reference
Lactating dairy cows	Fresh grass, concentrate	RC vs. SF_6	(Grainger et al., 2007)
Whether	Alfalfa silage	RC vs. SF_6	(Pinares-Patiño et al., 2011)
Lactating dairy cows	Grass silage, concentrate	RC vs. SF_6	(Muñoz et al., 2012)
Lactating dairy cows	TMR	GF vs. SF_6	(Dorich et al., 2015)
Lactating dairy cows	TMR	GF vs. Sniffer method	(Huhtanen et al., 2015)
Lactating dairy cows	TMR	GF vs. SF_6	(Hristov et al., 2016)
Dairy heifers	Corn and grass silages, extruded flaxseed, pasture	RC vs. GF vs. SF_6	(Hammond et al., 2015)
Beef heifers	Alfalfa silage	RC vs. GF vs. SF_6	(Jonker et al., 2016)
Crossbred beef heifers	Barley silage, steam-rolled barley, molasses	RC vs. GF	(Alemu et al., 2017)
Dry dairy cows	Hay, concentrate	RC vs. GF vs. SF_6	(Doreau et al., 2018)

RC = respiration chamber; SF_6 = sulfur hexafluoride technique; GF = GreenFeed system.

these two methods. However, the CV was greater within cows between days with the SF_6 tracer technique (CV = 6.1%) vs. RC (CV = 4.3%). Furthermore, the variability among cows was greater for the SF_6 tracer technique (CV = 19.6%) than for RC (CV = 17.8%). Muñoz et al. (2012) also conducted simultaneous gaseous measurements using RC and the SF_6 tracer technique and observed similar enteric CH_4 production with both methods in the first three periods of the study (mean = 404 and 442 g/day for RC and SF_6, respectively). In contrast, CH_4 production was significantly greater when using SF_6 (mean = 447 g/day) than RC (mean = 369 g/day) in period four of the study, possibly because the SF_6 release rate of the permeation tubes decreased over time, leading to an overestimation in CH_4 output. Nevertheless, a regression using all data across the four periods revealed linear relationships between RC and SF_6 for CH_4 production (R^2 = 0.69) and CH_4 intensity (g/kg of milk yield; R^2 = 0.88), as well as a quadratic relationship for CH_4 yield (g/kg of DMI; R^2 = 0.64).

Both Dorich et al. (2015) and Hristov et al. (2016) compared the SF_6 tracer technique vs. the GF system using lactating dairy cows. Dorich et al. (2015) reported similar enteric CH_4 production comparing the SF_6 against the GF system in dairy cows fed total mixed ration ad libitum or restricted. However, the variability of the SF_6 tracer technique was substantially greater than that of GF when the whole (CV = 111% and 22.4%, respectively) or reduced (CV = 41.1% and 22.4%, respectively) enteric CH_4 emission data sets were used in the analysis. Similarly, Hristov et al. (2016) reported a greater variability for enteric CH_4 production with SF_6 than GF (38.6% and 25.8%, respectively). They also observed that, across experimental phases, CH_4 production was significantly greater for SF_6 compared with GF. Furthermore, the correlation and concordance coefficients between the two methods were relatively low, averaging 0.40 and 0.34, respectively (Hristov et al., 2016). Therefore, it can be concluded, based on results from Dorich et al. (2015) and Hristov et al. (2016), that under the conditions of these two experiments in which measurements were taken inside poorly ventilated barns, the GF method yielded enteric CH_4 production results with less variability relative to the SF_6 technique. Huhtanen et al. (2015) compared the GF system and the sniffer method across two sites (Umeå, Sweden and Michigan, United States) and observed that the between-cow CV for enteric CH_4 production was smaller (mean CV = 14.3%) with GF than the sniffer method (mean CV = 22.3%). They also showed that the GF system was more reliable in measuring enteric CH_4 production compared to the sniffer method (Huhtanen et al., 2015).

The choice of method used for measuring enteric CH_4 production in lactating dairy cows depends on the experimental design adopted, the type of facility available, study conditions (e.g. grazing, confinement), and equipment costs. Undoubtedly, RC, GF, and SF_6 have been the most common methods to measure enteric CH_4 emissions in lactating dairy cows (Tedeschi et al., 2022,

Niu et al., 2018). Comparisons among these three methods have been made in studies conducted with dairy heifers (Hammond et al., 2015), beef heifers (Jonker et al., 2016), and dry cows (Doreau et al., 2018). Hammond et al. (2015), using dairy heifers, concluded that the GF system was not able to detect treatment and animal differences in enteric CH_4 production as both RC and SF_6 were likely because of the limited number of gaseous collections and timing of measurements obtained. According to Jonker et al. (2016), average CH_4 yield in beef heifers did not significantly differ when comparing RC (mean = 24.5 g/kg of DM), SF_6 (mean = 22.8 g/kg of DM), and GF (mean = 26.2 g/kg of DM). Doreau et al. (2018) reported that CH_4 production and CH_4 yield in dry cows were significantly greater when using RC (means = 367 g/d and 27.7 g/kg of DM) than SF_6 (means = 310 g/d and 24.1 g/kg of DM) or GF (means = 319 g/d and 23.7 g/kg of DM) and that the individual correlations between methods were low to moderate, which may limit their interchangeability for quantifying gaseous emissions of individual animals. It should be noted that studies comparing RC, SF_6 tracer technique, and the GF system in the same experiment under similar conditions using lactating dairy cows are not available to date, indicating that research to address this knowledge gap is needed.

4.2 Group-level methods

It is well known that measuring enteric CH_4 emissions in commercial dairy farms is challenging because experimental conditions are difficult to control, and individual DMI measurements are usually impossible. However, technologies currently available can be used for measuring enteric CH_4 emissions from individual cows and whole farms, thus providing opportunities for developing dietary and management strategies to mitigate enteric CH_4 emissions in different dairy systems.

The GF system is one of the few technologies that can be used for individual measurements of CH_4 production in commercial dairy farms because the units can be placed inside the barns or outside in pastures with minimal interference in animal behavior. Note that the most recent GF units come with automatic calibrations, thus minimizing labor. Nevertheless, the GF units require routine maintenance, animal training, and data management, which may discourage their use in commercial dairies. In addition, the monetary investment is substantial, which can further limit farmers' adoption of GF systems unless carbon credits offset some or all of the purchase costs. We are reporting the data below from studies conducted on working dairy farms whereby technologies such as the GF system and open-path laser were used for individual animal and whole-farm measurements.

Reyes et al. (2023) conducted a study in a dairy operation in Freeport, Maine, United States (Wolfe's Neck Center for Agriculture and the Environment) in which lactating dairy cows were fed diets supplemented or not with the red seaweed *Chondrus crispus*. Specifically, cows (n = 11/treatment) were assigned to a control diet without *C. crispus* or with 6% of *C. crispus* on a DM basis, with the experiment lasting 12 weeks, including a 2-week for baseline data collection (i.e. covariate period). Enteric CH_4 production was measured using one GF unit located inside a composted bedded-pack barn, and it was 14% lower in cows fed C. *crispus* than the control diet (means = 466 and 401 g/d, respectively). However, CH_4 intensity (g/kg of energy-corrected milk) did not differ between treatments. In this study, the seaweed was individually fed when cows were restrained in head-locks, with the remaining dietary ingredients (baleage and concentrate) group fed. Therefore, total DMI was not quantified for individual cows, which is a limitation commonly found during on-farm studies. Nevertheless, the Reyes et al. (2023) study is an example of using a method (i.e. GF system) with the potential to be applied to commercial dairies for supporting dietary strategies to reduce enteric CH_4 emissions and improve energy utilization in dairy cows.

Eckert et al. (2018) investigated the effects of feeding systems (partial mixed ration or partial mixed ration plus grazing) on the diurnal concentration of enteric CH_4 in individual cows on 12 commercial dairy farms in the United Kingdom. Enteric CH_4 measurements were quantified for a minimum of seven days from individual cows' breath using CH_4 analyzers installed in robotic milking stations. Results showed that CH_4 emissions followed a consistent diurnal pattern and did not differ between the feeding systems studied. However, the response in enteric CH_4 emissions to increasing DMI was greater for cows fed partial mixed ration plus grazing than partial mixed ration alone. Eckert et al. (2018) concluded that repeated spot measurements of CH_4 emissions during milking can be used to assess the effects of feeding systems and potentially benchmark farms on their emission levels.

Pereira et al. (2023) conducted a longitudinal study over six months using a laser CH_4 detector to record enteric CH_4 emissions from Holstein and Jersey cows in an experimental dairy farm in the Azores, Portugal. The laser sensor was pointed to the cows' right nostrils at a fixed distance of 1.2 m, with 10 measurements recorded individually over time in dry and lactating dairy cows. Their results showed that the concentration of CH_4 associated with eructation was greater in Holstein (mean = 344 ppm) than in Jersey (mean = 295 ppm) cows and that CH_4 production (g/d) tended to increase in Holsteins vs. Jerseys, respectively. However, CH_4 intensity (g/kg of milk yield) did not differ significantly between breeds. Furthermore, CH_4 concentration was lower in early than mid or late lactation. Pereira et al. (2023) concluded that the laser CH_4 method is minimally invasive and relatively cheap, and it was sensitive to detect breed and lactation stage differences, thus having the potential to be used in commercial dairy farms to support management decisions to reduce enteric CH_4 emissions at the farm level.

Gao et al. (2011) conducted a study on a commercial dairy farm in China (Baoding City; North China Plain) to quantify whole-farm CH_4 emissions during the winter, spring, summer, and fall using open-path lasers in combination with an IDM. Their results showed that CH_4 emissions peaked after feeding and that daily CH_4 production averaged 0.28, 0.32, 0.33, and 0.30 kg/head for the winter, spring, summer, and fall, respectively. They also reported that the annual CH_4 production averaged 112.4 kg/head and included enteric emissions from lactating and nonlactating cows, heifers, and calves (3–12 months), as well as emissions from manure. Methane emission measurements from Gao et al. (2011) obtained with open-path lasers appear to be sensitive to feeding schedules and seasonality, suggesting that this methodology may be helpful for on-farm application and support strategies to mitigate enteric CH_4 emissions in dairy operations.

Arndt et al. (2018) estimated CH_4 emissions of 2 dairy farms in California, United States, using open-path measurements with IDM, vehicle measurements with the tracer flux ratio method, and aircraft measurements with the closed-path method. These three techniques were used to estimate whole-facility CH_4 emissions for 3–6 d per farm in the summer of year 1. During year 2, the open-path technique was used to estimate CH_4 emissions for 13–14 d per farm in the winter. Results showed that whole-facility CH_4 estimates were similar among measurement techniques, and no seasonality was detected for CH_4 emissions from animal housing. In contrast, CH_4 emissions from liquid manure storage were 3 to 6 times greater in the summer than in winter. They also observed that manure CH_4 emissions contributed 69% to 79% and 26% to 47% of whole-facility CH_4 emissions during the summer and winter measurement periods, respectively. Furthermore, CH_4 yield from animal housing was similar between farms, averaging 20.9 g of/kg of DMI. Arndt et al. (2018) concluded that whole-facility CH_4 emissions need to be measured throughout the year to estimate and evaluate annual inventories.

5 Technologies and practices to reduce methane

The primary objective in reducing CH_4 emissions must always be considered in the context of dairy productivity, specifically CH_4 emissions relative to the production of milk solids (i.e. water-free milk). While various technologies and dietary interventions can effectively lower CH_4 emissions, assessing any potential impacts on feed intake is crucial, as it may reduce milk yield, particularly in high-genetic merit cows. Therefore, strategies such as fat supplementation or feed additives should be evaluated for their ability to mitigate CH_4 concurrently to their effects on animal performance, ensuring that environmental benefits do not come at the cost of reduced milk production or overall farm efficiency (i.e. financial sustainability).

Several innovations and best practices can help reduce CH_4 emissions, such as dietary supplements (e.g. feed additives like seaweed), breeding for lower CH_4-producing cows, and improved manure management techniques. The adoption of these technologies may not only reduce CH_4 yield (g/kg of DMI) and CH_4 intensity (g/kg of energy-corrected milk) but also improve farm productivity and sustainability in confined or grazing conditions (Brito et al., 2022). Reducing CH_4 emissions from dairy farms is a critical goal in the quest to mitigate climate change. Various technologies and practices have been developed to address the two primary sources of CH_4 emissions: enteric fermentation and manure management. However, implementing these technologies and practices requires a holistic approach to farm management. Farmers must consider the specific conditions and constraints of their operations to choose the most effective strategies (Figure 11). As discussed below, financial incentives and technical support from government agencies and industry organizations can also play a significant role in encouraging the adoption of CH_4-reducing technologies.

5.1 Diet management

One of the most effective strategies for reducing CH_4 emissions from enteric fermentation involves dietary manipulation. By altering the composition of a cow's diet, farmers can reduce the amount of CH_4 produced during digestion. Incorporating more easily digestible feeds, such as high-quality forages and grains, can reduce CH_4 production. Additionally, certain feed additives have been found to inhibit CH_4-producing microbes in the rumen. For example, compounds like 3-NOP specifically target and reduce methanogenesis. Other additives, such as fats and oils, can also lower CH_4 emissions by altering the microbial population in the rumen.

Feeding management that targets reducing feed overconsumption has been shown to lower CH_4 emissions and enhance feed efficiency, particularly in the least efficient lactating dairy cows, according to Fischer et al. (2020). These researchers explored the impact of precision feed restriction on feed efficiency and CH_4 emissions in lactating Holstein cows, focusing on those that were the least efficient in their feed utilization. The trial included 68 cows, and during 92 days of feed restriction, the least efficient cows had their daily DMI reduced by 2.6 kg compared to their intake during ad libitum feeding. This reduction led to a significant decrease in CH_4 emissions by 49.2 g/day for these cows. Despite the feed restriction, there were no adverse effects on milk production, body weight, or body weight change. The study also found that the variability in feed and milk energy efficiencies among the cows was reduced, narrowing the differences between the most and least efficient cows. Notably, the most efficient

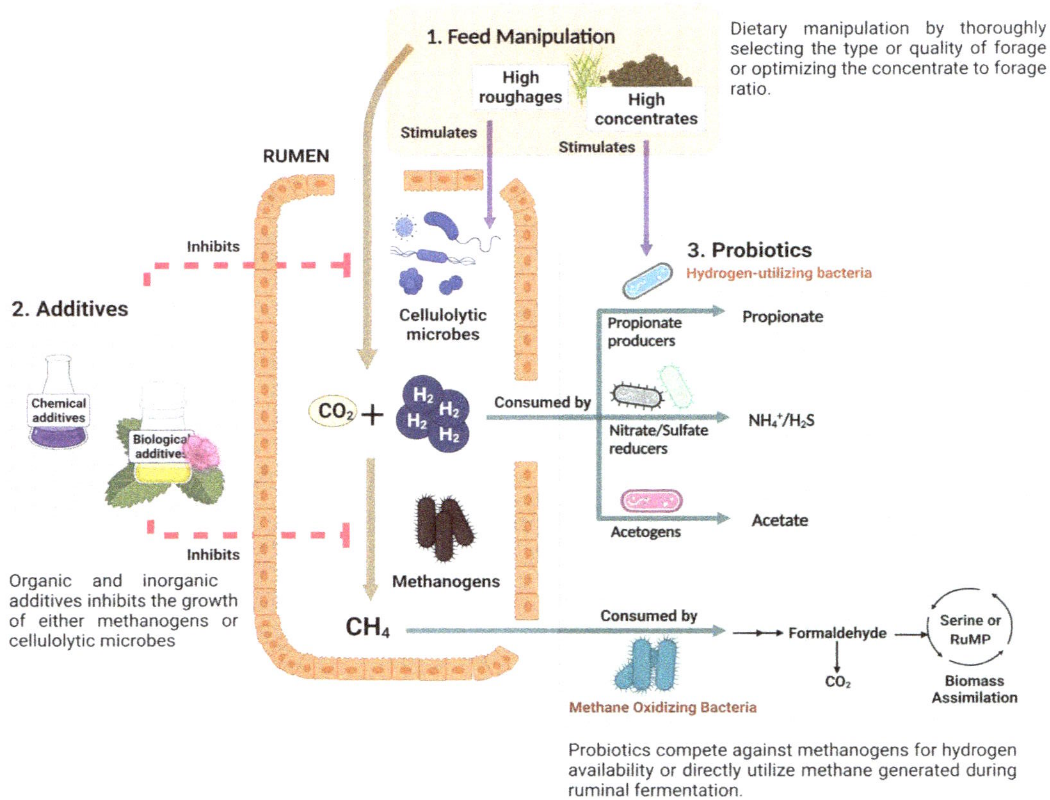

Figure 11 Feed manipulation, supplementation of additives, and probiotics. Light brown line (flow of rumen fermentation), pink line (inhibition), purple line (stimulation), and green line (consumption). Reprinted under Creative Commons Attribution 4.0 license (CC BY; https://creativecommons.org/licenses/by/4.0/) from Tseten et al. (2022).

cows under ad libitum feeding remained efficient even under feed restriction. This suggests that overconsumption of feed is a significant contributor to feed inefficiency and that managing feed intake can enhance efficiency and reduce CH_4 emissions without compromising the overall performance of the cows and heifers (Boadi and Wittenberg, 2002).

5.1.1 Forage versus concentrate

Dietary manipulation remains a potent and cost-effective approach to mitigating enteric CH_4 emissions from ruminants. Strategies primarily involve adjusting the type and quality of forage and the forage-to-concentrate ratio in the feed (Haque, 2018, Kebreab et al., 2010). High-quality forage, characterized by younger plants with higher fermentable carbohydrates and lower non-digestible fiber, promotes higher digestibility and passage rate, which shifts rumen fermentation toward propionate, an alternative hydrogen sink that reduces CH_4 production (Table 3) (Benchaar et al., 2001, Mosier et al., 1998). Supplementing forage with concentrates, which contain less cell wall material and more readily fermentable carbohydrates such as starch and sugar, further decreases CH_4 emissions by favoring propionic acid production (Johnson and Johnson, 1995). However, excessive concentrate supplementation, particularly beyond 80%, can lead to health issues like subacute ruminal acidosis,

Table 3 The effects of forage quality on intake and methane production of heifers fed ad libitum[1]

	Diet[3]			
Items[2]	High (n = 9)	Medium (n = 9)	Low (n = 9)	*P* value
Intake				
DMI, kg d^{-1}	9.7 ± 0.23[a]	8.9 ± 23[b]	6.3 ± 0.23[c]	<0.01
DOMI, kg d^{-1}	6.0 ± 0.14[a]	4.5 ± 0.14[b]	2.4 ± 0.14[c]	<0.01
GEI, kJ d^{-1}	185.6 ± 0.56[a]	159.3 ± 0.56[b]	116.2 ± 0.56[c]	<0.01
CH_4 production				
CH_4, L d^{-1}	285.2 ± 13.35[a]	289.8 ± 13.35[a]	203.5 ± 13.35[b]	<0.01
CH_4, L kg^{-1} DM	29.4 ± 1.83	32.5 ± 1.83	32.0 ± 1.83	0.46
CH_4, L kg^{-1} DOM	47.8 ± 4.02[a]	63.7 ± 4.02[b]	83.2 ± 4.02[c]	<0.01
CH_4, % GE	6.0 ± 0.38	7.1 ± 0.38	6.9 ± 0.98	0.14

[1]Values are least-square means ± standard deviations.
[2]DMI: dry matter intake; DOMI: digestible organic matter intake; GEI: gross energy intake; and CH_4: methane.
[3]The diet high-, medium-, and low-quality forages were based on in vitro organic matter digestibility 61.5%, 50.7%, and 38.5%, respectively. Note that the CH_4 (L d^{-1}) of the high-quality group (285.2) was computed as 9.7 (kg d^{-1}) x 29.4 (L kg^{-1} DM).
Adapted from Boadi and Wittenberg (2002).

highlighting the need for balanced dietary interventions (Ogata et al., 2019, Owens et al., 1998). For instance, replacing grass silage with maize silage, which provides higher starch content and improves DMI and lactating cows' performance, significantly reduces CH_4 emissions (Beauchemin et al., 2008, Hassanat et al., 2013). This combined approach of enhancing forage quality and strategically supplementing with concentrates forms a comprehensive dietary strategy to curb CH_4 emissions while maintaining animal productivity (Agle et al., 2010, Jiao et al., 2014).

The study by van Gastelen et al. (2024) examined the impact of diets, with and without 3-NOP supplementation, on CH_4 emissions from dairy cows. The diets differed according to the lactation stage: early lactation included 44.5%

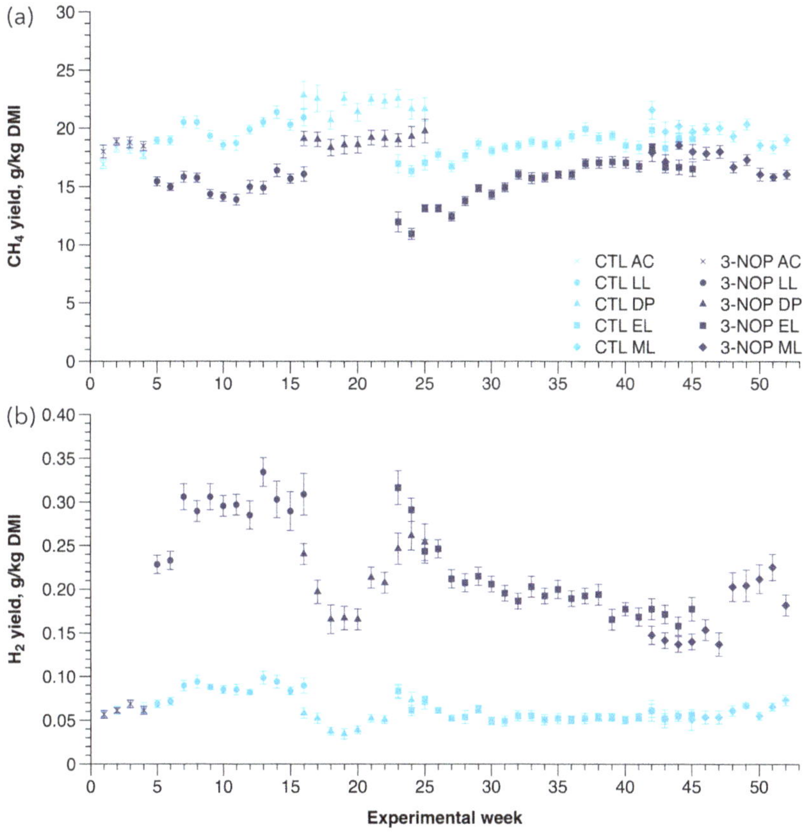

Figure 12 Methane (CH_4) yield (panel A) and H_2 yield (panel B) for the complete experimental period per diet (AC = adaptation + covariate period, LL = late lactation, DP = dry period, EL = early lactation, or ML = mid-lactation) of cows receiving a diet supplemented with a placebo (CTL) or with 3-nitrooxypropanol (3-NOP; mean ± SEM). Adapted from van Gastelen et al. (2024).

concentrate, mid-lactation had 34.8% concentrate, late lactation contained 24.1% concentrate, and the dry period diet had 18.7% concentrate. The results indicated that CH_4 reduction was influenced by diet composition, particularly the concentrate amount and dietary neutral detergent fiber. Diets with higher levels of concentrate generally led to a more substantial reduction in CH_4 emissions (g/kg DMI), regardless of 3-NOP supplementation (Figure 12). These findings suggest that diets with a higher proportion of concentrate levels and a lower proportion of neutral detergent fiber content can improve the effectiveness of CH_4 mitigation strategies in dairy cows.

5.1.2 Fat supplementation

Adding fat to ruminant diets has emerged as a promising strategy for mitigating CH_4 emissions, traditionally used to enhance dietary energy content for high-producing dairy cows. When carbohydrates in the diet are replaced with fats, there is a reduction in fermentation and, consequently, CH_4 production, primarily due to the reduction of organic matter fermentation and fiber digestibility, along with the direct inhibition of methanogens via hydrogenation of unsaturated fatty acids (Johnson and Johnson, 1995). Unsaturated fatty acids, particularly medium-chain fatty acids (C8: C14) from sources such as coconut and palm oil, are most effective, acting as hydrogen sinks through dehydrogenation processes (Boadi et al., 2004). Although biohydrogenation only minimizes hydrogen utilization (1–2%) (Giger-Reverdin et al., 2003), it remains a critical mechanism when methanogenesis is inhibited. Studies have shown that low levels of lipid supplementation (<4% of dietary DMI) can reduce CH_4 production by up to 20%, though results vary. A meta-analysis indicates a decrease of 1–5% in CH_4 emissions per 10 g/kg of dry matter (DM) of dietary fat (Grainger and Beauchemin, 2011, Patra, 2012). However, lipid supplementation is costly and can negatively impact fiber digestibility, DMI, rumen fermentation, milk fat synthesis, and fatty acid composition in products (Doreau and Chilliard, 1997, Grainger and Beauchemin, 2011). Despite its immediate applicability on commercial farms, the overall potential for CH_4 mitigation through lipid supplementation is limited due to cost and possible adverse effects (e.g. Table 2 in Patra (2013)) on animal productivity and product quality.

5.1.3 Ionophores

Ionophores, such as monensin (Rumensin), lasalocid (Bovatec), salinomycin (Bio-cox, Sacox), and laidlomycin (Cattlyst), have been widely used since their approval by the United States FDA in 1975 (Russell and Strobel, 1989). These compounds enhance energy metabolism efficiency and improve ruminal nitrogen metabolism while reducing the risk of bloating and acidosis (Bergen and Bates, 1984, Richardson et al., 1976) as well as decreasing the deamination of amino acids in the rumen (Tedeschi et al., 2003, 2011). Ionophores modulate the ratio of propionic to acetic acid production, improving feed efficiency and resulting in body weight gain (Odongo et al., 2007, Singh and Mohini, 1999, Thivend and Jouany, 1983). Their antimicrobial action disrupts ion concentration gradients in microbial membranes, inhibiting gram-positive bacteria that produce lactate, acetate, butyrate, formate, and hydrogen, thereby reducing hydrogen availability for methanogens (Marques and Cooke, 2021, McGuffey et al., 2001, Russell and Houlihan, 2003). Supplementation with ionophores has been shown to significantly decrease the ciliate protozoal population and lower CH_4 production, with reported reductions of up to 30% in beef cattle and over 9% in dairy cows (Guan et al., 2006, Intergovernmental Panel on Climate Change, 2007, McGuffey et al., 2001, Odongo et al., 2007). The results of a meta-analysis (Appuhamy et al., 2013) indicated that the effects of monensin on CH_4 production in dairy cows were inconsistent, with an equal number of studies showing positive and negative results (Figure 13), likely because dairy cows usually receive lower monensin doses (21 mg/kg DM) over a longer duration compared to beef steers, which receive higher doses (32 mg/kg DM) over shorter periods (Appuhamy et al., 2013). However, their efficacy diminishes over time due to microbial adaptation and the development of resistance (Appuhamy et al., 2013, Guan et al., 2006, Patra et al., 2017, Tedeschi et al., 2003). Additionally, public pressure to reduce antimicrobial feed additives limits their long-term use for CH_4 mitigation in ruminants (Beauchemin et al., 2008, Guan et al., 2006, Johnson and Johnson, 1995).

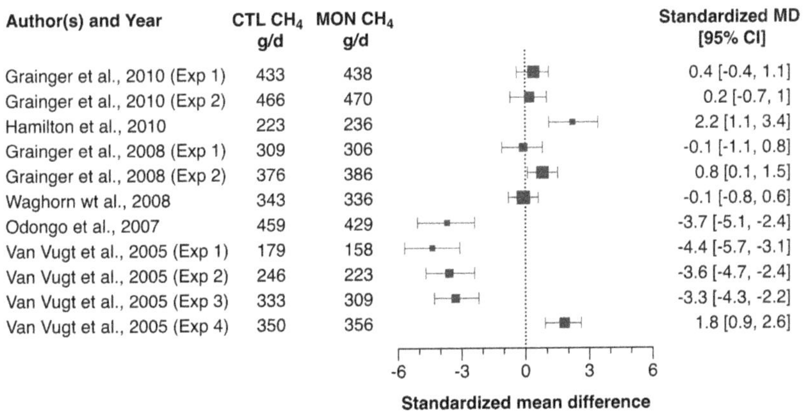

Figure 13 Forest plot showing mean methane production (g/d) in control (CTL CH_4) and monensin treatment (Mon CH_4) groups along with standardized mean difference (MD) and its 95% confidence interval (CI) for dairy cow studies. The dotted line represents a 0 standardized mean difference. Adapted from Appuhamy et al. (2013).

5.1.4 Phytocompounds

The potential of phytocompounds or phytochemicals to mitigate CH_4 production in ruminants has garnered significant research interest (Tedeschi et al., 2021), particularly following the European Union's 2006 ban on antibiotics as growth promoters (https://www.abdn.ac.uk). Various secondary plant compounds, such as essential oils, tannins, and saponins, have demonstrated promising anti-methanogenic properties (Tedeschi et al., 2021). Essential oils, which are volatile and aromatic oily liquids extracted from diverse plant materials, have shown broad-spectrum antimicrobial properties. A meta-analysis by Belanche et al. (2020) revealed that supplementing dairy cows with 1 g/d of Agolin® Ruminant (a specific essential oils blend) increased milk yield by 3.6% and decreased CH_4 production per day by 8.8% without changing the rumen fermentation pattern. Their efficacy is primarily attributed to their chemical constituents and functional groups, such as terpenoids and phenols, which interact with microbial cell membranes, thereby reducing microbial populations, including methanogens. Studies have shown that essential oils from plants like garlic, thyme, and eucalyptus can significantly decrease CH_4 production in vitro, although long-term in vivo effects remain extensively evaluated (Cobellis et al., 2016).

Additionally, tannins, polyphenolic compounds found in various plants, have shown variable effects on CH_4 reduction depending on their source, type, and molecular weight. A meta-analysis indicated that increasing levels of tannins decreased CH_4 production, with in vitro studies reasonably predicting in vivo responses up to a particular concentration (Figure 14) (Jayanegara et al., 2012). Furthermore, saponins have also been effective, with studies reporting up to a 29% reduction in CH_4 emissions in vitro (Guyader et al., 2017). Despite the promising results, the variability in response and potential adverse effects on animal performance necessitate further research to optimize these phytocompounds in practical livestock farming.

5.1.5 Other organic additives

Macroalgae and yeast-based probiotics are promising phytocompounds for reducing CH_4 production in the rumen of ruminants. Macroalgae, such as *Dictyota* and *Asparagopsis taxiformis* (AT), have demonstrated significant potential in inhibiting methanogenesis. Studies by Machado et al. (2014) and Kinley et al. (2016) have shown that AT, when supplemented at 20 g/kg of forage, can nearly eliminate CH_4 production without affecting forage digestibility. Li et al. (2018) further reported that supplementing sheep diets with up to 3% AT can reduce CH_4 emissions by 80% over 72 days without compromising body mass gain. However, some studies have shown a decrease in intake when AT

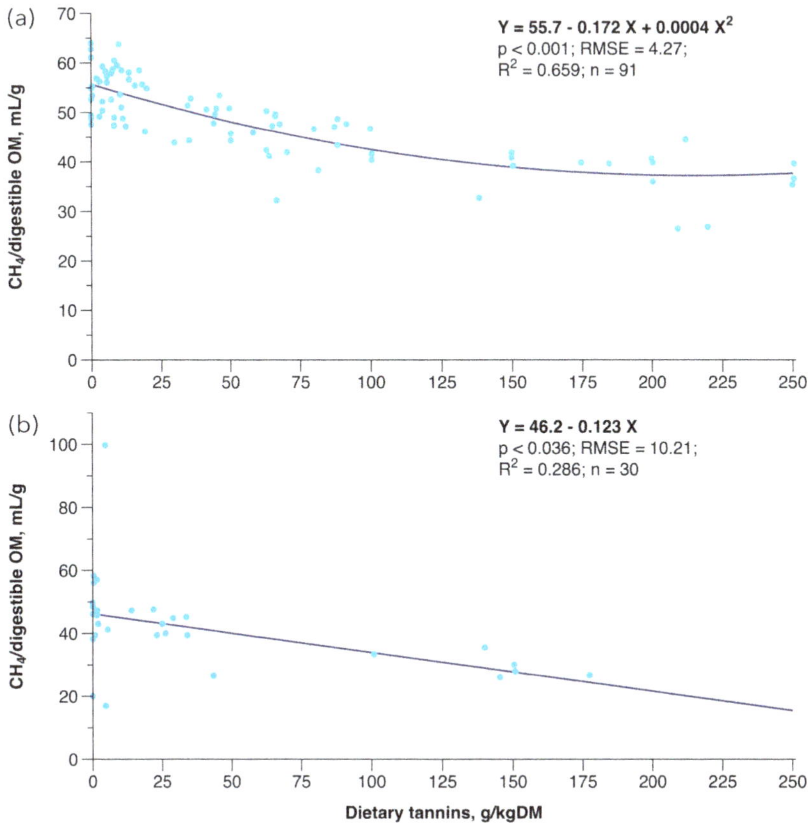

Figure 14 Relationships between dietary tannins (g/kg dry matter) and methane (CH_4) per digestible organic matter (ml/g) from (a) in vitro batch culture and (b) in vivo experiments. Adapted from Jayanegara et al. (2012).

was used at concentrations higher than 0.5% of DM (Min et al., 2021), suggesting that further research is necessary before making a recommendation. Also, the safety of bromoform, a halogenated compound found in these algae, must be assessed due to its environmental and health risks. Additionally, a life cycle assessment is necessary to evaluate the carbon footprint associated with algae production and transportation.

In the study by Stefenoni et al. (2021), the use of AT as a dietary supplement in lactating dairy cows, when included at 0.5% of DM, AT reduced daily enteric CH_4 emissions by up to 65% and CH_4 yield by 55% during the initial experimental periods. However, the effectiveness of AT declined in later periods, likely due to a substantial decrease (up to 84% over four months) in the concentration of bromoform ($CHBr_3$; Figure 15), an essential compound responsible for its anti-methanogenic properties. Despite its efficacy in mitigating CH_4 emissions,

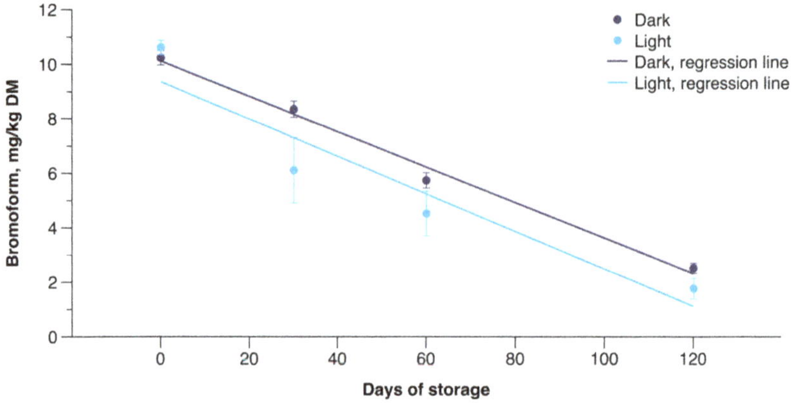

Figure 15 Effect of storage conditions and storage time on concentration of bromoform ($CHBr_3$) in *Asparagopsis taxiformis*. Light = seaweed material exposed to continuous fluorescent light; Dark = seaweed material stored in brown glass bottles with plastic lids. Adapted from Stefenoni et al. (2021).

the study also found that AT supplementation reduced DMI and milk yield. These findings highlight the promise and the challenges of using AT as a CH_4 mitigation strategy.

Yeast-based probiotics, particularly those containing *Saccharomyces cerevisiae*, are also gaining traction for improving rumen fermentation, DMI, and milk yield while reducing CH_4 emissions. These probiotics modify hydrogen production and the rumen microbial community, thereby inhibiting methanogenesis (Vallejo-Hernández et al., 2018, Zanferari et al., 2018). Despite their potential and relatively low cost, further research is needed to optimize these products for CH_4 abatement. In addition to algae and yeast, biochar, prebiotics such as chitosan and inulin, and other seaweeds have shown anti-methanogenic properties, though their use in ruminant diets remains limited and warrants further investigation (Sun et al., 2021). Collectively, these phytocompounds offer promising avenues for reducing enteric CH_4 emissions, but their widespread adoption will require thorough safety evaluations and further refinement to maximize their efficacy and acceptance in ruminant production systems.

5.1.6 Methanogenesis inhibitors

Methyl-coenzyme M reductase (McR) is pivotal in anaerobic methanogenesis, catalyzing the final step where CH_4 is released (Shima et al., 2012). Disrupting this process via halogenated and nitro-derivatives of hydrocarbons has shown promise. Bromoethane sulfonate and bromopropanesulfonic acid, which mimic coenzyme M, reduce CH_4 emissions by up to 80% without impacting organic matter digestibility and volatile fatty acid concentrations (Dong et al., 1999, Hwang et al., 2012). Chloroform and other derivatives affect specific steps in the methanogenesis pathway, notably decreasing CH_4 production by 30% in cattle (Martinez-Fernandez et al., 2018). Notably, 3-NOP, a nitro derivative, binds to McR's active site, inactivating it through oxidation, leading to a 20–60% reduction in CH_4 emissions depending on supplementation duration and method (Jayanegara et al., 2018, Zhang et al., 2018). Despite its inhibitory success, 3-NOP requires low dosages (1–2 g/day), demonstrates high specificity, and breaks down to safe byproducts, making it a promising candidate for large-scale use (Thiel et al., 2019a, 2019b). However, long-term CH_4 reduction can result in hydrogen accumulation in the rumen, with unknown effects, and microbial resistance may emerge over time (Mitsumori et al., 2012). However, the study by van Gastelen et al. (2024) explored the long-term effects of 3-NOP on CH_4 emissions and milk production in dairy cows over a full year, encompassing all stages of lactation. The results demonstrated that 3-NOP supplementation led to significant reductions in CH_4 emissions, with an overall decrease of 21% in total CH_4 production, 20% in CH_4 yield per kilogram of DMI, and 27% in CH_4 intensity per kilogram of fat- and protein-corrected milk. Additionally, the study found that 3-NOP improved milk production characteristics, including a 6.5% increase in energy-corrected milk yield and enhanced feed efficiency. However, the efficacy of 3-NOP was influenced by factors such as lactation stage, diet type, and composition, with varying levels of CH_4 reduction observed at different points in the trial. The study concluded that while 3-NOP effectively reduces CH_4 emissions and improves milk production, its long-term efficacy is closely tied to dietary factors, necessitating further research to optimize its use in varying nutritional conditions (Figure 12).

While these inhibitors have demonstrated potential for CH_4 mitigation, their widespread adoption will hinge on balancing efficacy, safety, cost, and obtaining regulatory approval. Additionally, the high costs and stringent regulatory processes present significant challenges for their practical application (Boadi et al., 2004, Mitsumori et al., 2012).

5.1.7 Nanoclays and other nano-particles

The application of feed additives, such as natural zeolite clay, in ruminant diets is a promising solution to improve animal performance by manipulating ruminal fermentation patterns and enhancing nutrient utilization. Zeolite, composed of crystalline aluminosilicates, has high cation exchange capacity and sorbent properties (Baglieri et al., 2013, Jaynes et al., 2007). Zeolite's high cation exchange capacity and sorbent properties enable it to modify ruminal fluid viscosity and bind NH_3-N, thereby improving nutrient utilization (Baglieri et al., 2013, Jaynes et al., 2007). Studies have shown that dietary supplementation with zeolite, ranging from 1% to 9% of DM, positively affects nutrient digestion and animal health, with nano-zeolite enhancing the adsorption and absorption properties (Mahler et al., 2012, Trckova et al., 2004). A comprehensive review by Valpotić et al. (2018) summarizes the potential benefits of clinoptilolite, a specific type of natural zeolite, in dairy cattle, including its ability to improve energy status, reproductive efficiency, and milk production in dairy cows. Although zeolite is safe for long-term use, with no adverse effects on trace element bioavailability or hematological parameters, it is unlikely to replace conventional antibiotic growth promoters completely, and further research is needed to understand its potential and limitations fully. Additionally, El-Nile et al. (2021) found that both natural and nano forms of zeolite reduced CH_4 production by 9.9%. This finding indicates that as a feed additive, zeolite effectively enhances the rumen environment, underscoring its potential as a promising modifier for reducing enteric CH_4 emissions. However, due to the limited number of studies on this topic, we advise caution in recommending natural clays specifically for CH_4 reduction until more research is available.

5.2 Cow breeding

Selective breeding is another promising approach to reducing enteric CH_4 emissions. By identifying and breeding cows that naturally produce less CH_4, farmers can gradually decrease the overall CH_4 output of their herds. Advances in genomic technologies have made identifying genetic markers associated with lower CH_4 production possible, allowing for more targeted and effective breeding programs. Several studies have investigated the potential of selective breeding to reduce CH_4 emissions in cattle. These studies generally focus on identifying genetic traits and host-microbiome interactions that contribute to lower CH_4 production. One significant approach involves incorporating CH_4 emissions as a breeding goal. This method has been shown to potentially reduce CH_4 intensity by up to 24% by 2050, highlighting the effectiveness of genetic improvement as a mitigation strategy (González-Recio et al., 2020). Others have identified specific genetic markers and host-microbiome interactions that influence CH_4 emissions. For instance, some companies have claimed to use phenotypic and metagenomic data from over 1000 dairy cows across two breeds to identify genetic traits associated with lower CH_4 production. Other studies have explored the phenotypic and genetic relationships between CH_4 emissions recorded at different life stages and under various physiological conditions, emphasizing the importance of comprehensive recording strategies to optimize breeding for reduced CH_4 production. Overall, these studies suggest that selective breeding is a viable and promising strategy for reducing CH_4 emissions from cattle, contributing to more sustainable livestock farming practices in the long term.

The heritability of absolute CH_4 production in cattle and sheep has been found to be moderate, with higher heritability observed in sheep compared to dairy cows (Manzanilla-Pech et al., 2021). This implies that selective breeding for low CH_4 production could lead to permanent and cumulative gains in reducing CH_4 emissions, as improvements associated with host genetics are lasting (de Haas et al., 2021).

Selective breeding programs have shown promising results in reducing CH_4 emissions and improving productivity. For example, over ten years, a breeding program that selected ewes for lower CH_4 yield achieved a 12% reduction in CH_4 emissions. Interestingly, these ewes also weaned heavier and leaner lambs that produced more wool, indicating that selection for low CH_4 production did not compromise productivity (Rowe et al., 2019). These findings suggest that with the proper selection criteria, it is possible to breed environmentally efficient and productive animals.

Studies on residual CH_4 production further support the potential of selective breeding as a CH_4 mitigation strategy. Research involving 15 320 records from 2990 dairy cows across multiple countries found that residual CH_4 metrics are more appropriate for inclusion in selection indices than absolute CH_4 production metrics (Manzanilla-Pech et al., 2021). This approach allows for more efficient genetic selection, as it accounts for variables such as DMI, daily body mass gain, and milk production, making it a more comprehensive tool for reducing CH_4 emissions in cattle.

However, the implementation of genetic selection for low CH_4 production faces practical challenges, particularly in measuring CH_4 emissions on a large scale in commercial farms. While technologies such as sniffers for measuring CH_4 concentration during feeding or milking have shown some success, the need for continuous measurements over weeks and the requirement of thousands of data points for a genetic selection program make it challenging (Løvendahl et al., 2018, de Haas et al., 2017). Nonetheless, developing reliable and practical proxies for CH_4 production, such as feed intake behavior and rumen volatile fatty acid concentration, could facilitate the broader application of this mitigation strategy in the future (Beauchemin et al., 2020).

In addition to the promising results of selective breeding for reduced CH_4 emissions, it would be helpful to recognize some limitations in the current research. Many studies focus on empirical associations between genetic data and performance outcomes, often neglecting a deeper analysis of the biological processes driving CH_4 emissions. For example, while breeding programs may identify genetic markers or traits associated with lower CH_4 production, the lack of integration with fundamental biological concepts such as the nature of ruminal fermentation, appetite regulation, and milk yield can limit the relevance of these findings. Understanding the trade-offs between CH_4 production and physiological functions, such as energy partitioning for milk production vs.

metabolic demand, is critical for developing more comprehensive breeding strategies. One area that requires further exploration is the relationship between the efficiency of rumen fermentation and its impact on CH_4 emissions. A deeper mechanistic analysis of how selective breeding influences microbial populations, fermentation end-products, and feed intake behavior would provide valuable insights beyond empirical, correlational analyses. Additionally, breeding strategies that consider residual CH_4 production have shown potential in accounting for factors like DMI and productivity. However, this approach still relies heavily on empirical data, and more work is needed to connect these traits with underlying biological pathways to improve predictive accuracy.

In light of these challenges, future breeding programs should prioritize integrating a mechanistic understanding of rumen microbiology and animal physiology with genomic and phenotypic data. Such an approach would ensure that selective breeding for low methane emissions does not inadvertently compromise important traits like milk yield or appetite regulation, ultimately leading to more sustainable and productive livestock systems.

5.3 Cow health

Improving overall animal health and productivity also plays a significant role in reducing CH_4 emissions. Healthier cows with higher milk yields produce less CH_4 per liter of milk. Ensuring optimal nutrition, health care, and management practices can lead to more efficient production and lower emissions. Technologies such as precision feeding systems, which tailor feed rations to the specific needs of each cow, can enhance feed efficiency and reduce CH_4 production. Poor health in cows, including diseases and suboptimal environmental conditions, leads to decreased DMI and compromised performance, which in turn increases CH_4 emission intensity. Healthy cows with higher productivity tend to have lower CH_4 emissions per unit of milk produced. For instance, cows with multiple lactations show up to a 40% reduction in CH_4 emission intensity compared to those with only one lactation, highlighting the importance of longevity and health in reducing emissions (Knapp et al., 2014, Kebreab et al., 2006). Maintaining good cow health through proper nutrition, genetic selection, and effective health management practices can significantly reduce CH_4 emissions in cattle. This holistic approach not only benefits the environment but also enhances the productivity and sustainability of dairy farming.

5.4 Manure management

Manure management practices are equally crucial in the effort to reduce CH_4 emissions. One effective technology is the use of anaerobic digesters. These systems capture CH_4 from manure storage facilities and convert it into biogas, which can be used as a renewable energy source. This not only reduces CH_4 emissions but also provides a valuable energy resource for the farm. Anaerobic digesters can be particularly effective in large dairy operations where large amounts of manure are produced and stored. Another important practice is the frequent removal and proper handling of manure. By minimizing the time manure spends in anaerobic conditions, CH_4 production can be reduced. Immediate application of manure to fields or composting it under aerobic conditions can further decrease CH_4 emissions. Separating solids from liquids in manure management systems can also help, as solid manure decomposes more slowly and produces less CH_4. Innovative technologies are continuously being developed to enhance these practices. For instance, automated manure scraping systems can frequently remove manure from barns, reducing the potential for anaerobic decomposition. Advanced composting systems that optimize aerobic conditions can further reduce CH_4 emissions from manure.

There is emerging interest in using CH_4 inhibitors to reduce CH_4 emissions from manure or slurry. While CH_4 inhibitors have primarily been developed to target enteric CH_4 emissions in ruminants, recent research suggests they may also have potential in manure management systems. Biological inhibitors, such as microbial additives or chemical compounds, could influence the microbial communities responsible for methane production during anaerobic decomposition in manure storage facilities. However, the effectiveness of these inhibitors may vary due to the different fermentation processes occurring in stored manure compared to the rumen. On the economic side, the feasibility of applying methane inhibitors to manure management would depend on their cost, ease of application, and the potential savings from reducing methane emissions or enhancing biogas production. While promising, more research is needed to fully assess methane inhibitors' biological impact and economic viability in this context.

6 Policies, challenges, and opportunities

In addition to environmental impact, dairy farmers face a multiplicity of challenges, such as animal health, finances, and overall farm management. While CH_4 management is another layer of responsibility, it can be integrated into broader farm management practices that enhance efficiency and productivity. Tools and policies should support farmers in this transition rather than impose new burdens.

6.1 Policies

Given the challenges that dairy producers are facing, the development of supportive policies for dairy farmers is critical if we are to meet sustainable development goals established by the United Nations in 2015 (https://sdgs.un.org/goals) to address various social, economic, and environmental challenges to promote prosperity while protecting the planet, including, more specifically, zero hunger (i.e. food for all) and climate action (i.e. combat climate change and its impacts). Such supportive policies could include financial incentives for adopting green technologies, technical support, and research into more cost-effective methods of emission measurement and reduction. Effective policies and support mechanisms are essential for encouraging and enabling dairy farmers to measure and reduce CH_4 emissions. Governments, industry organizations, and research institutions all have roles to play in creating an environment that supports sustainable practices in dairy farming.

One of the most impactful forms of support is financial assistance. Governments can provide subsidies, grants, and low-interest loans to help farmers invest in CH_4 reduction technologies and practices. For example, the United States Department of Agriculture offers programs such as the Environmental Quality Incentives Program, which provides financial and technical assistance to farmers implementing conservation practices, including those that reduce CH_4 emissions. Similarly, the European Union's Common Agricultural Policy includes funding for sustainable farming practices. Financial incentives can significantly offset the high upfront costs associated with technologies like anaerobic digesters, precision feeding systems, and CH_4 measurement equipment. By reducing the financial barriers to adoption, these programs make it more feasible for farmers to invest in long-term sustainability.

Beyond financial assistance, technical support and education are crucial. Agricultural extension services and research institutions can provide farmers with the knowledge and skills needed to implement CH_4 reduction strategies effectively. This support can take the form of workshops, training programs, online resources, and one-on-one consultations. For instance, universities and agricultural research centers often run extension programs that offer practical advice and hands-on training in new technologies and best practices. These programs can help farmers understand the benefits of CH_4 reduction, navigate the complexities of implementation, and troubleshoot any issues that arise.

Continued research and development are vital for advancing CH_4 reduction technologies and practices. Governments and industry organizations can fund research projects that explore new methods for reducing CH_4 emissions, improve existing technologies, and develop cost-effective solutions. Collaboration between public research institutions and private companies can accelerate innovation and bring new products to market more quickly. For example, research into feed additives that reduce enteric CH_4 emissions, such as seaweed extracts and other dietary supplements, is ongoing and shows promising results. Funding for research and development can help bring these innovations from the lab to the farm, providing farmers with more effective tools for CH_4 mitigation.

Regulatory frameworks that recognize and reward sustainable practices can also drive the adoption of CH_4 reduction strategies. Governments can establish regulations that set emissions targets for the agricultural sector, including specific requirements for CH_4 reduction. These regulations can be coupled with incentives for compliance, such as tax breaks or access to additional funding. Moreover, creating certification programs for sustainably produced dairy products can help farmers who adopt CH_4 reduction practices differentiate themselves in the market. Certification schemes can provide a market advantage, allowing farmers to command higher prices for their products and appeal to environmentally conscious consumers.

Market-based mechanisms, such as carbon credits and emissions trading systems, offer another avenue for supporting CH_4 reduction. Farmers who reduce their CH_4 emissions can earn carbon credits, which can be sold to other businesses seeking to offset their emissions. This provides a financial incentive for farmers to invest in CH_4 reduction technologies and practices. For example, California's cap-and-trade program allows dairy farmers to generate carbon credits by installing anaerobic digesters. These credits can then be sold on the carbon market, providing a new revenue stream and encouraging the adoption of CH_4 reduction technologies.

Finally, collaboration between various stakeholders is essential for creating a supportive environment for CH_4 reduction. Governments, industry groups, research institutions, and non-governmental organizations can work together to develop comprehensive strategies that address the technical, financial, and educational needs of farmers. Collaborative initiatives, such as industry-wide sustainability programs, can set common goals and standards for CH_4 reduction. These initiatives can also facilitate the sharing of best practices and success stories, helping to build a culture of sustainability within the dairy industry.

6.2 Challenges

Dairy farmers face numerous challenges in their efforts to measure and reduce CH_4 emissions. One significant challenge is the financial burden associated with implementing new technologies and practices. Many CH_4 reduction strategies, such as anaerobic digesters or precision feeding systems, require a substantial upfront investment. Smaller farms, in particular, may struggle to afford these technologies without financial assistance or incentives.

Additionally, there is a complexity in accurately measuring CH_4 emissions on farms (Tedeschi and Beauchemin, 2023). Direct measurement techniques, while precise, can be labor-intensive and expensive. Indirect methods and models rely on accurate and comprehensive data, which can be difficult to obtain consistently. Variability in emissions due to factors such as diet, weather, and individual animal differences further complicates measurement efforts.

Another challenge is the need for continuous education and training. Farmers must stay informed about the latest research, technologies, and best practices for reducing CH_4 emissions. This requires ongoing support from agricultural extension services, research institutions, and industry organizations. However, the availability and accessibility of such resources can vary widely. Farmers also have to balance CH_4 reduction efforts with other critical aspects of farm management. They must ensure that new practices do not negatively impact animal health, milk production, or overall farm profitability. Integrating CH_4 reduction into the broader context of farm operations can be complex and requires careful planning and management.

6.3 Opportunities

Despite these challenges, there are significant opportunities for the dairy industry to advance CH_4 reduction efforts and achieve greater sustainability. One promising area is the development and adoption of new technologies. Innovations in CH_4 measurement, such as continuous monitoring systems and drone-based sensors, can provide more accurate and comprehensive data, facilitating better management decisions. Emerging feed additives and dietary supplements also offer the potential for substantial CH_4 reductions. Research into compounds like seaweed extracts, which have been shown to reduce CH_4 emissions from enteric fermentation, is ongoing. As these additives become more effective and affordable, they can be widely adopted by dairy farms.

Another opportunity lies in the integration of CH_4 reduction strategies with other sustainability practices. For example, anaerobic digesters not only reduce CH_4 emissions but also produce biogas, which can be used as a renewable energy source. This can help farms become more energy self-sufficient and reduce their reliance on fossil fuels. Additionally, improved manure management practices can enhance soil health and fertility, providing agronomic benefits alongside environmental ones.

Collaboration and knowledge-sharing within the industry can also drive progress. Farmer cooperatives, industry groups, and research networks can facilitate the dissemination of best practices and innovations. Pilot projects and demonstration farms can serve as valuable case studies, showcasing successful implementations and providing practical insights for other farmers.

Policy and regulatory support will be crucial in overcoming financial and technical barriers. Governments can provide subsidies, grants, and low-interest loans to help farmers invest in CH_4 reduction technologies. Regulatory frameworks that recognize and reward sustainable practices can also incentivize farmers to adopt new methods.

Furthermore, consumer demand for sustainably produced dairy products presents an opportunity for farms to differentiate themselves in the market. As awareness of environmental issues grows, consumers are increasingly willing to pay a premium for products that are produced using environmentally friendly practices. By marketing their sustainability efforts, dairy farms can enhance their brand value and potentially increase profitability.

7 Conclusion

Measuring and managing CH_4 emissions in dairy farms is vital in the context of global environmental health and the sustainability of the dairy industry. However, adding the responsibility of managing CH_4 emissions to the already heavy load of dairy farmers can be challenging. In hindsight, if approached through integrated farm management practices and supported by appropriate technologies and policies, it can lead to a more sustainable, efficient, and potentially more profitable farming operation. These measures must not be viewed simply as additional burdens but as part of a holistic approach to modern farming that benefits the farmers, the environment, and society at large. Therefore, a balanced approach that considers the economic and operational realities of dairy farming must be followed. Measuring CH_4 emissions from dairy farms requires a combination of direct and indirect methods to obtain a comprehensive understanding of emissions. By leveraging the strengths of each method and incorporating new technologies, it is possible to develop more accurate and efficient strategies for monitoring and mitigating CH_4 emissions in the dairy industry. Reducing CH_4 emissions from dairy farms involves a multifaceted approach that includes dietary adjustments, selective breeding, improved animal health and productivity, and advanced manure management practices. By integrating these strategies and leveraging innovative technologies, the dairy industry can make substantial progress in mitigating its environmental impact while maintaining productivity and profitability. While the challenges of measuring and reducing CH_4 emissions on dairy farms are significant, there are also numerous opportunities for progress. Advances in technology, innovative dietary solutions, integrated sustainability practices, and supportive policies can all contribute to a more sustainable and resilient dairy industry. By embracing these opportunities and addressing the challenges, dairy farmers can play a vital role in mitigating climate change and promoting environmental stewardship. Policy and support mechanisms play a crucial role in enabling dairy farmers to measure and reduce CH_4 emissions. Financial incentives, technical support, research funding, regulatory frameworks, market-based mechanisms, and collaborative approaches all contribute to creating a sustainable and resilient dairy industry. By leveraging these tools, policymakers and industry leaders can help farmers adopt effective CH_4 reduction strategies, ultimately contributing to global efforts to mitigate climate change.

8 Abbreviations

3-NOP = 3-nitrooxypropanol; AT = *Asparagopsis taxiformis*; CH_4 = methane; CO_2 = carbon dioxide; CV = coefficient of variation; DM = dry matter; DMI = dry matter intake; DNA = deoxyribonucleic acid; EC = Eddy covariance; FAO = Food and Agriculture Organization; FM = face mask; GF = GreenFeed; GQS = open-circuit gas quantification system; IDM = inverse distribution method; McR = methyl-coenzyme M reductase; RC = respiration chamber; RNA = ribonucleic acid; rRNA = ribosomal ribonucleic acid; SF_6 = sulfur hexafluoride; SR = stocking rate; and UAV = unmanned (unperson) aerial vehicle.

9 References

Agle, M., et al. 2010. Effect of dietary concentrate on rumen fermentation, digestibility, and nitrogen losses in dairy cows. *Journal of Dairy Science*, 93, 4211–4222. https://doi.org/10.3168/jds.2009-2977.

Alemu, A. W., et al. 2017. Enteric methane emissions from low- and high-residual feed intake beef heifers measured using GreenFeed and respiration chamber techniques. *Journal of Animal Science*, 95, 3727–3737. https://doi.org/10.2527/jas.2017.1501.

Amon, B., et al. 2006. Methane, nitrous oxide and ammonia emissions during storage and after application of dairy cattle slurry and influence of slurry treatment. *Agriculture, Ecosystems & Environment,* 112, 153–162. https://doi.org/10.1016/j.agee.2005.08.030.

Andersen, T., et al. 2018. A UAV-based active AirCore system for measurements of greenhouse gases. *Atmospheric Measurement Techniques*, 11, 2683–2699. https://doi.org/10.5194/amt-11-2683-2018.

Appuhamy, J. A., et al. 2013. Anti-methanogenic effects of monensin in dairy and beef cattle: a meta-analysis. *Journal of Dairy Science*, 96, 5161–5173. https://doi.org/10.3168/jds.2012-5923.

Arndt, C., et al. 2018. Short-term methane emissions from 2 dairy farms in California estimated by different measurement techniques and us Environmental Protection Agency inventory methodology: a case study. *Journal of Dairy Science*, 101, 11361–11479. https://doi.org/10.3168/jds.2017-13881.

Baglieri, A., et al. 2013. Organically modified clays as binders of fumonisins in feedstocks. *Journal of Environmental Science and Health, Part B*, 48, 776–783. https://doi.org/10.1080/03601234.2013.780941.

Beauchemin, K. A. 2018. Invited review: current perspectives on eating and rumination activity in dairy cows. *Journal of Dairy Science*, 101, 4762–4784. https://doi.org/10.3168/jds.2017-13706.

Beauchemin, K. A., et al. 2008. Nutritional management for enteric methane abatement: a review. *Australian Journal of Experimental Agriculture*, 48, 21–27. https://doi.org/10.1071/EA07199.

Beauchemin, K. A., et al. 2020. Review: fifty years of research on rumen methanogenesis: lessons learned and future challenges for mitigation. *Animal*, 14, s2–s16. https://doi.org/10.1017/S1751731119003100.

Bekele, W., et al. 2022. Contemporary methods of measuring and estimating methane emission from ruminants. *Methane* [Online], 1. https://doi.org/10.3390/methane1020008.

Belanche, A., et al. 2020. A meta-analysis describing the effects of the essential oils blend agolin ruminant on performance, rumen fermentation and methane emissions in dairy cows. *Animals (Basel)*, 10. https://doi.org/10.3390/ani10040620.

Benchaar, C., et al. 2001. Evaluation of dietary strategies to reduce methane production in ruminants: a modelling approach. *Canadian Journal of Animal Science*, 81, 563–574. https://doi.org/10.4141/A00-119.

Bergen, W. G. & Bates, D. B. 1984. Ionophores: their effect on production efficiency and mode of action. *Journal of Animal Science*, 58, 1465–1483. https://doi.org/10.2527/jas1984.5861465x.

Boadi, D., et al. 2004. Mitigation strategies to reduce enteric methane emissions from dairy cows: update review. *Canadian Journal of Animal Science*, 84, 319–335. https://doi.org/10.4141/A03-109.

Boadi, D. A. & Wittenberg, K. M. 2002. Methane production from dairy and beef heifers fed forages differing in nutrient density using the sulphur hexafluoride (SF_6) tracer gas technique. *Canadian Journal of Animal Science*, 82, 201–206. https://doi.org/10.4141/A01-017.

Borhan, M. & Khanaum, M. 2022. Sensors and methods for measuring greenhouse gas emissions from different components of livestock production facilities. *Journal of Geoscience and Environment Protection*, 10, 242–272. https://doi.org/10.4236/gep.2022.1012014.

Borhan, M. S., et al. 2011a. Determining seasonal greenhouse gas emissions from ground-level area sources in a dairy operation in central Texas. *Journal of the Air & Waste Management Association*, 61, 786–795. https://doi.org/10.3155/1047-3289.61.7.786.

Borhan, M. S., et al. 2011b. Greenhouse gas emissions from ground level area sources in dairy and cattle feedyard operations. *Atmosphere*, 2, 303. https://doi.org/10.3390/atmos2030303.

Brito, A. F., et al. 2022. Production performance, nutrient use efficiency, and predicted enteric methane emissions in dairy cows under confinement or grazing management system. *Translational Animal Science*, 6, txac028. https://doi.org/10.1093/tas/txac028.

Bühler, M., et al. 2021. Assessment of the inverse dispersion method for the determination of methane emissions from a dairy housing. *Agricultural and Forest Meteorology*, 307, 108501. https://doi.org/10.1016/j.agrformet.2021.108501.

Byskov, M. V., et al. 2015. Variations in automatically recorded rumination time as explained by variations in intake of dietary fractions and milk production, and between-cow variation. *Journal of Dairy Science*, 98, 3926–3937. https://doi.org/10.3168/jds.2014-8012.

Calderón-Chagoya, R., et al. 2019. Genome-wide association studies for methane production in dairy cattle. *Genes (Basel)*, 10. https://doi.org/10.3390/genes10120995.
Capper, J. L., et al. 2009. The environmental impact of dairy production: 1944 compared with 2007. *Journal of Animal Science*, 87, 2160–2167. https://doi.org/10.2527/jas.2009-1781.
Capuzzello, G., et al. 2023. Evaluation of an indwelling bolus equipped with a triaxial accelerometer for the characterisation of the diurnal pattern of bovine reticuloruminal contractions. *Journal of Dairy Research*, 90(1), 9–15. https://doi.org/10.1017/s0022029923000134.
Chagunda, M. G. G., et al. 2009. On the use of a laser methane detector in dairy cows. *Computers and Electronics in Agriculture*, 68, 157–160. https://doi.org/10.1016/j.compag.2009.05.008.
Cobellis, G., et al. 2016. Critical evaluation of essential oils as rumen modifiers in ruminant nutrition: a review. *Science of the Total Environment*, 545–546, 556–568. https://doi.org/10.1016/j.scitotenv.2015.12.103.
Crompton, L. A., et al. 2011. Fluctuations in methane emission in response to feeding pattern in lactating dairy cows. *In:* Sauvant, D., et al. (eds.) *Modelling Nutrient Digestion and Utilisation in Farm Animals*. Wageningen: Wageningen Academic Publishers. https://doi.org/10.3920/978-90-8686-712-7_19.
Da Cunha, L. L., et al. 2023. Characterization of rumen microbiome and metabolome from oro-esophageal tubing and rumen cannula in Holstein dairy cows. *Scientific Reports*, 13, 5854. https://doi.org/10.1038/s41598-023-33067-5.
De Haas, Y., et al. 2011. Genetic parameters for predicted methane production and potential for reducing enteric emissions through genomic selection. *Journal of Dairy Science*, 94, 6122–6134. https://doi.org/10.3168/jds.2011-4439.
De Haas, Y., et al. 2017. Invited review: Phenotypes to genetically reduce greenhouse gas emissions in dairying. *Journal of Dairy Science*, 100, 855–870. https://doi.org/10.3168/jds.2016-11246.
De Haas, Y., et al. 2021. Selective breeding as a mitigation tool for methane emissions from dairy cattle. *Animal*, 15, 100294. https://doi.org/10.1016/j.animal.2021.100294.
Dehareng, F., et al. 2012. Potential use of milk mid-infrared spectra to predict individual methane emission of dairy cows. *Animal*, 6, 1694–1701. https://doi.org/10.1017/S1751731112000456.
Deighton, M. H., et al. 2014. A modified sulphur hexafluoride tracer technique enables accurate determination of enteric methane emissions from ruminants. *Animal Feed Science and Technology*, 197, 47–63. https://doi.org/10.1016/j.anifeedsci.2014.08.003.
Della Rosa, M. M., et al. 2021. A review of technical variations and protocols used to measure methane emissions from ruminants using respiration chambers, SF_6 tracer technique and GreenFeed, to facilitate global integration of published data. *Animal Feed Science and Technology*, 279, 115018. https://doi.org/10.1016/j.anifeedsci.2021.115018.
Difford, G. F., et al. 2018. Host genetics and the rumen microbiome jointly associate with methane emissions in dairy cows. *PLoS Genetics*, 14, e1007580. https://doi.org/10.1371/journal.pgen.1007580.

Dong, Y., et al. 1999. Effects of exogenous fibrolytic enzymes, α-bromoethanesulfonate and monensin on fermentation in a rumen simulation (RUSITEC) system. *Canadian Journal of Animal Science*, 79, 491–498. https://doi.org/10.4141/A99-024.

Doreau, M. & Chilliard, Y. 1997. Digestion and metabolism of dietary fat in farm animals. *British Journal of Nutrition*, 78, S15–S35. https://doi.org/10.1079/BJN19970132.

Doreau, M., et al. 2018. Comparison of 3 methods for estimating enteric methane and carbon dioxide emission in nonlactating cows. *Journal of Animal Science*, 96, 1559–1569. https://doi.org/10.1093/jas/sky033.

Dorich, C. D., et al. 2015. *Short communication:* Use of a portable, automated, open-circuit gas quantification system and the sulfur hexafluoride tracer technique for measuring enteric methane emissions in Holstein cows fed ad libitum or restricted. *Journal of Dairy Science*, 98, 2676–2681. https://doi.org/10.3168/jds.2014-8348.

Duffield, T., et al. 2004. Comparison of techniques for measurement of rumen pH in lactating dairy cows. *Journal of Dairy Science*, 87, 59–66. https://doi.org/10.3168/jds.S0022-0302(04)73142-2.

Eckelkamp, E. A. 2019. Invited Review: Current state of wearable precision dairy technologies in disease detection. *Applied Animal Science*, 35, 209–220. https://doi.org/10.15232/aas.2018-01801.

Eckert, M., et al. 2018. Effect of feeding system on enteric methane emissions from individual dairy cows on commercial farms. *Land* [Online], 7. https://doi.org/10.3390/land7010026.

El-Nile, A., et al. 2021. In vitro and in vivo assessment of dietary supplementation of both natural or nano-zeolite in goat diets: effects on ruminal fermentation and nutrients digestibility. *Animals* [Online], 11. https://doi.org/10.3390/ani11082215.

Fischer, A., et al. 2020. Precision feed restriction improves feed and milk efficiencies and reduces methane emissions of less efficient lactating Holstein cows without impairing their performance. *Journal of Dairy Science*, 103, 4408–4422. https://doi.org/10.3168/jds.2019-17654.

Flesch, T. K., et al. 2004. Deducing ground-to-air emissions from observed trace gas concentrations: ā field trial. *Journal of Applied Meteorology*, 43, 487–502. https://doi.org/10.1175/1520-0450(2004)043<0487:DGEFOT>2.0.CO;2.

Food And Agriculture Organization. 2006. *Livestock's Long Shadow: Environmental Issues and Options*. Rome, Italy: Food and Agriculture Organization of the United Nations.

Gao, Z., et al. 2011. Diurnal and seasonal patterns of methane emissions from a dairy operation in north china plain. *Advances in Meteorology*, 2011, 190234. https://doi.org/10.1155/2011/190234.

Garnsworthy, P. C., et al. 2012. On-farm methane measurements during milking correlate with total methane production by individual dairy cows. *Journal of Dairy Science*, 95, 3166–3180. https://doi.org/10.3168/jds.2011-4605.

Garnsworthy, P. C., et al. 2019. Comparison of methods to measure methane for use in genetic evaluation of dairy cattle. *Animals (Basel)*, 9, 837. https://doi.org/10.3390/ani9100837.

Gerber, P. J., et al. 2013. *Tackling Climate Change Through Livestock - A Global Assessment of Emissions and Mitigation Opportunities*. Rome, Italy: Food and Agriculture Organization of the United Nations (FAO).

Giger-Reverdin, S., et al. 2003. Literature survey of the influence of dietary fat composition on methane production in dairy cattle. *Livestock Production Science*, 82, 73–79. https://doi.org/10.1016/S0301-6226(03)00002-2.

Goff, J. P. 2015. Ruminant digestive physiology and intestinal microbiology. *In:* Reece, W. O., et al. (eds.) *Dukes' Physiology of Domestic Animals.* 13th ed. Hoboken: John WIley.

González-Recio, O., et al. 2020. Mitigation of greenhouse gases in dairy cattle via genetic selection: 2. Incorporating methane emissions into the breeding goal. *Journal of Dairy Science*, 103, 7210–7221. https://doi.org/10.3168/jds.2019-17598.

Grainger, C. & Beauchemin, K. A. 2011. Can enteric methane emissions from ruminants be lowered without lowering their production? *Animal Feed Science and Technology*, 166–167, 308–320. https://doi.org/10.1016/j.anifeedsci.2011.04.021.

Grainger, C., et al. 2007. Methane emissions from dairy cows measured using the sulfur hexafluoride (SF_6) tracer and chamber techniques. *Journal of Dairy Science*, 90, 2755–2766. https://doi.org/10.3168/jds.2006-697.

Guan, H., et al. 2006. Efficacy of ionophores in cattle diets for mitigation of enteric methane. *Journal of Animal Science*, 84, 1896–1906. https://doi.org/10.2527/jas.2005-652.

Guyader, J., et al. 2017. Tea saponin reduced methanogenesis in vitro but increased methane yield in lactating dairy cows. *Journal of Dairy Science*, 100, 1845–1855. https://doi.org/10.3168/jds.2016-11644.

Hammond, K. J., et al. 2015. Methane emissions from cattle: estimates from short-term measurements using a GreenFeed system compared with measurements obtained using respiration chambers or sulphur hexafluoride tracer. *Animal Feed Science and Technology*, 203, 41–52. https://doi.org/10.1016/j.anifeedsci.2015.02.008.

Haque, M. N. 2018. Dietary manipulation: a sustainable way to mitigate methane emissions from ruminants. *Journal of Animal Science and Technology*, 60, 15. https://doi.org/10.1186/s40781-018-0175-7.

Hassanat, F., et al. 2013. Replacing alfalfa silage with corn silage in dairy cow diets: effects on enteric methane production, ruminal fermentation, digestion, N balance, and milk production. *Journal of Dairy Science*, 96, 4553–4567. https://doi.org/10.3168/jds.2012-6480.

Hegarty, R. S. 2013. Applicability of short-term emission measurements for on-farm quantification of enteric methane. *Animal*, 7(Suppl 2), 401–408. https://doi.org/10.1017/s1751731113000839.

Henderson, G., et al. 2013. Effect of DNA extraction methods and sampling techniques on the apparent structure of cow and sheep rumen microbial communities. *PLoS One*, 8, e74787. https://doi.org/10.1371/journal.pone.0074787.

Hernández Medina, R., et al. 2022. Machine learning and deep learning applications in microbiome research. *ISME Communications*, 2, 98. https://doi.org/10.1038/s43705-022-00182-9.

Hofmann, R. R. 1988. Anatomy of the gastro-intestinal tract. *In:* Church, D. C. (ed.) *The Ruminant Animal: Digestive Physiology and Nutrition.* Englewood Cliffs: Prentice Hall.

Holm-Nielsen, J. B., et al. 2009. The future of anaerobic digestion and biogas utilization. *Bioresource Technology*, 100, 5478–5484. https://doi.org/10.1016/j.biortech.2008.12.046.

Hristov, A. N., et al. 2013a. Special topics – Mitigation of methane and nitrous oxide emissions from animal operations: I. A review of enteric methane mitigation options. *Journal of Animal Science*, 91, 5045–5069. https://doi.org/10.2527/jas.2013-6583.

Hristov, A. N., et al. 2013b. *Mitigation of Greenhouse Gas Emissions in Livestock Production; A Review of Technical Options for Non-CO_2 Emissions*. Rome, Italy: Food and Agriculture Organization.

Hristov, A. N., et al. 2016. *Short communication:* comparison of the GreenFeed system with the sulfur hexafluoride tracer technique for measuring enteric methane emissions from dairy cows. *Journal of Dairy Science*, 99, 5461–5465. https://doi.org/10.3168/jds.2016-10897.

Huhtanen, P., et al. 2015. Comparison of methods to determine methane emissions from dairy cows in farm conditions. *Journal of Dairy Science*, 98, 3394–3409. https://doi.org/10.3168/jds.2014-9118.

Huhtanen, P., et al. 2019. Enteric methane emission can be reliably measured by the GreenFeed monitoring unit. *Livestock Science*, 222, 31–40. https://doi.org/10.1016/j.livsci.2019.01.017.

Hwang, H.-S., et al. 2012. Effects of halogenated compounds on in vitro fermentation characteristics in the rumen and methane emissions. *Journal of Life Science*, 22, 1187–1193. https://doi.org/10.5352/JLS.2012.22.9.1187.

Intergovernmental Panel On Climate Change. 2007. Climate change 2007: The physical science basis. Contribution of working group I to the fourth assessment report of the intergovernmental panel on climate change. *In:* Solomon, S., et al. (eds.) Cambridge and New York: Cambridge University Press.

Jackson, R. B., et al. 2020. Increasing anthropogenic methane emissions arise equally from agricultural and fossil fuel sources. *Environmental Research Letters*, 15, 071002. https://doi.org/10.1088/1748-9326/ab9ed2.

Jayanegara, A., et al. 2012. Meta-analysis of the relationship between dietary tannin level and methane formation in ruminants from in vivo and in vitro experiments. *Journal of Animal Physiology and Animal Nutrition*, 96, 365–375. https://doi.org/10.1111/j.1439-0396.2011.01172.x.

Jayanegara, A., et al. 2018. Use of 3-nitrooxypropanol as feed additive for mitigating enteric methane emissions from ruminants: a meta-analysis. *Italian Journal of Animal Science*, 17, 650–656. https://doi.org/10.1080/1828051X.2017.1404945.

Jaynes, W. F., et al. 2007. Aflatoxin B1 adsorption by clays from water and corn meal. *Applied Clay Science*, 36, 197–205. https://doi.org/10.1016/j.clay.2006.06.012.

Jiao, H. P., et al. 2014. Effect of concentrate feed level on methane emissions from grazing dairy cows. *Journal of Dairy Science*, 97, 7043–7053. https://doi.org/10.3168/jds.2014-7979.

Johnson, K., et al. 1994. Measurement of methane emissions from ruminant livestock using a sulfur hexafluoride tracer technique. *Environmental Science & Technology*, 28, 359–362. https://doi.org/10.1021/es00051a025.

Johnson, K. A. & Johnson, D. E. 1995. Methane emissions from cattle. *Journal of Animal Science*, 73, 2483–2492. https://doi.org/10.2527/1995.7382483x.

Jonker, A., et al. 2016. Enteric methane and carbon dioxide emissions measured using respiration chambers, the sulfur hexafluoride tracer technique, and a GreenFeed head-chamber system from beef heifers fed alfalfa silage at three allowances and four feeding frequencies. *Journal of Animal Science*, 94, 4326–4337. https://doi.org/10.2527/jas.2016-0646.

Kaharabata, S. K., et al. 2000. Estimating methane emissions from dairy cattle housed in a barn and feedlot using an atmospheric tracer. *Environmental Science & Technology*, 34, 3296–3302. https://doi.org/10.1021/es990578c.

Kebreab, E., et al. 2023. A meta-analysis of effects of 3-nitrooxypropanol on methane production, yield, and intensity in dairy cattle. *Journal of Dairy Science*, 106, 927–936. https://doi.org/10.3168/jds.2022-22211.

Kebreab, E., et al. 2006. Methane and nitrous oxide emissions from Canadian animal agriculture: a review. *Canadian Journal of Animal Science*, 86, 135–157. https://doi.org/10.4141/A05-010.

Kebreab, E., et al. 2010. Impact of dietary manipulation on nutrient flows and greenhouse gas emissions in cattle. *Revista Brasileira de Zootecnia*, 39, 458–464. https://doi.org/10.1590/S1516-35982010001300050.

Kinley, R. D., et al. 2016. The red macroalgae *Asparagopsis taxiformis* is a potent natural antimethanogenic that reduces methane production during in vitro fermentation with rumen fluid. *Animal Production Science*, 56, 282–289. https://doi.org/10.1071/AN15576.

Knapp, J. R., et al. 2014. Invited review: enteric methane in dairy cattle production: quantifying the opportunities and impact of reducing emissions. *Journal of Dairy Science*, 97, 3231–3261. https://doi.org/10.3168/jds.2013-7234.

Lassen, J., et al. 2012. Accuracy of noninvasive breath methane measurements using Fourier transform infrared methods on individual cows. *Journal of Dairy Science*, 95, 890–898. https://doi.org/10.3168/jds.2011-4544.

Lassey, K. R. & Hegarty, R. S. 2020. *Overview of the SF_6 Tracer Technique and Its Evolution*. New Zealand: New Zealand Agricultural Greenhouse Gas Research Centre.

Li, X., et al. 2018. *Asparagopsis taxiformis* decreases enteric methane production from sheep. *Animal Production Science*, 58, 681–688. https://doi.org/10.1071/AN15883.

Livingston, G. P. & Hutchinson, G. L. 1995. Enclosure-based measurement of trace gas exchange: applications and sources of error. *In:* Matson, P. A. & Harris, R. C. (eds.) *Biogenic Trace Gases: Measuring Emissions from Soil and Water.* New York: Wiley-Blackwell.

Lopes, L. S. F., et al. 2024. Estimates of genetic parameters for rumination time, feed efficiency, and methane production traits in first-lactation Holstein cows. *Journal of Dairy Science*, 107, 4704–4713. https://doi.org/10.3168/jds.2023-23751.

Løvendahl, P., et al. 2018. Review: selecting for improved feed efficiency and reduced methane emissions in dairy cattle. *Animal*, 12, s336–s349. https://doi.org/10.1017/s1751731118002276.

Ma, X., et al. 2024. Evaluating GreenFeed and respiration chambers for daily and intraday measurements of enteric gaseous exchange in dairy cows housed in tie-stalls. *Journal of Dairy Science*, 107(12), 10913–10931. https://doi.org/10.3168/jds.2024-25246.

Machado, L., et al. 2014. Effects of marine and freshwater macroalgae on in vitro total gas and methane production. *PLoS One*, 9, e85289. https://doi.org/10.1371/journal.pone.0085289.

Mahler, G. J., et al. 2012. Oral exposure to polystyrene nanoparticles affects iron absorption. *Nature Nanotechnology*, 7, 264–271. https://doi.org/10.1038/nnano.2012.3.

Manriquez, D., et al. 2024. Comparison of group-scale emissions in the milking parlor and a tunnel-ventilated housing barn in a commercial dairy farm. *Journal of Dairy Science*, 107, 77.

Manzanilla-Pech, C. I. V., et al. 2021. Breeding for reduced methane emission and feed-efficient Holstein cows: an international response. *Journal of Dairy Science*, 104, 8983–9001. https://doi.org/10.3168/jds.2020-19889.

Marques, R. D. S. & Cooke, R. F. 2021. Effects of ionophores on ruminal function of beef cattle. *Animals (Basel)*, 11, 2871. https://doi.org/10.3390/ani11102871.

Martinez-Fernandez, G., et al. 2018. 3-NOP vs. Halogenated compound: methane production, ruminal fermentation and microbial community response in forage fed cattle. *Frontiers in Microbiology*, 9, 1582. https://doi.org/10.3389/fmicb.2018.01582.

Massaro, S., et al. 2024. Review of equations to predict methane emissions in dairy cows from milk fatty acid profiles and their application to commercial dairy farms. *Journal of Dairy Science* (in press). https://doi.org/10.3168/jds.2024-24814.

McGuffey, R. K., et al. 2001. Ionophores for dairy cattle: current status and future outlook. *Journal of Dairy Science*, 84, E194–E203. https://doi.org/10.3168/jds.S0022-0302(01)70218-4.

Mikuła, R., et al. 2022. The effect of rumination time on milk performance and methane emission of dairy cows fed partial mixed ration based on maize silage. *Animals* [Online], 12. https://doi.org/10.3390/ani12010050.

Min, B. R., et al. 2021. The role of seaweed as a potential dietary supplementation for enteric methane mitigation in ruminants: challenges and opportunities. *Animal Nutrition*, 7, 1371–1387. https://doi.org/10.1016/j.aninu.2021.10.003.

Mitsumori, M., et al. 2012. Responses in digestion, rumen fermentation and microbial populations to inhibition of methane formation by a halogenated methane analogue. *British Journal of Nutrition*, 108, 482–491. https://doi.org/10.1017/s0007114511005794.

Mohamad Salleh, S., et al. 2023. Is it possible to predict the methane emission intensity of Swedish dairy cows from milk spectra? *Smart Agricultural Technology*, 5, 100286. https://doi.org/10.1016/j.atech.2023.100286.

Møller, H. B., et al. 2004. Methane productivity of manure, straw and solid fractions of manure. *Biomass and Bioenergy*, 26, 485–495. https://doi.org/10.1016/j.biombioe.2003.08.008.

Mønster, J. G., et al. 2014. Quantifying methane emission from fugitive sources by combining tracer release and downwind measurements – A sensitivity analysis based on multiple field surveys. *Waste Management*, 34, 1416–1428. https://doi.org/10.1016/j.wasman.2014.03.025.

Montes, F., et al. 2013. Special topics – Mitigation of methane and nitrous oxide emissions from animal operations: II. A review of manure management mitigation options. *Journal of Animal Science*, 91, 5070–5094. https://doi.org/10.2527/jas.2013-6584.

Mosier, A. R., et al. 1998. Mitigating agricultural emissions of methane. *Climatic Change*, 40, 39–80. https://doi.org/10.1023/A:1005338731269.

Moss, A. R., et al. 2000. Methane production by ruminants: its contribution to global warming. *Annales de Zootechnie*, 49, 231–253. https://doi.org/10.1051/animres:200011.hal-00889894.

Muñoz, C., et al. 2012. Comparison of the sulfur hexafluoride tracer and respiration chamber techniques for estimating methane emissions and correction for rectum methane output from dairy cows. *Journal of Dairy Science*, 95, 3139–3148. https://doi.org/10.3168/jds.2011-4298.

Nagaraja, T. G. 2016. Microbiology of the rumen. *In:* Millen, D. D., et al. (eds.) *Rumenology*. Cham: Springer International Publishing. https://doi.org/10.1007/978-3-319-30533-2_2.

Negussie, E., et al. 2017. Non-invasive individual methane measurement in dairy cows. *Animal*, 11, 890–899. https://doi.org/10.1017/S1751731116002718.

Niu, M., et al. 2018. Prediction of enteric methane production, yield, and intensity in dairy cattle using an intercontinental database. *Global Change Biology*, 24, 3368–3389. https://doi.org/10.1111/gcb.14094.

Norris, A. B., et al. 2020. Effect of quebracho condensed tannin extract on fecal gas flux in steers. *Journal of Environmental Quality*, 49, 1225–1235. https://doi.org/10.1002/jeq2.20110.

Odongo, N. E., et al. 2007. Long-term effects of feeding monensin on methane production in lactating dairy cows. *Journal of Dairy Science*, 90, 1781–1788. https://doi.org/10.3168/jds.2006-708.

Ogata, T., et al. 2019. Effects of an increased concentrate diet on rumen pH and the bacterial community in Japanese Black beef cattle at different fattening stages. *The Journal of Veterinary Medical Science*, 81, 968–974. https://doi.org/10.1292/jvms.19-0077.

Oss, D. B., et al. 2016. An evaluation of the face mask system based on short-term measurements compared with the sulfur hexafluoride (SF_6) tracer, and respiration chamber techniques for measuring CH_4 emissions. *Animal Feed Science and Technology*, 216, 49–57. https://doi.org/10.1016/j.anifeedsci.2016.03.008.

Owens, F. N. & Basalan, M. 2016. Ruminal fermentation. *In:* Millen, D. D., et al. (eds.) *Rumenology*. Cham: Springer International Publishing. https://doi.org/10.1007/978-3-319-30533-2_3.

Owens, F. N., et al. 1998. Acidosis in cattle: a review. *Journal of Animal Science*, 76, 275–286. https://doi.org/10.2527/1998.761275x.

Pathak, N., et al. 2024. Comparing rumen fluid collection methods on fermentation profile and microbial population in lactating dairy cows. *JDS Communications*, 6(1), 34–38 https://doi.org/10.3168/jdsc.2024-0566.

Patra, A. K. 2012. Enteric methane mitigation technologies for ruminant livestock: a synthesis of current research and future directions. *Environmental Monitoring and Assessment*, 184, 1929–1952. https://doi.org/10.1007/s10661-011-2090-y.

Patra, A. K. 2013. The effect of dietary fats on methane emissions, and its other effects on digestibility, rumen fermentation and lactation performance in cattle: a meta-analysis. *Livestock Science*, 155, 244–254. https://doi.org/10.1016/j.livsci.2013.05.023.

Patra, A. K., et al. 2017. Rumen methanogens and mitigation of methane emission by anti-methanogenic compounds and substances. *Journal of Animal Science and Biotechnology*, 8, 13. https://doi.org/10.1186/s40104-017-0145-9.

Paz, H. A., et al. 2016. Rumen bacterial community composition in holstein and jersey cows is different under same dietary condition and is not affected by sampling method. *Frontiers in Microbiology*, 7, 1206. https://doi.org/10.3389/fmicb.2016.01206.

Peltola, O., et al. 2014. Evaluating the performance of commonly used gas analysers for methane eddy covariance flux measurements: the InGOS inter-comparison field experiment. *Biogeosciences*, 11, 3163–3186. https://doi.org/10.5194/bg-11-3163-2014.

Peng, C., et al. 2023. Unveiling microbial biomarkers of ruminant methane emission through machine learning. *Frontiers in Microbiology*, 14, 1308363. https://doi.org/10.3389/fmicb.2023.1308363.

Pereira, A. M., et al. 2023. A longitudinal study with a laser methane detector (LMD) highlighting lactation cycle-related differences in methane emissions from dairy cows. *Animals* [Online], 13. https://doi.org/10.3390/ani13060974.

Pereira, G. M., et al. 2018. Technical note: Validation of an ear-tag accelerometer sensor to determine rumination, eating, and activity behaviors of grazing dairy cattle. *Journal of Dairy Science*, 101, 2492–2495. https://doi.org/10.3168/jds.2016-12534.

Pickering, N. K., et al. 2015. Animal board invited review: genetic possibilities to reduce enteric methane emissions from ruminants. *Animal*, 9, 1431–1440. https://doi.org/10.1017/s1751731115000968.

Pinares-Patiño, C. S., et al. 2011. Assessment of the sulphur hexafluoride (SF_6) tracer technique using respiration chambers for estimation of methane emissions from sheep. *Animal Feed Science and Technology*, 166–167, 201–209. https://doi.org/10.1016/j.anifeedsci.2011.04.067.

Place, S. E. & Mitloehner, F. M. 2010. Invited review: contemporary environmental issues: a review of the dairy industry's role in climate change and air quality and the potential of mitigation through improved production efficiency. *Journal of Dairy Science*, 93, 3407–3416. https://doi.org/10.3168/jds.2009-2719.

Powers, W. & Capelari, M. 2016. Analytical methods for quantifying greenhouse gas flux in animal production systems. *Journal of Animal Science*, 94, 3139–3146. https://doi.org/10.2527/jas.2015-0017.

Ramin, M. & Huhtanen, P. 2013. Development of equations for predicting methane emissions from ruminants. *Journal of Dairy Science*, 96, 2476–2493. https://doi.org/10.3168/jds.2012-6095.

Reyes, D. C., et al. 2023. Maine organic dairy producers' receptiveness to seaweed supplementation and effect of Chondrus crispus on enteric methane emissions in lactating cows. *Frontiers in Veterinary Science*, 10. https://doi.org/10.3389/fvets.2023.1153097.

Richardson, L. F., et al. 1976. Effect of monensin on rumen fermentation in vitro and in vivo. *Journal of Animal Science*, 43, 657–664. https://doi.org/10.2527/jas1976.433657x.

Rochette, P. 2011. Towards a standard non-steady-state chamber methodology for measuring soil N_2O emissions. *Animal Feed Science and Technology*, 166–167, 141–146. https://doi.org/10.1016/j.anifeedsci.2011.04.063.

Roessler, R., et al. 2018. Using a portable laser methane detector in goats to assess diurnal, diet- and position-dependent variations in enteric methane emissions. *Computers and Electronics in Agriculture*, 150, 110–117. https://doi.org/10.1016/j.compag.2018.04.010.

Ross, E. M. & Hayes, B. J. 2022. Metagenomic predictions: a review 10 years on. *Frontiers in Genetics*, 13, 865765. https://doi.org/10.3389/fgene.2022.865765.

Ross, E. M., et al. 2013. Metagenomic predictions: from microbiome to complex health and environmental phenotypes in humans and cattle. *PLoS One*, 8, e73056. https://doi.org/10.1371/journal.pone.0073056.

Ross, E. M., et al. 2020. Genomic predictions for enteric methane production are improved by metabolome and microbiome data in sheep (*Ovis aries*). *Journal of Animal Science*, 98. https://doi.org/10.1093/jas/skaa262.

Rotz, C. A. 2004. Management to reduce nitrogen losses in animal production. *Journal of Animal Science*, 82, E119–E137. https://doi.org/10.2527/2004.8213_supplE119x.

Rowe, S. J., et al. 2019. Selection for divergent methane yield in New Zealand sheep - a ten-year perspective. *Proceedings of the 23rd Conference of the Association for the Advancement of Animal Breeding and Genetics (AAABG).* Armidale, New South Wales, Australia: Association for the Advancement of Animal Breeding and Genetics.

Russell, J. B. & Houlihan, A. J. 2003. Ionophore resistance of ruminal bacteria and its potential impact on human health. *FEMS Microbiology Reviews*, 27, 65–74. https://doi.org/10.1016/s0168-6445(03)00019-6.

Russell, J. B. & Strobel, H. J. 1989. Effect of ionophores on ruminal fermentation. *Applied and Environmental Microbiology*, 55, 1–6. https://doi.org/10.1128/aem.55.1.1-6.1989.

Schirmann, K., et al. 2009. Technical note: validation of a system for monitoring rumination in dairy cows. *Journal of Dairy Science*, 92, 6052–6055. https://doi.org/10.3168/jds.2009-2361.

Sepulveda, B. J., et al. 2022. Eating time as a genetic indicator of methane emissions and feed efficiency in australian maternal composite sheep. *Frontiers in Genetics*, 13. https://doi.org/10.3389/fgene.2022.883520.

Shima, S., et al. 2012. Structure of a methyl-coenzyme M reductase from Black Sea mats that oxidize methane anaerobically. *Nature*, 481, 98–101. https://doi.org/10.1038/nature10663.

Silveira, S. R., et al. 2019. Replacement of soybean meal with soybean cake reduces methane emissions in dairy cows and an assessment of a face-mask technique for methane measurement. *Frontiers in Veterinary Science*, 6, 295. https://doi.org/10.3389/fvets.2019.00295.

Singh, G. P. & Mohini, M. 1999. Effect of different levels of rumensin in diet on rumen fermentation, nutrient digestibility and methane production in cattle. *Asian-Australas Journal of Animal Science*, 12, 1215–1221. https://doi.org/10.5713/ajas.1999.1215.

Smith, P., et al. 2021. Agricultural methane emissions and the potential formitigation. *Philosophical Transactions of the Royal Society A: Mathematical, Physical and Engineering Sciences*, 379, 20200451. https://doi.org/10.1098/rsta.2020.0451.

Sorg, D., et al. 2017. The agreement between two next-generation laser methane detectors and respiration chamber facilities in recording methane concentrations in the spent air produced by dairy cows. *Computers and Electronics in Agriculture*, 143, 262–272. https://doi.org/10.1016/j.compag.2017.10.024.

Stefenoni, H. A., et al. 2021. Effects of the macroalga *Asparagopsis taxiformis* and oregano leaves on methane emission, rumen fermentation, and lactational performance of dairy cows. *Journal of Dairy Science*, 104, 4157–4173. https://doi.org/10.3168/jds.2020-19686.

Sun, K., et al. 2021. Research progress on the application of feed additives in ruminal methane emission reduction: a review. *PeerJ*, 9, e11151. https://doi.org/10.7717/peerj.11151.

Sypniewski, M., et al. 2019. Technical note: interchangeability and comparison of methane measurements in dairy cows with 2 noninvasive infrared systems. *Journal of Dairy Science*, 102, 9512–9517. https://doi.org/10.3168/jds.2019-16258.

Taylor, A. M., et al. 2017. Direct whole-farm greenhouse gas flux measurements for a beef cattle operation. *Agriculture, Ecosystems & Environment*, 239, 65–79. https://doi.org/10.1016/j.agee.2017.01.011.

Tedeschi, L. O., et al. 2022. Quantification of methane emitted by ruminants: a review of methods. *Journal of Animal Science*, 100, 1–22. https://doi.org/10.1093/jas/skac197.

Tedeschi, L. O. & Beauchemin, K. A. 2023. Galyean appreciation club review: a holistic perspective of the societal relevance of beef production and its impacts on climate change. *Journal of Animal Science*, 101, 1–19. https://doi.org/10.1093/jas/skad024.

Tedeschi, L. O. & Fox, D. G. 2020. *The Ruminant Nutrition System: Volume I - An Applied Model for Predicting Nutrient Requirements and Feed Utilization in Ruminants.* Ann Arbor: XanEdu.

Tedeschi, L. O., et al. 2003. Potential environmental benefits of ionophores in ruminant diets. *Journal of Environmental Quality*, 32, 1591–1602. https://doi.org/10.2134/jeq2003.1591.

Tedeschi, L. O., et al. 2011. Potential environmental benefits of feed additives and other strategies for ruminant production. *Revista Brasileira de Zootecnia*, 40, 291–309.

Tedeschi, L. O., et al. 2021. Nutritional aspects of ecologically relevant phytochemicals in ruminant production. *Frontiers in Veterinary Science*, 8, 1–24. https://doi.org/10.3389/fvets.2021.628445.

Thiel, A., et al. 2019a. 3-NOP: ADME studies in rats and ruminating animals. *Food and Chemical Toxicology*, 125, 528–539. https://doi.org/10.1016/j.fct.2019.02.002.

Thiel, A., et al. 2019b. 3-NOP: mutagenicity and genotoxicity assessment. *Food and Chemical Toxicology*, 123, 566–573. https://doi.org/10.1016/j.fct.2018.11.010.

Thivend, P. & Jouany, J. P. 1983. effect of lasalocid sodium on rumen fermentation and digestion in sheep. *Reproduction Nutrition Development (1980)*, 23, 817–828. https://doi.org/10.1051/rnd:19830602.

Trckova, M., et al. 2004. Kaolin, bentonite, and zeolites as feed supplements for animals: health advantages and risks. *Veterinarni Medicina*, 49, 389–399. https://doi.org/10.17221/5728-VETMED.

Tseten, T., et al. 2022. Strategies to mitigate enteric methane emissions from ruminant animals. *Journal of Microbiology and Biotechnology*, 32, 269–277. https://doi.org/10.4014/jmb.2202.02019.

Uddin, M. E., et al. 2022. Impact of nitrate and 3-nitrooxypropanol on the carbon footprints of milk from cattle produced in confined-feeding systems across regions in the United States: a life cycle analysis. *Journal of Dairy Science*, 105, 5074–5083. https://doi.org/10.3168/jds.2021-20988.

Vallejo-Hernández, L. H., et al. 2018. Environmental impact of yeast and exogenous xylanase on mitigating carbon dioxide and enteric methane production in ruminants. *Journal of Cleaner Production*, 189, 40–46. https://doi.org/10.1016/j.jclepro.2018.03.310.

Valpotić, H., et al. 2018. Zeolite clinoptilolite nanoporous feed additive for animals of veterinary importance: potentials and limitations. *Periodicum Biologorum*, 189, 40–46. https://doi.org/10.18054/pb.v119i3.5434.

Van Gastelen, S., et al. 2024. Long-term effects of 3-nitrooxypropanol on methane emission and milk production characteristics in Holstein-Friesian dairy cows. *Journal of Dairy Science*, 107, 5556–5573. https://doi.org/10.3168/jds.2023-24198.

Van Soest, P. J. 1994. *Nutritional Ecology of the Ruminant*. Ithaca: Comstock Publishing Associates.

Vanderzaag, A. C., et al. 2014. Measuring methane emissions from two dairy farms: seasonal and manure-management effects. *Agricultural and Forest Meteorology*, 194, 259–267. https://doi.org/10.1016/j.agrformet.2014.02.003.

Vanlierde, A., et al. 2016. Milk mid-infrared spectra enable prediction of lactation-stage-dependent methane emissions of dairy cattle within routine population-scale milk recording schemes. *Animal Production Science*, 56, 258–264. https://doi.org/10.1071/AN15590.

Vinković, K., et al. 2022. Evaluating the use of an unmanned aerial vehicle (UAV)-based active aircore system to quantify methane emissions from dairy cows. *Science of The Total Environment*, 831, 154898. https://doi.org/10.1016/j.scitotenv.2022.154898.

Waghorn, G. C., et al. 2016. Measuring methane from grazing dairy cows using GreenFeed. *Animal Production Science*, 56, 252–257. https://doi.org/10.1071/AN15491.

Wallace, R. J., et al. 2019. A heritable subset of the core rumen microbiome dictates dairy cow productivity and emissions. *Science Advances*, 5, eaav8391. https://doi.org/10.1126/sciadv.aav8391.

Weller, J. I. & Ezra, E. 2024. Genetic analysis of rumination time based on an analysis of 77,697 Israeli dairy cows. *Journal of Dairy Science*, 107, 4793–4803. https://doi.org/10.3168/jds.2023-24095.

Yu, G., et al. 2021. A review of 3-nitrooxypropanol for enteric methane mitigation from ruminant livestock. *Animals* [Online], 11. https://doi.org/10.3390/ani11123540.

Zanferari, F., et al. 2018. Effects of chitosan and whole raw soybeans on ruminal fermentation and bacterial populations, and milk fatty acid profile in dairy cows. *Journal of Dairy Science*, 101, 10939–10952. https://doi.org/10.3168/jds.2018-14675.

Zetouni, L., et al. 2018. Is rumination time an indicator of methane production in dairy cows? *Journal of Dairy Science*, 101, 11074–11085. https://doi.org/10.3168/jds.2017-14280.

Zhang, B., et al. 2023. Methane prediction equations including genera of rumen bacteria as predictor variables improve prediction accuracy. *Scientific Reports*, 13, 21305. https://doi.org/10.1038/s41598-023-48449-y.

Zhang, Z.-W., et al. 2018. Nitrocompounds as potential methanogenic inhibitors in ruminant animals: a review. *Animal Feed Science and Technology*, 236, 107–114. https://doi.org/10.1016/j.anifeedsci.2017.12.010.

Zimmerman, P. R. 1993. *System for Measuring Metabolic Gas Emissions From Animals*. United States Patent Application.

Part 2

Strategies to achieve net zero: supporting the dairy cow

Chapter 6

Genetics and breeding for reduced methane emissions

A. E. van Breukelen, Wageningen University & Research, The Netherlands

1 Introduction

Selective breeding is increasingly being recognised as an effective methane (CH_4) mitigation strategy for dairy farming as implementation costs are low, the results are cumulative, and the effect is permanent. In selective breeding, animals with desirable traits are selected to reproduce, aiming to pass those traits to the next generation. Over multiple generations, selective breeding improves favourable characteristics, as has been done to reach higher milk yield and improve fertility in dairy cows. By carefully selecting individuals with preferred traits, breeders can gradually change a population, making the animals suited for changes in agricultural environments driven by, for example, societal and economic demands, and the changing climate.

Previous research has shown that individual dairy cows differ in the amount of enteric CH_4 emissions they emit, and that part of this variation between animals is heritable and driven by genetics (Lassen and Difford, 2020; Pickering et al., 2015b). Thus, implicating that genetic selection could be an effective CH_4 mitigation strategy. Although CH_4 is produced by microorganisms primarily in the rumen, and not by the cow itself, variation in CH_4 emissions is not only modulated by the rumen microbiome but also by the host genotype (Difford et al., 2018; Saborío-Montero et al., 2019). Therefore,

http://dx.doi.org/10.19103/AS.2025.0157.13

also without information on the rumen microbiome, genetic selection for reduced CH_4 emissions is believed to be an effective CH_4 mitigation strategy. The implementation of CH_4 mitigation in breeding programmes is, however, still in its infancy.

A limitation to practical application has been the lack of information on recorded CH_4 emissions of individual cows. Information on thousands of individual cows is needed to estimate sufficiently reliable genetic parameters, which are required for informed breeding decisions (de Haas et al., 2017). However, recent innovations in methods to record enteric CH_4 emissions have accelerated the collection of CH_4 phenotypes. Over the past years, various countries have focused on collecting enteric CH_4 measurements on hundreds to thousands of individual cows to be able to make informed breeding decisions in the future (Manzanilla-Pech et al., 2021).

Now that the number of recorded CH_4 phenotypes is gradually increasing, researchers and industry work on developing statistical models that estimate genetic parameters and later the breeding values for CH_4 for commercial dairy bulls. The breeding value is an estimate of an individual's genetic potential to pass on desirable traits to its offspring (Falconer and Mackay, 1996). Appropriate statistical models are essential to estimate breeding values accurately, which, for example, correct for differences in management between farms, to be able to include CH_4 phenotypes in breeding programmes. Once breeding values have been estimated, so-called selection indices can be developed that encourage CH_4 mitigation while maintaining production and without negatively impacting reproduction, health and welfare, conformation, and feed intake (Gonzalez-Recio et al., 2020; Pryce and Haile-Mariam, 2020).

The innovations related to implementing selective breeding as a CH_4 mitigation strategy will be discussed in this chapter, where we discuss:

- Methods of recording CH_4 emissions of individual cows.
- Defining CH_4 traits to be used in breeding programmes.
- Genetic parameters estimated to date.
- The implementation of selection on CH_4 mitigation in new or existing breeding programmes.

2 Recording methane emissions of individual cows

To make informed breeding decisions, phenotypes for CH_4 need to be available for thousands of individual cows (de Haas et al., 2017). Over the past few decades, several methods have been developed to directly measure CH_4 emissions from individual cows (de Haas et al., 2017; Negussie et al., 2017; Zhao et al., 2020). The most common methods that have been used to measure dairy cows are summarised in this chapter.

2.1 Respiration chambers

Respiration chambers (RC) are generally considered the gold standard to accurately measure CH_4 emissions from individual animals (Hammond et al., 2016). A RC is an airtight chamber that is usually designed to house a single animal, where all incoming and outgoing gases are continuously measured. By measuring the airflow within the RC, a direct measurement of the total CH_4 emitted by the animal can be obtained. Respiration chambers are able to measure total CH_4 emissions from the animal coming from oral, nasal, and anal routes.

Measuring individual dairy cows in RC is time-consuming, labour-intensive, and expensive. This makes the technique unsuitable for phenotyping large numbers of individual cows, required for the estimation of breeding values. In addition, RC only provides short-term measurements while the cow is in the RC, which is generally for an hour to a few days, and this does not provide information on CH_4 emissions during the full production cycle of the cow. Furthermore, separating the cow from its day-to-day environment and housing it in a RC may change feeding behaviour, which could potentially influence CH_4 emissions (Hammond et al., 2016). To overcome some of the challenges, alternative methods have been developed that can be operated long-term and on a larger scale, maximising the number of animals that can be phenotyped. Each of the methods has its advantages and disadvantages for large-scale application for breeding programmes (Garnsworthy et al., 2019).

2.2 GreenFeed

A system that was designed to measure CH_4 emissions of individual cows while remaining in their day-to-day environment is the GreenFeed (GF) (C-lock Inc. Rapid City, SD, US; Zimmerman (2011). The GF is a standalone unit which provides a cow concentrate as bait and measures CH_4 and CO_2 concentrations while the cow is eating the concentrate during spot-sample visits that generally last about three to ten minutes several times a day (Hammond et al., 2015; Huhtanen et al., 2015; Jonker et al., 2020) (Fig. 1).

The GF registers when the cow has its head in the feed bin, and measurements taken while the cows' head is not in the feed bin are discarded. Similar to a RC, the GF uses an air flux method with a strong pump at 1560–2250 L/min to be able to capture all emissions from breath. The spot sample of recorded concentrations is thereby extrapolated to total emissions in g/day (Huhtanen et al., 2015). Because GF units are a spot sample method, they are not able to precisely measure the total animal emissions during a full day or longer.

Figure 1 A GreenFeed unit, used to measure individual cows in the barn or in pasture.

Before using the CH_4 measurements recorded by a GF unit, animals need a period to get accustomed to using the unit (Garnsworthy et al., 2019), which can be challenging for animals that are not used to eating from e.g. concentrate feeders. Furthermore, the method is classified as medium throughput, resulting in relatively high costs in proportion to the number of animals that can be recorded (generally around 25 animals per recording period). Nonetheless, high phenotypic correlations with RC measurements have been reported of 0.85 and 0.96 (0.96 is transformed from R squared (R^2)) (Hristov et al., 2018; Velazco et al., 2016). Since the GF system has a higher throughput than RCs, its measurements are potentially valuable for benchmarking other direct or indirect methods of measuring CH_4 emissions from individual animals.

2.3 Sniffers

Sniffers are used to measure CH_4 and CO_2 concentrations from the breath of cattle and are often placed near an automatic milking system (AMS) (Fig. 2). They are installed with a tube leading from the feed bin of the AMS to the sniffer that is pumping air from the feed bin (Madsen et al., 2010; Teye et al., 2009). Thereby, breath concentrations of an individual cow are measured while the cow is eating the pellets provided in the feed bin of the AMS during milking. The devices are relatively inexpensive compared to other techniques to measure CH_4 and can easily be integrated with existing facilities, such as an AMS, where typically 40 to 70 cows can be measured multiple times during the day (Garnsworthy et al., 2019). To date, sniffers have mostly been used in research projects, on both research and commercial dairy farms (Breider, 2018; Huhtanen et al., 2015; Lassen and Lovendahl, 2013; Lopez-Paredes et al., 2020; Madsen et al., 2010; Pszczola et al., 2017; van Engelen et al., 2018).

Figure 2 A cow standing in an automated milking system (AMS), where breath is sampled from the feed bin to record methane emissions with a sniffer which is installed near the AMS at the top left (left). The sniffer records a value of 215 ppm methane and 2795 ppm carbon dioxide (right).

Different research groups have used various types of sniffers, which have often been developed by the research group together with companies with expertise in sensors for recording emissions in various types of industries, inside and outside of agriculture (an overview of available sniffer devices is given in Jonker et al. (2020)). As a result, large differences exist in hardware, the recording set-up, and the handling of raw measurements. Understanding these differences is essential to draw conclusions from the data and to correctly apply the phenotypes in industry breeding programmes in the near future.

In general, sniffers use a low-capacity pump of 0.6 to 4 L/min to transport breath samples of the cow to the CH_4 sensor, through the sampling tube that is placed in the feed bin of the AMS. Because of the pump's low capacity, the total breath of the cow is not captured and measured and thus the total CH_4 production cannot be measured (Huhtanen et al., 2015; Wu et al., 2018). However, methods have been developed to extrapolate short-term CH_4 concentration measurements to total emissions in g/day. For example, total CH_4 production can be approximated using recovery rates (Garnsworthy et al., 2012) or by using predicted CO_2 production as a tracer gas (based on e.g. body weight (BW), milk production) (Madsen et al., 2010). On the other hand, also without knowing the total CH_4 production in g/day, the CH_4 concentration measurements could potentially be used to rank cows from low to high emitters, which would be sufficient for breeding purposes (Garnsworthy et al., 2019).

Next to the low capture rate, sniffer measurements are faced with inaccuracies due to environmental conditions, such as wind and draft going through the AMS, movement of the cows, and the design and volume of the feed bin of the AMS. Also, sniffers can only be used to record short spot samples during the time that the cow is visiting the AMS. However, long-term recording of spot samples is possible as long as the cow is lactating. Statistical modelling and taking repeated measurements over a longer period of time and averaging these measurements is expected to leverage some of the measurement errors (Bovenhuis et al., 2018; Falconer and Mackay, 1996). Nonetheless, repeated recording and modelling cannot resolve all measurement error, and some inaccuracy will remain (Wu et al., 2018).

For sniffers, several studies report correlation estimates with RC measurements; however, most of the comparisons are difficult to interpret. Interpretation is made difficult because of differences in trait definitions (e.g. derived from peaks in CH_4 concentrations, Garnsworthy et al. (2012)), random measurement noise in single samples, or because sniffers are placed directly in a RC and these measurements do not correspond to on-farm conditions, as was shown by Difford (2018). One study reported correlations between CH_4 concentration measured by on-farm sniffers and CH_4 production measured in RCs and reported a moderate phenotypic correlation of 0.34 ± 0.22 and a high individual level correlation of 0.75 ± 0.20 (Difford et al., 2019). Another study reported a moderate Pearson's correlation between sniffer and GF measurements (0.30, transformed from R^2 (Huhtanen et al., 2015)). Furthermore, a high genetic correlation was estimated between sniffer-recorded CH_4 concentration (ppm) and GF-recorded CH_4 production (g/day) of 0.76 (van Breukelen et al., 2023).

The reported correlation estimates indicate that sniffer concentration measurements can provide a good indicator to reduce CH_4 production in breeding programmes, assuming that the GF and RC phenotypes are equal to true total emissions. Due to the lower accuracy, sniffers have the potential to serve as an indicator for total CH_4 production; however, even with suitable modelling and inference of the data, it is challenging to form conclusions on the true CH_4 production of cows. Nonetheless, because of the relatively low phenotyping costs, sniffers are a popular method for phenotyping individual cows, which allow recording full production cycles on thousands of individual cows for breeding programmes, and most results reported in the literature related to breeding for low CH_4 emitting cows have been based on the sniffer method (Garnsworthy et al., 2019).

2.4 Laser methane detectors

The laser methane detector (LMD) is a handheld device that is pointed to the cows' nostrils to measure CH_4 along the length of the laser beam in ppm per meter (Chagunda et al., 2009; Jonker et al., 2020). The LMD has been used in several studies that aimed to record CH_4 emissions from individual animals, largely on small groups of animals (as reviewed by Sorg (2022)). Advantages of the LMD are that the cost of the method is low, it is flexible, portable, easy to handle, and does not require an external power source. However, disadvantages are that the method requires intensive labour to record a large number of cows and the repeatability is low because the method is strongly affected by environmental conditions. The LMD is often handheld and requires a cow to be immobile for five minutes to maintain a stable measuring distance and angle. The method is also more prone to human error as the cows' identification numbers are not registered automatically.

The LMD has been compared to other techniques such as RC, GF, sniffers, and the SF_6 technique (as reviewed by Sorg (2022)). On-farm studies revealed strong phenotypic correlations with GF (r = 0.66) and sniffers (r = 0.57 and 0.60) (Sorg et al., 2018). Other studies that compared LMD to RC or sniffers reported moderate to high correlations between the methods; however, not all studies are directly comparable to situations and the environment on-farm. This highlights the need for standardised comparisons to better characterise the LMD's capabilities for large scale phenotyping.

Similar to sniffers, LMD measure concentrations and do not directly allow for the quantification of an animal's total CH_4 production, as is possible with GF and RC (Sorg, 2022). Attempts have been made to estimate daily CH_4 production from LMD concentration measurements; however, the equations proved to be unsuitable for accurate quantification of individual animal CH_4 production and were only useful to give a rough estimate of the total CH_4 production of a group (Chagunda et al., 2009; Sorg et al., 2017).

Currently, there is no consensus on a uniform measurement and evaluation protocol, so results between studies using the LMD are poorly comparable (Sorg, 2022). These factors make LMD challenging to use for large-scale phenotyping for genetic evaluation programmes, and therefore they have to date not been applied in large-scale commercial settings. Nonetheless, similar to the use of sniffers, the concentrations can be used to rank animals from low to high emitters, making it a promising tool for low-cost recording of CH_4 emissions of dairy cows. But before large-scale application, further research is needed to develop a joint protocol for measurement and data analysis.

2.5 SF_6 technique

The SF_6 technique measures enteric CH_4 emissions by using sulfur hexafluoride as a tracer gas. With the SF_6 technique, air is sampled 24 hours a day near the cow's nostrils through a tube attached to a halter and connected to a canister worn around the neck or on the back of the cow (Johnson et al., 1994; Zimmerman, 1993). A permeation tube containing SF_6 is placed into the rumen of the cow. The pre-determined release rate of SF_6 is multiplied by the ratio of CH_4/SF_6 concentrations in the canister to calculate the CH_4 emission rate.

An advantage of the SF_6 technique is that cows can be recorded during grazing. A linear relationship between SF_6 and RC measurements has been reported in the literature (R^2 = 0.69, Munoz et al. (2012)). However, the SF_6 technique is invasive and labour-intensive, making it unsuitable for large-scale phenotyping. In addition, concerns have arisen about the technique's accuracy in specific conditions as the SF_6 method shows greater day-to-day variability (Berends et al., 2014; Lassey et al., 2011). To improve reliability, further research is recommended, focusing on experimental designs to investigate these issues.

2.6 Other methods

This section discussed the most common methods in recording CH_4 emissions of individual dairy cows. In addition to the methods mentioned above, more methods to record individual CH_4 emissions are being developed. A method that has been extensively used in sheep and goats is portable accumulation chambers (PAC) (Jonker et al., 2018, 2020). Portable Accumulation Chambers are boxes that can fit one animal at a time in an airtight environment. During measurement sessions, which typically last between 30 minutes and 1 hour, the animal is placed inside the chamber, allowing exhaled gases to accumulate. The concentration of CH_4 increases over time, and this accumulation is measured to estimate the animal's CH_4 production rate. The method is a practical and cost-effective alternative to RCs and can be used to visit farms to record animals, and developments are ongoing to extend the use of PACs to cattle (AgResearch, 2023). Other alternative methods apply, for example, a hood system that can be placed above a feeding bin or lying cubicle (Troy et al., 2016; Wu, 2016). These systems have to date only been tested on small groups of cows; however, they show potential for further application.

3 Trait definition

Before a new phenotype can be implemented in a breeding programme, the trait should be clearly defined, recordable, affordable, have phenotypic variation, be heritable, and the genetic correlations with other selection index traits need to be known (Hazel, 1943). Worldwide, research groups and breeding organisations are working on including CH_4 mitigation in breeding programmes. Often, different methods of phenotyping are used, with each having their own defined traits and units of measurement.

Although several traits have been previously suggested (de Haas et al., 2017; Lassen and Difford, 2020), currently there is no consensus on which direct CH_4 traits are the most promising for breeding programmes that aim to mitigate CH_4 emissions. To answer these questions, more research will be required that aims to standardise traits, or additional estimates of phenotypic or genetic correlations can provide valuable knowledge into the relationships between different CH_4 recording methods and different phenotypes defined from one recording method.

3.1 Mean emissions

The most reported traits for CH_4 emissions are mean CH_4 concentration (ppm) and mean CH_4 production (g/day). Between phenotypes based on mean CH_4, differences exist in trait definition. For example, for mean traits defined from sniffers, studies by van Engelen et al. (2018) and van Breukelen et al. (2022) analysed the average CH_4 concentration per AMS visit, studies by Pszczola et al. (2017) and Sypniewski et al. (2021) averaged CH_4 concentration and CH_4 production per day, studies by Breider et al. (2019), van Breukelen et al. (2022), Difford et al. (2020), Lassen and Lovendahl (2016), Manzanilla-Pech et al. (2022b), and Zetouni et al. (2018) analysed average CH_4 concentration and CH_4 production per week, whereas a study by Lassen et al. (2016) analysed a single weekly average of CH_4 production. The mean concentration and production traits are popular, as the traits are easy to compute, explain and interpret. In addition, the high genetic correlation estimated between sniffer-recorded CH_4 concentration (ppm) and GF-recorded CH_4 production (g/day) of 0.76 suggests that selection for such traits will result in a decrease in enteric CH_4 emissions (van Breukelen et al., 2023). Nonetheless, genetic correlations with RC phenotypes need to be estimated in the future, to investigate the relationship with total CH_4 production recorded over longer periods of time.

3.2 Methane to carbon dioxide ratio

One of the underlying traits in estimating sniffer CH_4 production is the ratio between CH_4 and CO_2 (Madsen et al., 2010). The ratio is of interest because energetically it would be optimal to metabolise all feed carbon to the production of milk, BW gain, pregnancy, and CO_2, and the carbon extracted as CH_4 reflects the energy that is lost. Therefore, the CH_4/CO_2 ratio describes the proportion of excreted carbon that is not metabolised as CO_2 and a lower ratio reflects a more efficient cow. By determining the CH_4/CO_2 ratio and combining it with the estimated CO_2 production, the CH_4 production from the animal can be estimated.

For breeding programmes, using ratio traits is generally discouraged as a multi-trait approach is likely to result in higher genetic gain than by selecting for ratio traits. In addition, responses in ratio traits can be hard to interpret, as it is not immediately apparent if the response is due to a change in the numerator or the denominator trait (Veerkamp, 2002; Veerkamp et al., 2013; Zetouni et al.,

2017). In addition, when using the CH_4 to CO_2 ratio it is assumed that efficiency of energy utilisation from feed for maintenance and production is constant. However, in practice variation in diets and efficiency of cows may exist, which can result in overestimating the CH_4 production of high-efficiency cows and favour inefficient cows (Huhtanen et al., 2020).

3.3 Peaks of methane

Other traits of interest for breeding, commonly reported in the literature by studies using sniffers, are traits based on peaks in CH_4 emissions. Cows are known to belch or eruct gases from the rumen at intervals of around one minute (Garnsworthy et al., 2012). Only considering peaks in CH_4, as opposed to an average of all concentrations recorded during a full milking robot visit, ensures that cows' belches are analysed with possibly less disturbance from background CH_4 concentrations in between belches. However, questions remain about how peaks from belches reflect total CH_4 production. Only analysing the peaks in CH_4 emissions might miss the emissions that are present in breath. The relationship between mean CH_4 emissions and CH_4 peaks has not been extensively analysed and remains unclear.

Although traits based on peaks in emissions are common in the literature, the definition of a peak can differ between studies. For example, peak traits for sniffers have been calculated from: (1) the area under the curve, (2) a maximum value, or (3) an average of multiple peaks (Bell et al., 2019; Garnsworthy et al., 2012; Lopez-Paredes et al., 2020; Rey et al., 2019). From LMD measurements, several peak traits have been defined using the mean, number of peaks, and sum of peaks, and separating respiration peaks and eructation peaks. Also, for LMD traits there is no consensus on which trait would be the most suitable for genetic evaluation programs (Sorg, 2022). Therefore, with new definitions of peak traits, it is always important to first investigate the genetic relationship with recorded CH_4 production (as is done with RC or GF) to ensure that genetic selection has the desired outcomes.

3.4 Methane efficiency and residual methane

Other traits that have been mentioned in the literature are CH_4 efficiency traits. Efficiency traits are usually defined as CH_4 as a ratio to production or intake, making them independent of the denominator trait. Well-known efficiency traits for dairy cows are CH_4 intensity (g CH_4/kg milk) and CH_4 yield (g CH_4/kg dry matter intake (DMI)), as reviewed by de Haas et al. (2017). However, as these traits are ratio traits, they are generally not preferred for inclusion in breeding programmes. The use of ratio traits persists in genetic research, largely because these traits are commonly used in other fields, making the ratio traits useful for cross-disciplinary comparisons.

An alternative to using ratio traits is using a residual CH_4 trait (Berry et al., 2015; de Haas et al., 2017; Manzanilla-Pech et al., 2021). In dairy breeding, residual CH_4 is CH_4 that is phenotypically or genetically adjusted for production, BW, and/or DMI. By doing so, animals that have a lower CH_4 emission regardless of their level of production, BW, and DMI can be more easily identified. Similar approaches are often applied to DMI data, to adjust DMI for production and other energy sinks, resulting in residual feed intake (RFI) (Koch et al., 1963; Pryce et al., 2014; Veerkamp et al., 1995). In these approaches, phenotypic RFI is generally modelled as DMI minus predicted DMI based on, for example, milk production and growth (Tempelman et al., 2015; Veerkamp and Emmans, 1995). Genetic RFI is generally modelled by performing a genotypic regression of DMI on milk production and other energy sinks (Kennedy et al., 1993). Similar methods have been used to make live weight independent of body condition score (Veerkamp and Brotherstone, 1997) and to adjust longevity for milk production (van der Linde et al., 2007).

A study by Manzanilla-Pech et al. (2021), showed that genetic correlations between residual CH_4 (adjusted for metabolic BW and energy corrected milk yield (ECM)) were not only zero with BW and ECM, but also with DMI. Thus, selecting for residual CH_4 adjusted for metabolic BW and ECM would also not have an impact on DMI. Furthermore, in the study by Manzanilla-Pech et al. (2021) residual CH_4 was positively correlated with RFI, indicating that lower CH_4 emitting animals were more efficient at converting feed. Thereby, residual CH_4 showed potential for inclusion in breeding programmes. However, when the traits by which CH_4 is adjusted are also included in the breeding goal, undesired responses in the traits can also be avoided by appropriate weighing in the total merit index. In addition, genetic correlations between sniffer-recorded CH_4 and other breeding goal traits on a large dataset on dairy cows in the Netherlands showed close to zero correlations between CH_4 and all other traits (van Breukelen, 2024). With close to zero correlations between CH_4 and the traits by which residual CH_4 is adjusted, there is no advantage in using residual CH_4 as a trait for breeding over traits with a more simple definition such as mean CH_4 emissions.

3.5 Proxies for methane

Some studies have suggested using proxies (i.e. indicator or indirect traits) for CH_4 to improve the accuracy of breeding value estimations for CH_4. Proxies are traits that are correlated to recorded CH_4 traits, but which are easy and relatively low-cost to record on a large scale. Using proxies helps to increase the number of cows with phenotypes in reference populations, as the number of cows with recorded CH_4 phenotypes is still limited. Suggested proxies for CH_4 are, for example: feed intake and diet composition (Difford et al., 2020), fatty acid profiles (Dijkstra et al., 2011), milk mid-infrared spectra (MIR) (Dehareng et al., 2012), rumination time (Lopes et al., 2022), or rumen microbial composition (González-Recio et al., 2023; Wallace et al., 2015; Zhang et al., 2020). Proxies are unlikely accurate enough to predict CH_4 by themselves, but combining proxies can increase the accuracy by 15 to 35% (Negussie et al., 2017). The use of proxies as predictor traits can be combined with datasets on recorded CH_4 emissions to enlarge genomic reference populations, improving the accuracy of genomic prediction. Some of the most suggested proxies are discussed here and a more extensive list of proxies can be found in the review by ICAR (2020).

3.5.1 Feed intake

Feed intake and differences in diet are important drivers of the amount of CH_4 production (Herliatika et al., 2024; Hristov et al., 2018). Johnson and Johnson (1995) initially estimated that CH_4 emissions could represent an energy loss of 2% to 12% of cows' gross energy intake, sparking interest in potential links between CH_4 emissions and feed efficiency. A connection between CH_4 emissions and feed efficiency would suggest a possible dual benefit: improvements in feed efficiency, which hold significant economic value, alongside reductions in CH_4 emissions. However, nutritionists and physiologists cautioned that lower CH_4 emissions could come with trade-offs, such as reduced fibre digestion and quicker passage through the rumen, making cows less effective at digesting feed (Huhtanen and Hristov, 2018).

Numerous studies have reported clear relationships between differences in diet, such as starch content, and the associated amount of CH_4 production (as reviewed by Herliatika et al. (2024)). Additionally, CH_4 production is influenced by feed intake levels, though this relationship is modulated by dietary factors and the time of measuring CH_4 relative to the time of feeding (Hristov et al., 2018; Hristov and Melgar, 2020). To account for variation in diet and rations across farms, mixed models that incorporate farm and temporal effects are used when estimating breeding values. These models largely correct for dietary differences, yet individual feed intake data can still serve as a potential proxy for estimating CH_4 production on an individual cow level (Hristov et al., 2018).

On an individual genetic level, associations between feed efficiency and CH_4 traits have been identified. Difford et al. (2020) assessed genetic correlations between feed efficiency traits and CH_4 concentration (ppm) across two Holstein populations. In Denmark, strong favourable genetic correlations (0.42–0.69) were found across various definitions of RFI. However, in the

Netherlands, genetic correlations varied widely (−0.69 to 0.46) depending on the RFI definition. Studies by Manzanilla-Pech et al. (2021) and Manzanilla-Pech et al. (2022c) reported that CH_4 production was positively correlated with RFI (0.38 ± 0.18 to 0.76 ± 0.09), indicating that lower CH_4 emitting animals were more efficient at converting feed. In these studies, different definitions for CH_4 traits and RFI were used and this had an influence on the magnitude of the genetic correlations. Thus, in determining the applicability of RFI as a predictor for CH_4, it is important to always re-evaluate the genetic correlations between newly recorded traits and newly defined trait definitions.

3.5.2 Milk mid-infrared spectra

Milk MIR spectroscopy data has emerged as a promising tool for predicting CH_4 emissions from ruminants (Dehareng et al., 2012; Vanlierde et al., 2015). MIR spectroscopy analyses the chemical composition of milk samples to indirectly estimate CH_4 production, based on correlations between specific MIR profiles and CH_4 recorded with RC, SF_6 or GF. Using predictions from MIR spectra is advantageous because it is non-invasive, cost-effective, and can be integrated with existing milk analysis procedures.

Previous studies have reported correlations between MIR CH_4 and recorded CH_4 ranging from 0.23 to 0.75, and R^2 values ranging from 0.26 to 0.87 (Coppa et al., 2022; Dehareng et al., 2012, 2019; McParland et al., 2024; Shadpour et al., 2022; Vanlierde et al., 2015, 2018). Though promising, the accuracy of MIR predictions can vary greatly depending on the sample type and the calibration model used. Ongoing research focuses on improving prediction models by refining calibration techniques and ensuring consistency across different breeds, diets, and environmental conditions. For example, the study by Coppa et al. (2022) showed that predictions were poor for herds where the CH_4 inhibitor 3NOP was fed. This suggests that specific models would be required for MIR prediction when cows receive chemical inhibitors of CH_4 emissions that do not affect milk composition.

Although work is ongoing to refine MIR prediction models, a first breeding value for CH_4 for commercial use in Canada has been published based on MIR predictions since April 2023 (Oliveira et al., 2024). When launching the breeding value, the model was based on a limited sample size from 2 research stations. Nonetheless, work is ongoing to enlarge the quantity of data used for the prediction and to further validate the prediction. This will improve the accuracy of the genomic predictions for CH_4 in Canada, to guarantee sufficient response to selection.

3.5.3 Rumination time

Rumination has been proposed as a possible indicator of CH_4 emissions (Negussie et al., 2017). It is measured using acoustic tags attached to the collars of dairy cows, which record the time spent chewing. Chewing is linked to CH_4 emissions through the digestion and fermentation of fibre. A study where CH_4 was recorded with GF units reported an initial moderate genetic correlation estimate of −0.53 ± 0.24 between rumination time and CH_4 production (Lopes et al., 2024).

3.5.4 Rumen microbiome

The rumen microbiome has received considerable attention as a predictor for breeding low CH_4 emitting cows (as reviewed by González-Recio et al. (2023)), which is a logical link as the rumen microbiome is what produces the CH_4 during fermentation of feed in the rumen. Methane emissions are primarily influenced by methanogens, specific types of archaea within the microbiome, alongside other microbial communities that contribute to hydrogen production and consumption. It has been shown that across animals, also under different environments, similar bacteria and archaea dominate the rumen microbiome (Henderson et al., 2015). Especially methanogenic archaea, which have been extensively characterised, have been shown to be highly conserved across the world. The dominance of a few methanogens could enable the development of strategies to target these key species, potentially reducing methane emissions.

The rumen microbiome composition has been shown to be partially controlled by host genomics, with heritabilities ranging from 0.10 to 0.40 (González-Recio et al., 2023). A study by Martinez-Alvaro et al. (2022) observed strong genetic correlations with CH_4 emissions among the heritable part of the rumen microbiome (0.59–0.93). However, the genetic parameters need to be verified as the estimates were based on a small population (n = 359). Nonetheless, in beef cattle, it has been shown that more similar archaea:bacteria ratios were obtained across sire progeny groups, adjusted for diet, RC, and randomised block effects (Roehe et al., 2016). Other studies have shown differences in microbial composition between dairy breeds (Holstein, Jersey and Brown Swiss) for cows reared under the same conditions with the same diets (Gonzalez-Recio et al., 2018; King et al., 2011). Thus, microbial composition or specific microbial traits are partially under genetic control and can be included in genetic evaluation models.

Different approaches have been suggested for the statistical treatment of microbial information in animal breeding, targeting either the microbiome as a trait or the microbiome as a source of phenotypic variability. Specific strategies to include microbiome information in breeding programmes that aim to reduce CH_4 emissions have yet to be applied in practice, as the inclusion of metagenomic information still poses significant challenges in modelling and metagenomic information is still expensive to obtain due to sampling and sequencing costs (see the review by González-Recio et al. (2023)). However, most likely the microbiome will provide a major information source that helps to reduce CH_4 emissions of cattle in the future in the field of animal breeding and other fields with the development of new technologies to collect and sequence metagenomic data.

4 Genetic parameters

Previous research has shown that individual dairy cows differ in the amount of enteric CH_4 emissions they emit, and that part of this variation between animals is heritable (Lassen and Difford, 2020). However, differences in parameter estimates exist between trait definitions and choice of model, which have important implications for setting up breeding programmes.

4.1 Heritability

Several studies have estimated heritabilities for different CH_4 phenotypes in cattle, where the estimates in dairy cattle ranged from 0 to 0.53 ± 0.06 (Table 1). The average heritability of the estimates reported in the table is 0.19 and the estimates provide confidence that there is a genetic component to enteric CH_4 emissions and thus that genetic selection for CH_4 mitigation is possible. To date, not all methods of recording CH_4 production have resulted in heritability estimates for dairy cattle, with most genetic research in dairy cattle currently being conducted with sniffer sensors due to the ability to record large numbers of individual cows cost-effectively.

4.2 Recording strategy

In general, for traits with higher heritability and repeatability, accurate breeding values can be estimated with fewer records. As mentioned before, diverse studies used different trait definitions, with some studies averaging CH_4 emissions per AMS visit, whereas other studies average CH_4 emissions from AMS visits per day or week. Weekly averages reported in the literature generally are associated with higher heritability and repeatability estimates, which would imply that averaging per week would be preferred for breeding value estimation over visit or daily averages. However, this is not necessarily the case when considering that an average of seven visits can be recorded per week and because of the larger number of repeated records the two reliabilities

Table 1 Summary of heritability ($h^2 \pm SE$) estimates for recorded methane emissions in dairy cattle, recorded by GreenFeed units (GF), Laser Methane Detectors (LMD), the SF_6 tracer gas technique (SF_6), and sniffers (transformed to production in g/day or L/day, as concentrations in ppm)

Method of recording	Mean ± SD	h^2 ± SE	Reference
Combined dataset[1] (g/day)	392 ± 86	0.21 ± 0.04	Manzanilla-Pech et al. (2021)
GF (g/day)	464 ± 115	0.16 ± 0.10	Kamalanathan et al. (2023)
GF (g/day)[2]	436 ± 120 and 435 ± 98	0.19 ± 0.02 and 0.33 ± 0.04	van Breukelen et al. (2023)
GF (g/day)	492 ± 81	0.36 ± 0.12	Lopes et al. (2024)
LMD (mg/kg)	180 ± 47	0.05 ± 0.07	Pickering et al. (2015a)
SF_6 (g/day)	111 ± 34	0.23 ± 0.23	Manzanilla-Pech et al. (2016)
SF_6 (g/day)	469 ± 81	0.16 ± 0.11	Richardson et al. (2021b)
Sniffer (g/day)	395 ± 58	0.24 ± 0.15	Lassen et al. (2016)
Sniffer (g/day)	315 ± 36	0.21 ± 0.06	Lassen and Lovendahl (2016)
Sniffer (g/day)[3]	279 ± 64	0.23 ± 0.12 to 0.3 ± 0.08	Pszczola et al. (2017)
Sniffer (g/day)[3]		0.12 ± 0.16 to 0.45 ± 0.11	Breider et al. (2019)
Sniffer (g/day)	183 ± 67	0.12 ± 0.04	Lopez-Paredes et al. (2020)
Sniffer (g/day)[3]	396 ± 60 and 503 ± 93	0.13 to 0.26	Sypniewski et al. (2021)
Sniffer (g/day)[3,4]	307 ± 78 and 355 ± 82	0.11 ± 0.07 to 0.47 ± 0.13	Manzanilla-Pech et al. (2022a)
Sniffer (g/day)	355 ± 73	0.12 ± 0.03	Manzanilla-Pech et al. (2022b)
Sniffer (g/day)	338 ± 86	0.21 ± 0.05	Manzanilla-Pech et al. (2022c)
Sniffer (L/d)	380 ± 60	0.25 ± 0.07	Zetouni et al. (2018)
Sniffer (ppm)	254 ± 230	0.11 ± 0.02	van Engelen et al. (2018)
Sniffer (ppm)	853 ± 278	0.12 ± 0.01	Saborío-Montero et al. (2019)
Sniffer (ppm)		0.15 ± 0.15 and 0.26 ± 0.11	Difford et al. (2020)
Sniffer (ppm)	1288 ± 468	0.11 ± 0.03	Lopez-Paredes et al. (2020)
Sniffer (ppm)	511 ± 178	0 to 0.14	Sypniewski et al. (2021)
Sniffer (ppm)[3,4]	576 ± 54 and 573 ± 51	0.10 ± 0.06 to 0.53 ± 0.06	Manzanilla-Pech et al. (2022a)
Sniffer (ppm)	577 ± 46	0.15 ± 0.03	Manzanilla-Pech et al. (2022b)
Sniffer (ppm)	574 ± 52	0.20 ± 0.05	Manzanilla-Pech et al. (2022c)
Sniffer (ppm)[2]	328 ± 269	0.09 ± 0.01 to 0.32 ± 0.03	van Breukelen et al. (2022)

(Continued)

Table 1 *(Continued)*

Method of recording	Mean ± SD	h^2 ± SE	Reference
Sniffer (ppm)[2]	325 ± 251 and 331 ± 218	0.18 ± 0.01 and 0.32 ± 0.02	van Breukelen et al. (2023)
Sniffer (ppm)[3]	668 ± 299	0.10 ± 0.04 to 0.21 ± 0.03	van Breukelen et al. (2024)

[1]Including measurements from GF, SF_6, and sniffers, [2]For various trait definitions, [3]Using a random regression model over a lactation, [4]For lactation one and two separately.

would be approximately equal (seven visit records per week versus one weekly record) (van Breukelen et al., 2022).

Nonetheless, the number of records and thus the duration of CH_4 recording are important factors to determine the expected reliability in estimating breeding values. Longer recording periods generally give higher accuracies initially, because for CH_4 traits analysed with repeatability models the heritability is generally lower than the repeatability. However, the reliability will plateau at an ideal number of records which will depend on the trait definition. For example, a study by van Breukelen et al. (2022) showed that to achieve a minimum reliability of 50% for the estimated breeding value of a bull, 25 visit records on mean CH_4, measured on 10 different daughters, were sufficient. This is approximately equal to recording five weekly mean CH_4 records on 10 different daughters to achieve a minimum reliability of 50% for the estimated breeding value of a bull.

In addition, the genetic variance of CH_4 traits has been shown to vary during lactation for CH_4 recorded with sniffers, indicating that the moment of recording is also of importance (Figure 3) (Manzanilla-Pech et al., 2022a;

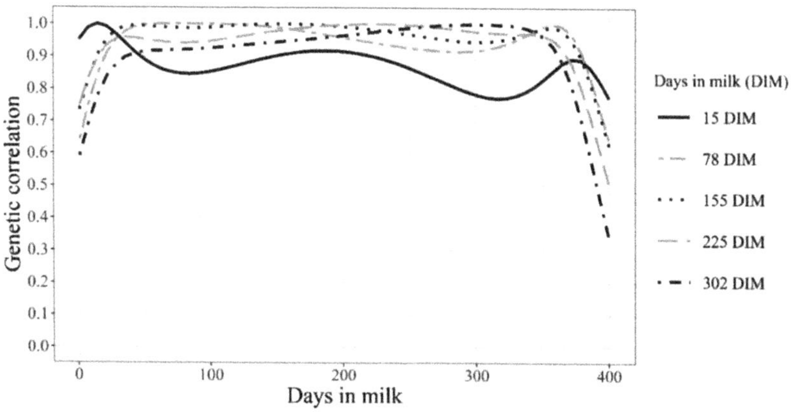

Figure 3 The genetic correlations between CH_4 concentration recorded on 15, 78, 155, 225, or 302 days in milk (DIM) and recorded on all other DIM, from van Breukelen (2024).

Pszczola et al., 2017; Sypniewski et al., 2021; van Breukelen et al., 2024). When sniffers are used to record CH_4 for short periods of time, recording CH_4 emissions in mid or late lactation will likely give a higher accuracy compared to early lactation, up to 25 days in milk, where the heritability and repeatability are lower.

Phenotyping strategy, trait definition, and choice of model can result in heterogeneity in heritability estimates, making the heritability by itself an unsuitable parameter to support decision making in which trait to include in breeding programmes. Deviations in parameter estimates can result in achieving lower genetic progress than what was estimated using a different model or trait. It is therefore important to re-estimate genetic parameters when making changes in phenotyping practices, trait definition, or when a new and largely unrelated population of animals is recorded (Falconer and Mackay, 1996). Furthermore, reflecting on phenotyping strategies, next to giving insight into the expected genetic gain, helps to apply phenotyping instruments in a cost-effective and time-efficient manner.

4.3 Genetic correlations

When selecting for reduced CH_4 emissions, it is important to know the genetic relationships between CH_4 traits and other breeding goal traits. Commonly, national breeding goals for dairy cattle include traits on milk production and components, feed efficiency, health, reproduction, and/or conformation. Relationships between traits can be estimated as genetic correlations, which are essential parameters to ensure that selecting for reduced CH_4 emissions does not negatively impact other traits in the breeding goal. For example, some studies indicate there is a positive genetic correlation between CH_4 traits and DMI, indicating that selecting for lower CH_4 emissions may have an impact on feed intake (Lopes et al., 2024; Richardson et al., 2021b). By using weighted breeding goals, selection will never be fully on one trait, but will always be part of a total merit index with a proper weight to ensure balanced breeding without undesired responses in other breeding goal traits.

To date, few genetic correlations have been estimated between CH_4 emissions and current breeding goal traits. For the estimates that are published, they are often associated with large standard errors probably due to a limited numbers of records (Hossein-Zadeh, 2022). Thus, the effect of genetic selection for lower CH_4 emissions on other breeding goal traits is still largely unknown, whereas they are required to make informed breeding decisions. Further analysis on the correlation structure with other traits is needed on larger datasets to validate these relationships. With greater certainty about these correlations, it will become more feasible to assign the appropriate weight to CH_4 emissions in breeding goals, without compromising milk production, fertility, or health.

Additionally, genetic correlations could be informative in deciding if other traits can be used as predictor traits for cows without CH_4 records (Negussie et al., 2017). Using predictor traits is useful when the accuracy of the estimated breeding value (EBV) for the predictor trait, multiplied by the genetic correlation, is higher than the accuracy for the EBV for the recorded CH_4 trait.

5 Implementation in breeding programmes

5.1 Incentives

Reducing CH_4 emissions from dairy farming is not yet economically advantageous for most farmers globally. Although carbon taxes are widespread, they have not been extended to agriculture. However, this is expected to change soon, as Denmark has announced plans to implement a carbon tax for agriculture starting in 2030 (Olsen, 2024). Meanwhile, societal and governmental pressures to lower greenhouse gas emissions in agriculture may encourage the sector to incorporate CH_4 emissions into breeding objectives with non-market value before carbon taxes become enforced (Boichard and Brochard, 2012; Brascamp, 1984).

The dairy industry itself could also incentivise lower CH_4 emissions. For instance, milk processing companies may offer premium prices for milk produced with a smaller carbon footprint, including herds with lower-than-average CH_4 emissions. This approach could align with industry programmes focused on reducing milk's carbon footprint. For example, a Danish-based company that collects milk from Denmark, Sweden, the United Kingdom, and Germany introduced a 0.04 eurocent milk payment increase related to greenhouse gas emissions in 2023 (Arla, 2023). Similarly, a company based in the Netherlands pays 0.05 eurocent per kg of milk for milk that is rewarded their sustainability label (FrieslandCampina, 2018). If this would be linked to breeding values, it can be a strong incentive for farmers to select for low CH_4 emitting cows.

5.2 Accelerating genetic progress

As the initial effects of climate change are already apparent, there is an urgency in reducing carbon emissions to prevent worsening (IPCC, 2021). Genetic selection has the advantage that its effect is cumulative and permanent; however, changing populations through genetic selection can be a slow process. Genetic improvement of quantitative traits becomes visible only over several generations. Nonetheless, several opportunities exist to accelerate genetic progress.

5.2.1 Indirect selection for a lower carbon footprint of dairy

Methane traits can be implemented in new or existing selection indices once the economic value or desired gain of CH_4 has been determined. As an alternative to having an index that only includes direct CH_4 traits, a new index can be developed that targets overall environmental sustainability. In such an index, next to reducing direct CH_4 emissions, other NVI traits can be optimised to minimise the carbon footprint.

An example is the Economic Breeding Index (EBI) created by the Irish Beef Cattle Federation. The EBI uses current breeding goal traits to indirectly reduce greenhouse gas emissions, without using direct CH_4 measurements. This index has achieved an economic gain of 20 euros net profit per cow per year, while leading to a 2% reduction in carbon footprint (ICBF, 2018). Similarly, in the Netherlands, it was shown that breeding for increased production in livestock has indirectly reduced the environmental impact by 1% a year per kg of product produced (Mollenhorst and De Haas, 2019). Studies in the United Kingdom, Ireland, and Canada similarly showed that the emission intensity will be reduced by 1% per year through their national breeding programmes (Amer et al., 2018; Bell et al., 2014; Richardson et al., 2021a). In New Zealand, research on eight breeding goal traits indicated that increased production efficiency through selecting for farm profitability indirectly helps to mitigate CH_4 emissions by −0.04 kg CO_2-equivalents/kg milk protein equivalent/cow/year (Zhang et al., 2019). Thus, developing an environmental sustainability index shows potential to support continued genetic progress in production, fertility, conformation, and health traits, while also indirectly targeting a reduction in CH_4 emissions.

Despite the passive genetic gains in CH_4 reduction achieved through current breeding strategies, as outlined in the studies above, CH_4 mitigation targets are often still out of reach. Combining indirect selection for a reduced carbon footprint with direct selection for traits related to enteric CH_4 emissions could be a powerful approach to accelerate genetic progress in lowering greenhouse gas emissions. When estimating the carbon footprint of various breeding goal traits, it is important to consider greenhouse gas emissions from all sources, such as feed production and manure management, through a life cycle assessment. Neglecting these emissions may otherwise lead to bias in the estimated emissions (van Middelaar et al., 2014). To address trade-offs, breeding indices aimed at optimising the reduction of greenhouse gas emissions from economically important traits should combine bio-economic models with life cycle assessments, as demonstrated by Shi et al. (2024).

5.2.2 Applications in other countries and possibilities for collaboration

Most research that targets the use of selective breeding to reduce CH_4 emissions from dairy cattle is a result of populations recorded in Europe, North America, and Oceania. However, new initiatives to reduce emissions should be implemented globally and should also include animal production systems in the Global South. Currently, such initiatives are being developed by, for example, the Global Genetics Program together with ICAR and other groups with experience, where CH_4 phenotypes and genotypes that are collected worldwide will be gathered in a shared database to use for genetic evaluations within and across countries (WUR, 2024). Shared databases with many records across countries can help countries with scarce recording of CH_4 to set up breeding for low CH_4 and can also help countries with CH_4 recording to accelerate breeding for CH_4 mitigation by improving the accuracy of estimated breeding values.

However, applying genetic selection to reduce CH_4 emissions across countries worldwide comes with unique challenges (Manzanilla-Pech et al., 2021; Richardson et al., 2022). Dairy production systems are generally very diverse over different climates, and many different dairy breeds are kept next to the widespread Holsteins. To overcome these challenges, a joint dataset with data from various countries is likely required to be much larger than within-country datasets to be able to obtain sufficiently high reliabilities of breeding values across countries (de Haas et al., 2017; Richardson et al., 2022). Due to the urgency in reducing enteric CH_4 emissions, phenotyping projects are commencing worldwide, and collating this data is expected to result in a sufficiently large dataset with CH_4 records on tens of thousands of cows. In addition, combining data across countries has previously been done successfully for other traits, for example, to increase the reliability of dry matter intake in countries with scarce or no recording (Berry et al., 2014; de Haas et al., 2015).

Several methods exist to integrate data from multiple countries, which can process phenotypes that are recorded by different systems and reported in different units. Most commonly, multi-trait models are applied, where the phenotypes are analysed as separate traits in an international evaluation, as is, for example, done by the MACE procedure (Palucci et al., 2023). Multi-trait models can simultaneously account for genotype-by-environment interaction, which may be induced by a large variation in management systems, feed, climates, and/or recording techniques across countries. However, also when using such models, it is essential to have a number of phenotypes recorded in each country, breed, and production system for which breeding values are estimated.

An alternative approach is by standardising similar CH_4 traits per country and analysing the data as a single trait. A study by Manzanilla-Pech et al. (2021) demonstrated that CH_4 production was highly correlated to genetically standardised CH_4, which was standardised using within and across country genetic standard deviation. However, the dataset was insufficiently large to estimate genetic correlations between CH_4 production across and within countries. These genetic correlations should first be estimated to judge if there are genotype-by-environment interactions between countries or if implementing a single across country CH_4 trait in international genetic evaluations is possible in specific cases.

5.2.3 Using genomic selection

Genetic progress for scarcely recorded traits within and across countries can be accelerated by applying genomic prediction (Manzanilla-Pech et al., 2021). Over the past two decades, genomic prediction has become an important tool in animal breeding, which adds information on the genomics of individual animals to common mixed models primarily through single nucleotide polymorphisms (SNPs). In genomic prediction, a reference population is used consisting of cows that have phenotypes and are genotyped, which is applied as a training dataset. From the training dataset, SNP effects are estimated, which are consequentially used in a prediction equation for cows or bulls without phenotypes but with genotype information (the selection candidates). Thereby, also for genotyped bulls without phenotyped daughters breeding values can be estimated.

Genomic selection is widely applied in dairy cattle breeding, including in international genetic evaluations (Interbull Centre, 2022; Palucci et al., 2023). The accuracy of genomic predictions can be improved by: (1) expanding the number of recorded animals in the reference population, (2) recording animals within families that are relevant for selection (e.g. from sires with many daughters), (3) increasing marker density, and (4) increasing the heritability by reducing environmental effects (de Haas et al., 2012). Likewise, it has been shown that sharing data on CH_4 phenotypes and genotypes across countries can be a powerful method to increase the reliability of genomic prediction within and across participating countries (de Haas et al., 2015; Lund et al., 2011). To implement genomic prediction for CH_4, routine genotyping of all CH_4 recorded cows on farm is crucial.

5.3 Expected genetic progress

An estimation of the expected genetic progress that can be achieved by selecting for lower CH_4 emissions, using CH_4 recording by sniffers, was published in a paper by de Haas et al. (2021). Through the selection index calculations, the impact of genetic selection for CH_4 intensity (CH_4/kg milk) was estimated to be a reduction of 24% by 2050 compared to 2018 levels (de Haas et al., 2021). Because CH_4 is a scarcely recorded trait, recording on 100 daughters per sire will initially not be realistic. Using genomic prediction is often mentioned as a practical and more realistic alternative. The study by de Haas et al. (2021) showed that when using genomic prediction, the genetic improvement in CH_4 intensity was only slightly lower (21% for genomic prediction vs 24% for 100 recorded daughters per bull).

Larger reductions in CH_4 emissions are possible when more selective weight is put on CH_4, such as when assigning economic values of zero to all other breeding goal traits and putting all weight on CH_4. However, although putting all weight on CH_4 results in a larger reduction in CH_4, it will also likely result in reduced milk production and possible undesired responses in other traits. Thus, selection merely for lower CH_4 comes at a cost in genetic improvement in the other breeding goal traits, as was also pointed by de Haas et al. (2021) and by Gonzalez-Recio et al. (2020). Genetic correlation estimates between CH_4 emissions and current breeding goal traits are often associated with large standard errors probably due to a limited numbers of records (Hossein-Zadeh, 2022). When genetic correlations differ from what was previously assumed, the weight that can be put on CH_4 in breeding goals may change. In addition, the optimal method to balance between CH_4 and current breeding goal traits in selection indices in practical breeding programmes is yet to be determined. Together, these factors will significantly influence the level of genetic progress that will be realised in practice and should be further investigated.

6 Conclusion

This chapter reflects on the role of selective breeding in dairy production as a tool to reduce enteric CH_4 emissions to support climate-smart agriculture. Methane mitigation through selective breeding is cost-effective, cumulative, and permanent, making it an appealing strategy.

Several innovations in CH_4 emission measurement of individual cows were highlighted which have enabled large-scale phenotyping. Methods such as GF systems, sniffers, and LMDs provide innovative ways to measure CH_4 emissions at the individual cow level. Although these methods do not measure total emissions as precisely as RCs, they correlate with more accurate methods and are more practical for large-scale application. This allows breeders to rank cows based on CH_4 emissions, facilitating the selection of low-emission animals.

These methods enable the collection of CH_4 data across thousands of cows, supporting the integration of CH_4 traits into breeding programmes which can be further accelerated with the application of genomic selection. By combining CH_4 phenotypes with genomic data, breeding values can be accurately estimated even for animals without direct CH_4 measurements based on their DNA information. In addition, combining CH_4 phenotypes with genomic information from shared international databases supports global efforts to mitigate emissions and improve productivity across diverse agricultural systems.

Overall, selective breeding can serve as a powerful strategy to lower the environmental footprint of dairy farming. The inclusion of CH_4 traits in balanced breeding indices allows selection that reduces emissions while balancing productivity, fertility, health, and other important traits. Nonetheless, selective breeding for CH_4 has limited practical application to date, and standardising CH_4 measurement methods and trait definitions globally will be key to accelerate genetic selection worldwide.

7 Where to look for further information

For learning more about the principles of quantitative genetics:

- Falconer, D. S. & Mackay, T. F. C. (1996). *'Introduction to quantitative genetics'*. Essex, England: Longman.

The following articles provide a good overview of topics discussed in this chapter:

- de Haas, Y., et al. (2017). 'Invited review: phenotypes to genetically reduce greenhouse gas emissions in dairying'. *J Dairy Sci*, 100(2): 855–870.
- Lassen, J. & Difford, G. F. (2020). 'Review: genetic and genomic selection as a methane mitigation strategy in dairy cattle'. *Animal*, 14(S3): s473-s483.
- ICAR (2020). 'Section 20 - Recording dairy cattle methane emission for genetic evaluation'. https://www.icar.org/index.php/icar-recording-guidelines/ (online).
- MPI (2020), 'Guidelines for estimating methane emissions from individual ruminants using: GreenFeed, sniffers, hand-held laser detector and portable accumulation chambers'. ISBN No: 978-1-99-004336-9 (online).
- Sorg (2021), 'Measuring Livestock CH4 Emissions with the Laser Methane Detector: A Review'. Methane, 1: 38–57.
- González-Recio, O., et al. (2023). 'Invited review: Novel methods and perspectives for modulating the rumen microbiome through selective breeding as a means to improve complex traits: Implications for methane emissions in cattle'. Livest Sci, 269.

Key research in this area can be found at the following organisations:

- ICAR (www.icar.org) and the ICAR Feed and Gas Working Group (https://www.icar.org/index.php/technical-bodies/working-groups/working-groupsfeed-and-gas/).

Conferences that commonly highlight key research in this area:

- EAAP
- WCGALP
- ICAR
- GGAA
- ADSA

8 References

AgResearch. (2023). New tech to boost emissions reduction efforts. Retrieved from https://www.agresearch.co.nz/news/new-tech-to-boost-emissions-reduction-efforts/ Accessed on: 2024-12-12.

Amer, P. R., et al. 2018. A methodology framework for weighting genetic traits that impact greenhouse gas emission intensities in selection indexes. *Animal, 12*(1), 5–11. https://doi.org/10.1017/S1751731117001549.

Arla. (2023). Arla's sustainability incentive model: how Arla farmers are rewarded for their sustainability activities. Retrieved from https://www.arla.com/sustainability/the-farms/arlas-sustainability-incentive-model-qa/ Accessed on: 2024-04-18.

Bell, M. J., et al. 2014. Effects of changing cow production and fitness traits on profit and greenhouse gas emissions of UK dairy systems. *J Agric Sci, 153*(1), 138–151. https://doi.org/10.1017/s0021859614000847.

Bell, M. J., et al. 2019. Modified approach to estimating daily methane emissions of dairy cows by measuring filtered eructations during milking. *J Sustain Org Agric Syst, 69*(1), 47–56. https://doi.org/10.3220/LBF15841062570004.

Berends, H., et al. 2014. Evaluation of the SF6 tracer technique for estimating methane emission rates with reference to dairy cows using a mechanistic model. *J Theor Biol, 353*, 1–8. https://doi.org/10.1016/j.jtbi.2014.02.040.

Berry, D. P., et al. 2014. International genetic evaluations for feed intake in dairy cattle through the collation of data from multiple sources. *J Dairy Sci, 97*(6), 3894–3905. https://doi.org/10.3168/jds.2013-7548.

Berry, D. P., et al. 2015. *Residual feed intake and breeding approaches for enteric methane mitigation*. Retrieved from https://www.cabidigitallibrary.org/doi/book/10.1079/9781780644325.0000.

Boichard, D. & Brochard, M. 2012. New phenotypes for new breeding goals in dairy cattle. *Animal, 6*(4), 544–550. https://doi.org/10.1017/S1751731112000018.

Bovenhuis, H., et al. 2018. Letter to the editor: a response to. *J Dairy Sci, 101*(11), 9621–9622. https://doi.org/10.3168/jds.2018-15606.

Brascamp, E. W. 1984. Selection indices with constraints. *Anim Breed Abstr, 52*, 645–654.

Breider, I. S. 2018. *The genetic and genomic background of methane emissions and feed efficiency in dairy cattle*. University of Nottingham, Scotland's Rural College, Retrieved from https://eprints.nottingham.ac.uk/55321/. Retrieved from https://eprints.nottingham.ac.uk/55321/.

Breider, I. S., et al. 2019. Short communication: Heritability of methane production and genetic correlations with milk yield and body weight in Holstein-Friesian dairy cows. *J Dairy Sci, 102*(8), 7277–7281. https://doi.org/10.3168/jds.2018-15909.

Chagunda, M. G. G., et al. 2009. On the use of a laser methane detector in dairy cows. *Comput Electron Agric, 68*(2), 157–160. https://doi.org/10.1016/j.compag.2009.05.008.

Coppa, M., et al. 2022. Methodological guidelines: cow milk mid-infrared spectra to predict reference enteric methane data collected by an automated head-chamber system. *J Dairy Sci, 105*(11), 9271–9285. https://doi.org/10.3168/jds.2022-21890.

de Haas, Y., et al. 2012. Improved accuracy of genomic prediction for dry matter intake of dairy cattle from combined European and Australian data sets. *J Dairy Sci, 95*(10), 6103–6112. https://doi.org/10.3168/jds.2011–5280.

de Haas, Y., et al. 2015. Genomic prediction of dry matter intake in dairy cattle from an international data set consisting of research herds in Europe, North America, and Australasia. *J Dairy Sci, 98*(9), 6522–6534. https://doi.org/10.3168/jds.2014–9257.

de Haas, Y., et al. 2017. Invited review: Phenotypes to genetically reduce greenhouse gas emissions in dairying. *J Dairy Sci, 100*(2), 855–870. https://doi.org/10.3168/jds.2016-11246.

de Haas, Y., et al. 2021. Selective breeding as a mitigation tool for methane emissions from dairy cattle. *Animal, 15*, 100294. https://doi.org/10.1016/j.animal.2021.100294.

Dehareng, F., et al. 2012. Potential use of milk mid-infrared spectra to predict individual methane emission of dairy cows. *Animal, 6*(10), 1694–1701. https://doi.org/10.1017/S1751731112000456.

Denninger, T. M., et al. 2019. Persistence of differences between dairy cows categorized as low or high methane emitters, as estimated from milk mid-infrared spectra and measured by GreenFeed. *J Dairy Sci, 102*(12), 11751–11765. https://doi.org/10.3168/jds.2019-16804.

Difford, G. F. 2018. *Genetic control of methane emission, feed efficiency and metagenomics in dairy cattle*. Wageningen University, Wageningen. https://doi.org/10.18174/458036.

Difford, G. F., et al. 2018. Host genetics and the rumen microbiome jointly associate with methane emissions in dairy cows. *PLoS Genet, 14*(10), e1007580. https://doi.org/10.1371/journal.pgen.1007580.

Difford, G. F., et al. 2019. Ranking cows' methane emissions under commercial conditions with sniffers versus respiration chambers. *Acta Agric Scand A Anim Sci, 68*(1), 25–32. https://doi.org/10.1080/09064702.2019.1572784.

Difford, G. F., et al. 2020. Can greenhouse gases in breath be used to genetically improve feed efficiency of dairy cows? *J Dairy Sci, 103*(3), 2442–2459. https://doi.org/10.3168/jds.2019-16966.

Dijkstra, J., et al. 2011. Relationships between methane production and milk fatty acid profiles in dairy cattle. *Animal Feed Science and Technology, 166–167*, 590–595. https://doi.org/10.1016/j.anifeedsci.2011.04.042.

Falconer, D. S. & Mackay, T. F. C. 1996. *Introduction to quantitative genetics*. Essex, England: Longman.

FrieslandCampina. (2018). Eerste PlanetProof-melk nu verkrijgbaar. Retrieved from https://www.frieslandcampina.com/nl/nieuws/eerste-planetproof-melk-nu-verkrijgbaar/ Accessed on: 11-11-2024.

Garnsworthy, P. C., et al. 2012. On-farm methane measurements during milking correlate with total methane production by individual dairy cows. *J Dairy Sci, 95*(6), 3166–3180. https://doi.org/10.3168/jds.2011–4605.

Garnsworthy, P. C., et al. 2019. Comparison of methods to measure methane for use in genetic evaluation of dairy cattle. *Animals, 9*(10). https://doi.org/10.3390/ani9100837.

Gonzalez-Recio, O., et al. 2018. Short communication: Signs of host genetic regulation in the microbiome composition in 2 dairy breeds: Holstein and Brown Swiss. *J Dairy Sci, 101*(3), 2285–2292. https://doi.org/10.3168/jds.2017-13179.

Gonzalez-Recio, O., et al. 2020. Mitigation of greenhouse gases in dairy cattle via genetic selection: 2. Incorporating methane emissions into the breeding goal. *J Dairy Sci, 103*(8), 7210–7221. https://doi.org/10.3168/jds.2019-17598.

González-Recio, O., et al. 2023. Invited review: novel methods and perspectives for modulating the rumen microbiome through selective breeding as a means to improve complex traits: Implications for methane emissions in cattle. *Livest Sci, 269*. https://doi.org/10.1016/j.livsci.2023.105171.

Hammond, K. J., et al. 2015. Methane emissions from cattle: estimates from short-term measurements using a GreenFeed system compared with measurements obtained using respiration chambers or sulphur hexafluoride tracer. *Anim Feed Sci Technol, 203*, 41–52. https://doi.org/10.1016/j.anifeedsci.2015.02.008.

Hammond, K. J., et al. 2016. Review of current in vivo measurement techniques for quantifying enteric methane emission from ruminants. *Anim Feed Sci Technol, 219*, 13–30. https://doi.org/10.1016/j.anifeedsci.2016.05.018.

Hazel, L. N. 1943. The genetic basis for constructing selection indexes. *Genetics, 28*(6), 476–490.

Henderson, G., et al. 2015. Rumen microbial community composition varies with diet and host, but a core microbiome is found across a wide geographical range. *Sci Rep, 5*, 14567. https://doi.org/10.1038/srep14567.

Herliatika, A., et al. 2024. Meta-analysis of the relationship between dietary starch intake and enteric methane emissions in cattle from in vivo experiments. *J Adv Vet Anim Res, 11*(1), 212–230. https://doi.org/10.5455/javar.2024.k767.

Hossein-Zadeh, N. G. 2022. Estimates of the genetic contribution to methane emission in dairy cows: a meta-analysis. *Sci Rep, 12*(1), 12352. https://doi.org/10.1038/s41598-022-16778-z.

Hristov, A. N. & Melgar, A. 2020. Short communication: relationship of dry matter intake with enteric methane emission measured with the GreenFeed system in dairy cows receiving a diet without or with 3-nitrooxypropanol. *Animal, 14*(S3), s484–s490. https://doi.org/10.1017/S1751731120001731.

Hristov, A. N., et al. 2018. Symposium review: uncertainties in enteric methane inventories, measurement techniques, and prediction models. *J Dairy Sci, 101*(7), 6655–6674. https://doi.org/10.3168/jds.2017-13536.

Huhtanen, P. & Hristov, A. N. 2018. Letter to the editor: challenging one sensor method for screening dairy cows for reduced methane emissions. *J Dairy Sci, 101*(11), 9619–9620. https://doi.org/10.3168/jds.2018-14704.

Huhtanen, P., et al. 2015. Comparison of methods to determine methane emissions from dairy cows in farm conditions. *J Dairy Sci, 98*(5), 3394–3409. https://doi.org/10.3168/jds.2014-9118.

Huhtanen, P., et al. 2020. Short communication: variation in feed efficiency hampers use of carbon dioxide as a tracer gas in measuring methane emissions in on-farm conditions. *J Dairy Sci, 103*(10), 9090–9095. https://doi.org/10.3168/jds.2020-18559.

ICAR. 2020. Section 20: recording dairy cattle methane emission for genetic evaluation. Retrieved from https://www.icar.org/index.php/icar-recording-guidelines/ Accessed on: 2024-12-05.

ICBF, I. C. B. F. 2018. EBI - Breeding a sustainable future. *From news website:*. Retrieved from https://www.icbf.com/wp/?p=11176 Accessed on: 2024-04-16.

Interbull Centre. 2022. National genomic evaluation forms. Retrieved from https://wiki.interbull.org/public/cou_geno_forms Accessed on: 2024-04-24.

IPCC. 2021. Climate Change 2021: The Physical Science Basis. Contribution of Working Group I to the Sixth Assessment Report of the Intergovernmental Panel on Climate Change. Cambridge University Press: Cambridge, United Kingdom and New York, NY, USA. Accessed on: 03-09-2023 at https://www.ipcc.ch/report/ar6/wg1/.

Johnson, K., et al. 1994. Measurement of methane emissions from ruminant livestock using a sulfur hexafluoride tracer technique. *Environmental Science & Technology, 28*(2), 359–362. https://doi.org/10.1021/es00051a025.

Johnson, K. A. & Johnson, D. E. 1995. Methane emissions from cattle. *J Anim Sci, 73*(8), 2483–2492.

Jonker, A., et al. 2018. Genetic parameters of methane emissions determined using portable accumulation chambers in lambs and ewes grazing pasture and genetic correlations with emissions determined in respiration chambers. *J Anim Sci, 96*(8), 3031–3042. https://doi.org/10.1093/jas/sky187.

Jonker, A., et al. 2020. In: Guidelines for estimating methane emissions from individual ruminants using: GreenFeed, 'sniffers', hand-held laser detector and portable accumulation chambers. Retrieved from https://www.mpi.govt.nz/dmsdocument/43510/direct Accessed on: 2024-02-02.

Kamalanathan, S., et al. 2023. Genetic analysis of methane emission traits in holstein dairy cattle. *Animals, 13*(8). https://doi.org/10.3390/ani13081308.

Kennedy, B. W., et al. 1993. Genetic and statistical properties of residual feed intake. *J Anim Sci, 71*(12), 3239–3250. https://doi.org/10.2527/1993.71123239x.

King, E. E., et al. 2011. Differences in the rumen methanogen populations of lactating Jersey and Holstein dairy cows under the same diet regimen. *Appl Environ Microbiol, 77*(16), 5682–5687. https://doi.org/10.1128/AEM.05130-11.

Koch, R. M., et al. 1963. Efficiency of feed use in beef cattle. *J Anim Sci, 22*(2), 486–494. https://doi.org/10.2527/jas1963.222486x.

Lassen, J. & Difford, G. F. 2020. Review: genetic and genomic selection as a methane mitigation strategy in dairy cattle. *Animal, 14*(S3), s473–s483. https://doi.org/10.1017/S1751731120001561.

Lassen, J. & Lovendahl, P. 2013. Heritability for enteric methane emission from Danish Holstein cows using a non-invasive FTIR method. *Advances in the Biosciences, 4*(2), 280. https://doi.org/10.1017/S2040470013000083.

Lassen, J. & Lovendahl, P. 2016. Heritability estimates for enteric methane emissions from Holstein cattle measured using noninvasive methods. *J Dairy Sci, 99*(3), 1959–1967. https://doi.org/10.3168/jds.2015-10012.

Lassen, J., et al. 2016. Genetic and genomic relationship between methane production measured in breath and fatty acid content in milk samples from Danish Holsteins. *Anim. Prod. Sci., 56*(3). https://doi.org/10.1071/an15489.

Lassey, K. R., et al. 2011. Assessing the SF6 tracer technique as an estimator of methane emissions from ruminants New Zealand. Retrieved from https://www.mpi.govt.nz/dmsdocument/2937/direct Accessed on: 2024-11-13.

Lopes, L. S. F., et al. 2022. *Genetic parameters for rumination time and traits related to sustainable dairy production*. Presented in Proc. 12th World Congress on Genetics Applied to Livestock Production, Rotterdam, the Netherlands.

Lopes, L. S. F., et al. 2024. Estimates of genetic parameters for rumination time, feed efficiency, and methane production traits in first lactation Holstein cows. *J Dairy Sci*, ssue 7108, 7297–7309. https://doi.org/10.3168/jds.2023-23751.

Lopez-Paredes, J., et al. 2020. Mitigation of greenhouse gases in dairy cattle via genetic selection: 1. Genetic parameters of direct methane using noninvasive methods and proxies of methane. *J Dairy Sci, 103*(8), 7199–7209. https://doi.org/10.3168/jds.2019-17597.

Lund, M. S., et al. 2011. A common reference population from four European Holstein populations increases reliability of genomic predictions. *Gen Sel Evol, 43*(43). https://doi.org/10.1186/1297-9686-43-43.

Madsen, J., et al. 2010. Methane and carbon dioxide ratio in excreted air for quantification of the methane production from ruminants. *Livest Sci, 129*(1–3), 223–227. https://doi.org/10.1016/j.livsci.2010.01.001.

Manzanilla-Pech, C. I. V., et al. 2016. Genomewide association study of methane emissions in Angus beef cattle with validation in dairy cattle. *J. Anim. Sci., 94*, 4151–4166. https://doi.org/10.2527/jas2016-0431.

Manzanilla-Pech, C. I. V., et al. 2021. Breeding for reduced methane emission and feed-efficient Holstein cows: An international response. *J Dairy Sci, 104*(8), 8983–9001. https://doi.org/10.3168/jds.2020-19889.

Manzanilla-Pech, C. I. V., et al. 2022a. Genetic (co-)variation of methane emissions, efficiency, and production traits in Danish Holstein cattle along and across lactations. *J Dairy Sci, 105*(12), 9799–9809. https://doi.org/10.3168/jds.2022-22121.

Manzanilla-Pech, C. I. V., et al. 2022b. Genome-wide association study for methane emission traits in Danish Holstein cattle. *J Dairy Sci, 105*(2), 1357–1368. https://doi.org/10.3168/jds.2021-20410.

Manzanilla-Pech, C. I. V., et al. 2022c. Selecting for feed efficient cows will help to reduce methane gas emissions. *Front Genet, 13*, 885932. https://doi.org/10.3389/fgene.2022.885932.

Martinez-Alvaro, M., et al. 2022. Bovine host genome acts on rumen microbiome function linked to methane emissions. *Commun Biol, 5*(1), 350. https://doi.org/10.1038/s42003-022-03293-0.

McParland, S., et al. 2024. Predicting methane emissions of individual grazing dairy cows from spectral analyses of their milk samples. *J Dairy Sci, 107*(2), 978–991. https://doi.org/10.3168/jds.2023-23577.

Mollenhorst, H. & De Haas, Y. 2019. The contribution of breeding to reducing environmental impact of animal production Wageningen, the Netherlands. Accessed on: 2024-02-01 at https://edepot.wur.nl/472395.

Munoz, C., et al. 2012. Comparison of the sulfur hexafluoride tracer and respiration chamber techniques for estimating methane emissions and correction for rectum methane output from dairy cows. *J Dairy Sci, 95*(6), 3139–3148. https://doi.org/10.3168/jds.2011-4298.

Negussie, E., et al. 2017. Invited review: Large-scale indirect measurements for enteric methane emissions in dairy cattle: A review of proxies and their potential for use in management and breeding decisions. *J Dairy Sci, 100*(4), 2433–2453. https://doi.org/10.3168/jds.2016-12030.

Oliveira, H. R., et al. 2024. Symposium review: development of genomic evaluation for methane efficiency in Canadian Holsteins. *JDS Communications*. https://doi.org/10.3168/jdsc.2023-0431.

Olsen, J. M. 2024. Gassy cows and pigs will face a carbon tax in Denmark, a world first. *AP News*. Retrieved from https://apnews.com/article/denmark-cow-tax-greenhouse-gases-9a570518639e0a1990806fd1a05ac11a Accessed on: 2024-07-01.

Palucci, V., et al. 2023. Interbull centre activity report 2023. Retrieved from https://interbull.org/ Accessed on: 2024-04-24.

Pickering, N. K., et al. 2015a. Genetic parameters for predicted methane production and laser methane detector measurements. *J. Anim. Sci., 92*, 11–20. https://doi.org/10.2527/jas2014-8302.

Pickering, N. K., et al. 2015b. Animal board invited review: genetic possibilities to reduce enteric methane emissions from ruminants. *Animal, 9*(9), 1431–1440. https://doi.org/10.1017/S1751731115000968.

Pryce, J. E., et al. 2014. Genomic selection for feed efficiency in dairy cattle. *Animal, 8*(1), 1–10. https://doi.org/10.1017/S1751731113001687.

Pryce, J. E. & Haile-Mariam, M. 2020. Symposium review: genomic selection for reducing environmental impact and adapting to climate change. *J Dairy Sci, 103*(6), 5366–5375. https://doi.org/10.3168/jds.2019-17732.

Pszczola, M., et al. 2017. Heritability of methane emissions from dairy cows over a lactation measured on commercial farms. *J Anim Sci, 95*(11), 4813–4819. https://doi.org/10.2527/jas2017.1842.

Rey, J., et al. 2019. Comparison between non-invasive methane measurement techniques in Cattle. *Animals, 9*(8), 563. https://doi.org/10.3390/ani9080563.

Richardson, C. M., et al. 2021a. Estimating the environmental impact of dairy cattle breeding programs through emission intensity. *Animal, 15*(1), 100005. https://doi.org/10.1016/j.animal.2020.100005.

Richardson, C. M., et al. 2021b. Genetic parameters for methane emission traits in Australian dairy cows. *J Dairy Sci, 104*(1), 539–549. https://doi.org/10.3168/jds.2020-18565.

Richardson, C. M., et al. 2022. Reducing greenhouse gas emissions through genetic selection in the Australian dairy industry. *J Dairy Sci, 105*(5), 4272–4288. https://doi.org/10.3168/jds.2021-21277.

Roehe, R., et al. 2016. Bovine host genetic variation influences rumen microbial methane production with best selection criterion for low methane emitting and efficiently feed converting hosts based on metagenomic gene abundance. *PLoS Genet, 12*(2), e1005846. https://doi.org/10.1371/journal.pgen.1005846.

Saborío-Montero, A., et al. 2019. Structural equation models to disentangle the biological relationship between microbiota and complex traits: Methane production in dairy cattle as a case of study. *J Anim Breed Genet, 137*(1), 36–48. https://doi.org/10.1111/jbg.12444.

Shadpour, S., et al. 2022. Predicting methane emission in Canadian Holstein dairy cattle using milk mid-infrared reflectance spectroscopy and other commonly available predictors via artificial neural networks. *J Dairy Sci, 105*(10), 8272–8285. https://doi.org/10.3168/jds.2021-21176.

Shi, R., et al. 2024. Balancing farm profit and greenhouse gas emissions along the dairy production chain through breeding indices. *Journal of Cleaner Production, 451*. https://doi.org/10.1016/j.jclepro.2024.142099.

Sorg, D., et al. 2017, 28 August- 1 September. *Protocol for measuring CH4 concentrations with the laser methane detector in the breath of cows.* Proceedings of the 68th Annual Meeting of the European Federation of Animal Science, Tallinn, Estonia.

Sorg, D., et al. 2018. Comparison of a laser methane detector with the GreenFeed and two breath analysers for on-farm measurements of methane emissions from dairy cows. *Comput Electron Agric, 153*, 285–294. https://doi.org/10.1016/j.compag.2018.08.024.

Sorg, D. 2022. Measuring livestock CH4 emissions with the laser methane detector: a review. *Methane, 1*(1), 38–57. https://doi.org/10.3390/methane1010004.

Sypniewski, M., et al. 2021. Genetic variability of methane production and concentration measured in the breath of polish holstein-friesian Cattle. *Animals, 11*(11), 3175. https://doi.org/10.3390/ani11113175.

Tempelman, R. J., et al. 2015. Heterogeneity in genetic and nongenetic variation and energy sink relationships for residual feed intake across research stations and countries. *J Dairy Sci, 98*(3), 2013–2026. https://doi.org/10.3168/jds.2014.8510.

Teye, F. K., et al. 2009. Instrumentation, measurement and performance of three air quality measurement systems for dairy buildings. *Appl Eng Agric, 25*(2), 247–256. https://doi.org/10.13031/2013.26328.

Troy, S. M., et al. 2016. A comparison of methane emissions from beef cattle measured using methane hoods with those measured using respiration chambers. *Anim Feed Sci Technol, 211*, 227–240. https://doi.org/10.1016/j.anifeedsci.2015.12.005.

van Breukelen, A. E. 2024. *Breeding climate smart dairy cattle: from phenotyping to genetic selection for low methane emitting cattle*. Wageningen University & Research, Wageningen. https://doi.org/10.18174/671955.

van Breukelen, A. E., et al. 2022. Genetic parameters for repeatedly recorded enteric methane concentrations of dairy cows. *J Dairy Sci, 105*(5), 4256–4271. https://doi.org/10.3168/jds.2021-21420.

van Breukelen, A. E., et al. 2023. Heritability and genetic correlations between enteric methane production and concentration recorded by GreenFeed and sniffers on dairy cows. *J Dairy Sci, 106*(6), 4121–4132. https://doi.org/10.3168/jds.2022-22735.

van Breukelen, A. E., et al. 2024. Genetic parameter estimates for methane emission during lactation from breath and potential inaccuracies in reliabilities assuming a repeatability versus random regression model. *J Dairy Sci*. https://doi.org/10.3168/jds.2024-24285.

van der Linde, C., et al. 2007. *From functional to productive longevity in the Netherlands.* InterBull, Dublin.

van Engelen, S., et al. 2018. Genetic background of methane emission by Dutch Holstein Friesian cows measured with infrared sensors in automatic milking systems. *J Dairy Sci, 101*(3), 2226–2234. https://doi.org/10.3168/jds.2017-13441.

van Middelaar, C. E., et al. 2014. Methods to determine the relative value of genetic traits in dairy cows to reduce greenhouse gas emissions along the chain. *J Dairy Sci, 97*(8), 5191–5205. https://doi.org/10.3168/jds.2013-7413.

Vanlierde, A., et al. 2015. Hot topic: Innovative lactation-stage-dependent prediction of methane emissions from milk mid-infrared spectra. *J Dairy Sci, 98*(8), 5740–5747. https://doi.org/10.3168/jds.2014-8436.

Vanlierde, A., et al. 2018. Short communication: development of an equation for estimating methane emissions of dairy cows from milk Fourier transform mid-infrared spectra by using reference data obtained exclusively from respiration chambers. *J Dairy Sci, 101*(8), 7618–7624. https://doi.org/10.3168/jds.2018-14472.

Veerkamp, R. F. & Emmans, G. C. 1995. Sources of genetic variation in energetic efficiency of dairy cows. *Livest Prod Sci, 44*(2), 87–97. https://doi.org/10.1016/0301-6226(95)00065-0.

Veerkamp, R. F., et al. 1995. Variance components for residual feed intake in dairy cows *Livest Prod Sci, 41*(2), 111–120. https://doi.org/10.1016/0301-6226(94)00056-D.

Veerkamp, R. F. & Brotherstone, S. 1997. Genetic correlations between linear type traits, food intake, live weight and condition score in Holstein Friesian dairy cattle. *Anim Sci, 64*(3), 385–392. https://doi.org/10.1017/S1357729800015976.

Veerkamp, R. F. 2002. *Feed intake and energy balance in lactating animals.* Proceedings of the 7th World Congr Gene Appl Livest Prod, Montpellier, France.

Veerkamp, R. F., et al. 2013. *Selection on feed intake or feed efficiency: a position paper from gDMI Breeding Goal Discussions.* Proceedings of Interbull Meeting No. 47, Nantes, France.

Velazco, J. I., et al. 2016. Use of short-term breath measures to estimate daily methane production by cattle. *Animal, 10*(1), 25–33. https://doi.org/10.1017/S1751731115001603.

Wallace, R. J., et al. 2015. The rumen microbial metagenome associated with high methane production in cattle. *BMC Genomics, 16*, 839. https://doi.org/10.1186/s12864-015-2032-0.

Wu, L. 2016. *Measurement methods to assess methane production of individual dairy cows in a Barn.* Wageningen University and Research, Wageningen. Retrieved from https://edepot.wur.nl/370231.

Wu, L., et al. 2018. Uncertainty assessment of the breath methane concentration method to determine methane production of dairy cows. *J Dairy Sci, 101*(2), 1554–1564. https://doi.org/10.3168/jds.2017-12710.

WUR. 2024. Update on the global livestock genetics and genomics programme selecting for reduced methane emissions in ruminants. Retrieved from https://www.wur.nl/en/show/update-on-the-global-livestock-genetics-and-genomics-programme-selecting-for-reduced-methane-emissions-in-ruminants.htm Accessed on: 2024-05-29.

Zetouni, L., et al. 2017. Direct multitrait selection realizes the highest genetic response for ratio traits. *J Anim Sci, 95*(5). https://doi.org/10.2527/jas2016.1324.

Zetouni, L., et al. 2018. Genetic correlations between methane production and fertility, health, and body type traits in Danish Holstein cows. *J Dairy Sci, 101*(3), 2273–2280. https://doi.org/10.3168/jds.2017-13402.

Zhang, Q., et al. 2020. Bayesian modeling reveals host genetics associated with rumen microbiota jointly influence methane emission in dairy cows. *ISME J, 14*(8), 2019–2033. https://doi.org/10.1038/s41396-020-0663-x.

Zhang, X., et al. 2019. Prediction of effects of dairy selection indexes on methane emissions. *J Dairy Sci, 102*(12), 11153–11168. https://doi.org/10.3168/jds.2019-16943.

Zhao, Y., et al. 2020. A review of enteric methane emission measurement techniques in ruminants. *Animals, 10*(6). https://doi.org/10.3390/ani10061004.

Zimmerman, P. 2011. Method and system for monitoring and reducing ruminant methane production. USA Patent No. 7966971B2.

Zimmerman, P. R. 1993. System for measuring metabolic gas emissions from animals. U. S. P. a. T. Office.

Chapter 7

Improving feed conversion efficiency in dairy cattle

Robin R. White, Virginia Tech, USA

1 Introduction

Sustainable food systems seek to contribute to food security and nutrition for all in such a way that the economic, social, cultural, and environmental bases to generate food security and nutrition for future generations are safeguarded (1). Underpinning sustainable food systems are sustainable agricultural enterprises. These enterprises take a holistic approach to food production that aims to meet current societal needs with the future in mind, integrating environmental stewardship, economic viability, and social responsibility into farming practices. Key principles of sustainable agriculture include optimizing resource use, minimizing environmental impact, promoting biodiversity, and enhancing resilience to climate change (2).

Net-zero dairy farming takes sustainable agriculture a step further by targeting a balance between the greenhouse gas emissions produced and those removed from the atmosphere, effectively resulting in a net-zero warming footprint (3). This ambitious goal encompasses not only reducing emissions from dairy operations but also implementing strategies to sequester or offset carbon, such as reforestation, soil carbon enhancement, and renewable energy adoption. Despite the promise held by these emerging technologies, achieving net-zero status in dairy farming will likely require a multifaceted approach that addresses most aspects of farm management, including animal productivity and feed conversion efficiency.

http://dx.doi.org/10.19103/AS.2025.0157.15

Productivity in dairy farming has long been recognized as essential to sustainability (4, 5). Higher productivity means producing more milk or dairy products per unit of input (i.e. feed, water, land, etc.). By improving productivity, dairy farmers can achieve greater economic efficiency, reduce resource use per unit of output, and lower the environmental footprint of each liter of milk produced, largely through diluting maintenance costs of animal production (6). Additionally, increased productivity can enhance farm resilience to external challenges, such as climate variability and market fluctuations, contributing to the long-term sustainability of dairy operations (7). Thus, the pursuit of improved individual animal productivity and feed conversion efficiency is tightly linked to the overarching goals of sustainable agriculture and net-zero dairy farming.

2 Factors driving productivity and feed conversion efficiency

2.1 Feed conversion efficiency

Feed conversion efficiency (FCE) in dairy systems refers to the ability of dairy cows to convert feed inputs into milk output efficiently. It is a critical metric that reflects the economic and environmental performance of dairy operations. Feed conversion efficiency is typically measured as the amount of milk produced per unit of feed consumed (i.e. kilograms of milk per kilogram of dry matter intake); however, other metrics such as residual feed intake (8) are increasingly more commonly leveraged. Higher FCE values indicate greater efficiency in converting feed into product, resulting in lower feed costs per unit of product produced and reduced environmental impacts associated with feed production and digestion (9).

In dairy farming, FCE is influenced by various factors, including genetics, nutrition, management practices, and health status (10). Genetics plays a significant role in determining the inherent efficiency of dairy cows in converting feed into milk (11). Selective breeding programs can target traits related to FCE, such as improved digestibility, metabolic efficiency, and milk production potential. Nutrition also plays a crucial role, with balanced diets and optimal nutrient utilization contributing to higher FCE values. Additionally, effective management practices, such as proper feeding strategies, housing conditions, and herd health protocols, can further enhance FCE and overall farm productivity.

The concept of feed conversion efficiency is closely intertwined with the goals of net-zero dairy farming and sustainability. Efficient FCE directly contributes to net-zero dairy farming; improving feed conversion efficiency is a key strategy for achieving warming neutrality. By reducing the amount of feed required per unit of milk output, dairy farms can lower their overall carbon emissions and contribute to mitigating climate change. Sustainable intensification practices, such as precision nutrition, advanced feeding technologies, and genetic improvement, are instrumental in maximizing FCE while minimizing environmental impacts, thereby advancing the sustainability and net-zero objectives of dairy farming.

2.2 Selecting for more efficient animals and production systems

Genetics and breeding programs play a pivotal role in enhancing animal productivity, health, reproductive efficiency, and feed conversion in dairy farming. Selective breeding strategies aim to improve desirable traits in dairy cattle, leading to more efficient and sustainable production systems. Although the dairy industry focused on single-trait selection for milk production for several decades, the more recent diversification of breeding goals has contributed to maintaining year-over-year improvements in milk production while concurrently enhancing other production goals (12). Current breeding systems focus very effectively on enhancing the genetic potential of dairy cows for milk production while simultaneously addressing and maintaining the capacity to address health, fertility, and environmental considerations (13).

Reproductive efficiency is another critical target for farm-level productivity (14). Selecting for improved fertility traits, such as reduced calving difficulty, higher conception rates, and increased longevity, can significantly impact the reproductive performance of dairy herds (15). Efficient breeding programs target genetic markers associated with reproductive success, allowing farmers to breed cows that calve easily, have fewer reproductive disorders, and maintain optimal reproductive cycles throughout their productive lifespan. Each of these outcomes helps ensure cows are productive throughout their lifespans, supporting the farm-level dilution of the maintenance costs associated with raising each cow to maturity.

Feed conversion efficiency is also a key focus of genetic selection in dairy breeding programs (11). By identifying and breeding animals with superior feed efficiency traits, such as efficient nutrient utilization, lower maintenance requirements, and higher milk yield per unit of feed consumed, farmers can improve overall FCE at the herd level. Genetic selection for FCE not only reduces feed costs and environmental impacts but also contributes to achieving net-zero goals by optimizing resource utilization and minimizing waste in feed conversion processes.

2.3 Nutritional management

Although selection of more efficient animals is critical for the long-term advancement of productivity of dairy systems, the phenotypes expressed by cows are the product of genetics, management, environmental factors, technologies, and other inputs. Both because of its contribution to the expenses of dairy farming and to supporting animals in meeting their genetic potential, nutritional management is foundational to profitable and sustainable dairy farming. Meeting animals' energy and nutrient requirements is pivotal for optimizing milk production, reproductive performance, and overall health. A meticulously balanced diet that supplies adequate levels of energy, amino acids, vitamins, and minerals is essential for maximizing feed conversion efficiency and achieving optimal milk yield without compromising animal welfare.

Enhancing the feed conversion ratio of dairy cattle throughout nutritional management relies primarily on the use of traditional feed formulation approaches which integrate estimates of energy and protein or amino acid supplies with requirements. Ration formulation is typically completed by a nutritionist and rations are designed for groups of animals, rather than for individuals. Ration formulation systems are, in turn, derived from data obtained from the average of animal group responses. This strategy has largely evolved to reflect the housing systems used in dairy production, where animals have similar access to feeds within group-housing settings.

Enhancing feed conversion ratio in ration formulation exercises can be accomplished through improved digestibility of feedstuffs, or through formulating to higher concentrations of energy, amino acids, vitamins, and minerals. Despite the latter being an option, it is crucial to recognize that optimal production should not be equated solely with maximal production. While high milk yield is desirable, an excessive focus on maximizing production without considering overall herd health and resource efficiency can lead to detrimental outcomes. Striking the right balance between maximizing milk output and promoting efficient nutrient utilization is key to sustainable productivity and feed conversion in dairy farming.

Effective nutritional management requires precise understanding of both animal requirements and feed composition. Among numerous factors that can unexpectedly alter animal requirements, animal health is a critical focus of productive dairy systems. The immune system is energetically expensive and many dietary nutrients have direct impacts on immune function through nutraceutical action (16). Due to the tight linkages among food and health, maintaining optimal health in dairy cattle is paramount for supporting productivity and maximizing feed conversion efficiency (17). Disease prevention strategies, particularly those aimed at minimizing the likelihood of metabolic diseases like ketosis, acidosis, and fatty liver syndrome, play a crucial role in ensuring the well-being and performance of dairy cows (18). These metabolic disorders can significantly impact milk production, reproductive success, and overall farm profitability by disrupting normal metabolic processes and nutrient utilization.

3 The movement toward individual animal management

Transitioning from managing group feed conversion efficiency to understanding and managing individual animal feed conversion efficiency marks a shift toward a more personalized approach to dairy farming. While group-level metrics provide valuable insights into overall herd performance, focusing on individual dairy cows' feed conversion efficiency is essential for optimizing productivity and ensuring productive well-being.

Managing individual animal feed conversion efficiency involves tailoring nutrition, health, and management practices to meet the specific needs of each cow within the herd. This personalized approach recognizes that not all cows have the same nutritional requirements, health status, or metabolic efficiency. By understanding and addressing individual variations, dairy farmers can optimize feed utilization, reduce resource waste, and enhance overall farm profitability.

Key strategies for managing individual animal feed conversion efficiency and promoting productive well-being include:

- Implementing precision feeding programs that tailor diets to individual nutritional needs based on factors such as lactation stage, body condition, and milk yield;
- Monitoring and managing individual cow health through regular veterinary checks, health screenings, and proactive disease prevention measures;
- Providing comfortable and stress-free housing environments that support cow comfort, reduce heat stress, and promote natural behaviors; and
- Emphasizing cow welfare practices, such as proper handling, access to clean water and feed, and minimizing environmental stressors.

By adopting a personalized approach to managing individual animal feed conversion efficiency and prioritizing productive well-being, dairy farmers can optimize resource utilization, improve feed conversion efficiency, and sustainably enhance overall farm performance.

3.1 Opportunities in individual animal feeding

Transitioning from group animal feeding to individual animal feeding presents significant opportunities for improving feed efficiency and optimizing resource utilization in dairy farming. By tailoring diets and feeding strategies to meet the specific nutrient and energy needs of each cow, individual animal feeding systems offer several advantages that contribute to enhanced feed conversion efficiency and overall productivity (19).

Individual animal feeding allows for precise control over nutrient intake, ensuring that each cow receives the appropriate balance of proteins, carbohydrates, fats, vitamins, and minerals to support optimal health and productivity. In a gross example, cows in early lactation require higher energy and protein levels to support milk production, while dry cows have different nutrient requirements to support pregnancy and prepare for the next lactation. In a more specific application, cows with higher genetic merit may require different nutritional support than animals that have less capacity to produce milk. By customizing diets based on factors such as productivity, lactation stage, body condition score, and milk yield, farmers can maximize nutrient utilization and minimize waste, leading to improved feed conversion efficiency and reduced feed costs per unit of milk produced (20). Beyond formulating more specifically among individual animals, it is also feasible to focus on formulating rations more specifically through time. This can support adaptation across the production calendar, but also to shifts in weather or barn environments (21).

In traditional feeding systems, supplementation has been heavily leveraged as an approach to optimizing productivity, and it is likely that the importance of supplements will be extended in the transition to individual animal feeding. Supplements can be tailored to meet individual nutritional deficiencies, support metabolic health, enhance immune function, or alleviate stress, contributing to improved overall well-being and productivity. For example, cows experiencing suboptimal rumen function may benefit from targeted supplementation with buffers or probiotics to optimize digestion and nutrient absorption. Similarly, cows facing heat stress or environmental challenges may receive supplements containing electrolytes or antioxidants to support hydration and mitigate the effects of stress. Through this more personalized approach to supplemental nutrition support, farmers can mitigate the risk of metabolic disorders, reduce veterinary costs, and optimize milk production efficiency.

In conclusion, transitioning from group feeding to individual animal feeding has the potential to offer significant feed efficiency gains and productivity benefits in dairy farming. By meeting the nutrient and energy needs of individual animals, allowing for individual control or modulation of feed intake, and implementing targeted supplementation strategies, farmers can optimize resource utilization, enhance feed conversion efficiency, and promote sustainable and profitable dairy operations.

4 Technologies for individual animal monitoring and management

The above precision livestock feeding principles are predicated on the precise and accurate identification of individual animal needs, stress levels, feed intake, or other related factors, as well as the capacity to deliver individual feed amounts or mixes to cattle. Traditional animal feeding approaches have leveraged data representing groups of animals, and may not be appropriately suited to making decisions in support of individualized nutrition. As such, the emerging precision technologies available in the dairy industry likely will be essential in advancing precision feeding approaches.

4.1 Monitoring technologies for precision feeding

Individualized management approaches recognize that cows have unique behavioral, physiological, and environmental needs that can significantly impact their well-being, health, and productivity. Technologies supporting these outcomes include environmental condition monitoring systems, animal behavior monitoring systems, in-line milk sensor systems, and physiological sensing systems, among others not reviewed here.

Wearable and implantable technologies such as rumination sensors and activity monitors will likely be important in supporting individual animal nutrition (22, 23). These devices continuously track cow behavior and physiological parameters, providing real-time data on rumination patterns, activity levels, and rest periods. By analyzing these metrics, early signs of health issues such as rumen disorders or heat stress can be identified, allowing for timely interventions such as dietary adjustments or environmental modifications to mitigate risks and support cow well-being. Identifying how to effectively use these kinds of individual animal sensors in driving feeding decisions will be critical in advancing data-driven individual feeding strategies.

In-line milk sensors represent another innovative technology that contributes to precision management in dairy farming (24). These sensors measure milk composition, volume, and quality parameters directly during milking, providing real-time data on individual cow milk production, fat and protein content, and somatic cell counts. Milk sensor data can support precision feeding through identifying high-performing cows, adjusting feeding strategies to optimize nutrient intake, and detecting early signs of health issues such as mastitis or metabolic disorders. This data-driven approach enables optimization of milk yield, improved milk quality, and minimized waste, leading to enhanced feed conversion efficiency and sustainable production practices.

Automated weight and body condition scoring monitoring systems, such as body condition scoring cameras and automatic scale systems, offer additional opportunities for precision management and individualized care (25). Body condition scoring cameras capture images of cows to assess body condition scores, allowing for the adjustment of feeding recommendations in response to body condition changes over time. These adjustments are critical both for minimizing over-conditioning and under-conditioning. In addition, automated scales allow for precise monitoring of animal weight shifts on a daily basis, enabling assessment of long-term trends and monitoring the consistency of those trends. Feed adjustment in response to these body weight changes is also supportive of minimizing over- and under-conditioning due to diet.

4.2 Management technologies for precision feeding

In addition to requiring individual information necessary to understanding what kinds of nutritional intervention animals may require, precision feeding requires a means of delivering feed to individual animals. These individual animal feed delivery systems, such as automated feeders or milkers, have different capacities, and therefore, we must consider these limitations when designing precision feeding approaches.

Automatic milkers are sophisticated systems that automate the milking process and gather valuable data on individual cow milk production, milking frequency, milk quality parameters, and udder health indicators. By integrating milkers with data management software, farmers can track milk yield trends, detect deviations from normal patterns, and identify high-performing cows or those requiring additional attention. This data-driven approach allows for individualized milking protocols, such as adjusting milking frequency or optimizing milking settings based on each cow's unique needs, promoting udder health, milk quality, and overall productivity. Typically, automatic milkers provide a bait feed to animals while milking. As such, they provide an opportunity not only for data collection to support feed management, but also a means of providing individual supplements to animals.

Similarly, automatic feeders are advanced technologies that dispense feed rations based on individual cow requirements, feeding schedules, and nutritional targets. These feeders can be coupled with data from individual cow monitoring systems, such as rumination sensors or body condition scoring cameras, to customize feed formulations and portion sizes. Depending on the feeder system, automatic feeders can provide anywhere from a few pounds of feed per day to the entirety of the cow's ration. As such, automatic feeders have more flexibility in terms of how they can be incorporated into precision feeding systems. However, a major limitation of these systems is that they require substantial infrastructure changes, as cows must have access to a feeder to be able to consume feed daily.

The integration of automatic milkers and feeders with precision monitoring technologies creates a closed-loop system that continuously collects data, analyzes trends, and delivers individualized management interventions. For example, if a cow exhibits reduced rumination activity or abnormal behavior patterns indicative of health issues, the system can automatically adjust feeding protocols, notify farm personnel for further assessment, or schedule veterinary interventions as needed. This proactive approach ensures timely responses to individual cow needs, reduces stress, minimizes disruptions to productivity, and promotes sustainable management practices.

The future of individual animal management in dairy farming is poised for significant advancements driven by precision technologies and genetically informed strategies. These innovations will revolutionize how farmers monitor, assess, and meet the specific needs of each cow, leading to optimized feed conversion efficiency, improved health outcomes, and sustainable productivity levels. Looking ahead, continued innovation and adoption of precision technologies and genetically informed management strategies will further propel the dairy industry toward optimized individual animal management, sustainable production practices, and enhanced profitability. By leveraging these tools and insights, farmers can navigate challenges, maximize resource utilization, and cultivate resilient and productive dairy herds for the future.

4.3 Limitations in applying individual feeding strategies

While individual animal monitoring and management technologies offer significant benefits in optimizing dairy farming practices, there are several current limitations that need to be addressed before they can fully align with broader sustainability objectives in the dairy industry.

One of the primary limitations is the initial investment cost associated with implementing these technologies. The upfront expenses for purchasing and installing automatic milkers, feeders, and associated monitoring systems can be substantial, especially for smaller-scale dairy operations or farmers with limited financial resources. To ensure broader adoption and accessibility, there is a need for more cost-effective solutions, such as modular systems, leasing options, or government incentives to support technology adoption among dairy farmers.

Another challenge is the complexity of integrating multiple individual animal management technologies into existing farm infrastructure and management practices. Seamless data integration, compatibility between different systems, and user-friendly interfaces are crucial aspects that require improvement to facilitate efficient operation and data-driven decision-making. Standardization of data formats and protocols across platforms would also enhance interoperability and simplify data management for farmers.

Furthermore, current individual animal management technologies may have limitations in accurately predicting and addressing complex health and nutritional issues at the individual cow level. While these technologies can detect deviations from normal patterns and trigger automated responses, more advanced algorithms and predictive analytics are needed to anticipate and prevent health challenges proactively. Enhancements in data analytics, machine learning, and artificial intelligence algorithms would enable predictive modeling and early intervention strategies, improving overall cow health and welfare outcomes.

In terms of sustainability objectives, there is a growing emphasis on reducing environmental impact and resource utilization in dairy farming. While individual animal management technologies can contribute to optimizing feed conversion efficiency and minimizing waste, further improvements are needed to enhance energy efficiency, reduce water usage, and mitigate greenhouse gas emissions associated with technology operations. Developing energy-efficient systems, optimizing water recycling and conservation practices, and exploring alternative materials for equipment construction are areas that warrant attention for enhancing sustainability in dairy technology solutions.

In conclusion, while individual animal management technologies have shown promise in improving productivity and efficiency in dairy farming, addressing current limitations and making key improvements are essential steps toward their broader support of sustainability objectives. Collaborative efforts among technology developers, researchers, farmers, and policymakers are needed to drive innovation, address challenges, and ensure that these technologies align with the long-term sustainability goals of the dairy industry.

5 Conclusion

Major take-home messages from this chapter include:

- **Individual animal productivity and feed efficiency:**
    - Precision management at the individual level optimizes feed conversion efficiency and productivity.
    - Tailoring diets and care to individual needs maximizes nutrient utilization and minimizes waste.
    - Individual health management reduces metabolic disorders and supports long-term productivity.
- **Advantages over group management:**
    - Group-based management overlooks individual variations and may lead to suboptimal resource utilization.
    - Individualized care enhances cow well-being, reduces stress, and promotes sustainable productivity.

  - Precision technologies and data-driven decision-making enable targeted interventions and proactive management.
- **Future trends and improvements:**
  - Continued innovation in precision technologies, genetic profiling, and predictive analytics will further enhance individual animal management.
  - Addressing current limitations, such as integration complexities and predictive capabilities, is crucial for broader adoption and effectiveness.
  - Emphasis on energy efficiency, resource conservation, and environmental impact reduction drives sustainable technology development.

6 References

1. von Braun, J., et al. (2021). Food systems–definition, concept and application for the UN Food Systems Summit. Science and Innovations, 27, 1–7.
2. Velten, S., et al. (2015). What is sustainable agriculture? A systematic review. Sustainability, 7, 7833–7865.
3. Berry, J. (2022). Sustainability on the farm and the Net Zero Initiative [Report].
4. Capper, J. L. & Cady, R. A. (2020). The effects of improved performance in the US dairy cattle industry on environmental impacts between 2007 and 2017. Journal of Animal Science, 98, skz291. https://doi.org/10.1093/jas/skz291.
5. Capper, J. L., et al. (2009). The environmental impact of dairy production: 1944 compared with 2007. Journal of Animal Science, 87, 2160–2167. https://doi.org/10.2527/jas.2009-1781.
6. Capper, J. L. & Bauman, D. E. (2013). The role of productivity in improving the environmental sustainability of ruminant production systems. Annual Review of Animal Biosciences, 1, 469–489. https://doi.org/10.1146/annurev-animal-031412-103727.
7. Zawalińska, K., et al. (2022). A framework linking farming resilience with productivity: empirical validation from Poland in times of crises. Sustainability Science, 17, 81–103. https://doi.org/10.1007/s11625-021-00965-6.
8. Xi, Y., et al. (2016). Biological mechanisms related to differences in residual feed intake in dairy cows. Animal, 10, 1311–1318. https://doi.org/10.1017/S175173111600027X.
9. White, R. R. & Capper, J. L. (2013). An environmental, economic, and social assessment of improving cattle finishing weight or average daily gain within US beef production. Journal of Animal Science, 91, 5801–5812.https://doi.org/10.2527/jas.2013-6516.
10. Abd-El Hamed, A. M. & Kamel, E. R. (2021). Effect of some non-genetic factors on the productivity and profitability of Holstein Friesian dairy cows. Veterinary World, 14, 242–249. https://doi.org/10.14202/vetworld.2021.242-249.
11. VandeHaar, M. J., et al. (2016). Harnessing the genetics of the modern dairy cow to continue improvements in feed efficiency. Journal of Dairy Science, 99, 4941–4954. https://doi.org/10.3168/jds.2015-10352.

12. Miglior, F., et al. (2017). A 100-year review: identification and genetic selection of economically important traits in dairy cattle. Journal of Dairy Science, 100, 10251–10271.https://doi.org/10.3168/jds.2017-12968.
13. Brito, L. F., et al. (2021). Genetic selection of high-yielding dairy cattle toward sustainable farming systems in a rapidly changing world. Animal, 15, 100292. https://doi.org/10.1016/j.animal.2021.100292.
14. White, R. R., et al. (2015). Cow–calf reproductive, genetic, and nutritional management to improve the sustainability of whole beef production systems. Journal of Animal Science, 93, 3197–3211.https://doi.org/10.2527/jas.2014-8800.
15. Weigel, K. A. (2006). Prospects for improving reproductive performance through genetic selection. Animal Reproduction Science, 96, 323–330.https://doi.org/10.1016/j.anireprosci.2006.08.001.
16. Lopreiato, V., et al. (2020). Role of nutraceuticals during the transition period of dairy cows: a review. Journal of Animal Science and Biotechnology, 11, 1–18.https://doi.org/10.1186/s40104-020-00477-8.
17. Hoischen-Taubner, S., et al. (2021). The whole and the parts–A new perspective on production diseases and economic sustainability in dairy farming. Sustainability, 13, 9044. https://doi.org/10.3390/su13169044.
18. Donadeu, F. X., et al. (2020). Farmer and veterinary practices and opinions related to the diagnosis of mastitis and metabolic disease in UK dairy cows. Frontiers in Veterinary Science, 7, 127.https://doi.org/10.3389/fvets.2020.00127.
19. Maltz, E. (2020). Individual dairy cow management: achievements, obstacles and prospects. Journal of Dairy Research, 87, 145–157.https://doi.org/10.1017/S0022029920000372.
20. Souza, V., et al. (2022). Algorithm development for individualized precision feeding of supplemental top dresses to influence feed efficiency of dairy cattle. Journal of Dairy Science, 105, 4048–4063.https://doi.org/10.3168/jds.2021-20974.
21. White, R. R. & Capper, J. L. (2014). Precision diet formulation to improve performance and profitability across various climates. Journal of Dairy Science, 97, 1563–1577. https://doi.org/10.3168/jds.2013-7196.
22. Dijkstra, J., et al. (2020). Rumen sensors: data and interpretation for key rumen metabolic processes. Animal, 14, S176–S186.https://doi.org/10.1017/S1751731120000982.
23. Leso, L., et al. (2021). Validation of a commercial collar-based sensor for monitoring eating and ruminating behaviour of dairy cows. Animals, 11, 2852. https://doi.org/10.3390/ani11102852.
24. Zhu, Z. & Guo, W. (2021). Recent developments on rapid detection of main constituents in milk: a review. Critical Reviews in Food Science and Nutrition, 61, 312–324.https://doi.org/10.1080/10408398.2020.1717414.
25. Halachmi, I., et al. (2013). Automatic assessment of dairy cattle body condition score using thermal imaging. Computers and Electronics in Agriculture, 99, 35–40.https://doi.org/10.1016/j.compag.2013.08.012.

Chapter 8

Housing strategies to reduce the carbon footprint and improve animal welfare in dairy farming

Pol Llonch, Autonomous University of Barcelona, Spain; and Fernando Estellés, Polytechnic University of Valencia, Spain

1 Introduction

Livestock housing significantly influences the environmental impact of dairy farming. This chapter explores key housing strategies to mitigate greenhouse gas (GHG) emissions while maintaining animal welfare. Potential measures include managing pasture access, increasing stocking density, improving animal surveillance, enhancing diet quality, and optimising manure management.

The chapter also assesses the pros and cons of different strategies. Restricting pasture e.g. can reduce methane and nitrous oxide emissions

http://dx.doi.org/10.19103/AS.2025.0157.16

but must balance welfare concerns. Higher stocking densities improve land-use efficiency but require careful management to minimise stress. Precision livestock farming (PLF) technologies enhance herd monitoring, improving health, reproduction, and feed efficiency, reducing emissions per unit of product. Improving diet quality lowers methane emissions, though ingredient selection must consider both digestibility and environmental impact. Effective manure management, including frequent removal and optimised storage, mitigates methane and ammonia emissions. Implementing these strategies can improve dairy sustainability, but trade-offs between emissions reduction, welfare, and economic feasibility must be carefully managed. This chapter explores these different approaches.

2 Types of dairy production system

Dairy cattle farming varies widely around the world, with production systems adapted to environmental conditions, resource availability, cultural practices, economic goals, and market demands. These systems range from smallholder operations in developing countries to highly intensive, technology-driven farms in industrialised nations. Key factors such as feeding strategies, herd management, automation, and sustainability influence dairy productivity and efficiency. While some systems prioritise pasture-based or organic methods for environmental benefits, others focus on high-input, intensive approaches to maximise milk yields. A large variety of production systems have been described in the literature, mostly depending on their level of intensification, which are described below from less to more intensive.

Intensive systems emphasise high milk yields through technologies and controlled environments, such as free-stall or tie-stall barns, optimising feeding and health but potentially restricting natural behaviours. Semi-intensive systems blend indoor housing with grazing, balancing productivity with opportunities for natural behaviours while adapting to seasonal pasture availability. Extensive systems, relying mainly on grazing, prioritise low-input, sustainable practices closely aligned with natural cattle behaviours but yield lower productivity, influenced by forage quality and availability. The choice of housing system is influenced by factors such as farm size, location, economic constraints, and management priorities. Each system presents trade-offs in terms of productivity, animal welfare, and environmental impact, which need to be adapted to the local conditions.

Smallholder dairy farming is a low-input, labour-intensive system where small-scale farmers, typically owning a low number of cows (i.e. one to ten), which produce milk for household consumption and local markets. It is often integrated with crop farming, utilising crop residues as feed and manure for soil fertility. They rely primarily on family labour and have minimal mechanisation and depend on local resources. Smallholder dairy farming is common in developing countries, contributing to food security by supplying essential nutrients and economic opportunities to rural communities.

Pasture-based systems prioritise pasture use, which significantly influences economic performance. Factors such as pasture availability per hectare, grazing season length, and the proportion of purchased feed over the total feed quantity impact profitability. Efficient pasture utilisation is directly linked to higher gross and net profits. Similarly, silvopastoral systems, particularly in temperate and tropical areas of Asia and Latin America, integrate trees with pasture to enhance milk yield and reproductive efficiency. Improved silvopastoral systems demonstrate better productivity and demographic traits than traditional models, making them valuable for climate adaptation and sustainable dairy production.

Pasture-based or silvopastoral systems offer a more natural environment for dairy cattle, focusing on animal welfare and sustainability in detriment to productivity. Pasture-based systems allow cows to graze freely, supporting natural behaviours such as grazing and reducing housing costs. However, they depend on seasonal forage availability, which can impact production and expose cattle to weather extremes and predators. Hybrid systems that combine pasture access with indoor housing strike a balance between welfare and productivity. Cows can graze during the day and return to barns for feeding and milking, but this approach requires significant infrastructure and management effort.

In intensive systems, free-stall barns are popular choices. These barns provide individual stalls for cows to rest, with concrete floors typically covered by bedding materials like sand, sawdust, or rubber mats. Free-stall barns allow for efficient use of space and controlled environments, ensuring proper temperature and ventilation. However, if poorly managed, they can lead to increased risks of lameness and injuries. Another type of intensive housing is tie-stall barns, where cows are tethered in individual stalls for feeding, watering, and milking. While this system allows for close monitoring of individual animals, it limits movement and natural behaviours, potentially causing stress and health issues. Compost-bedded pack barns represent another less intensive option, providing open areas with deep bedding that is regularly aerated to manage waste and maintain cow comfort. This system promotes the free movement of cows and natural behaviours while supporting sustainable practices through composting, though it requires diligent management and higher labour inputs. Total mixed ration (TMR) feedlots, which keep cattle in confined areas with a nutritionally optimised mixed feed, are also widely used. Despite their efficiency, these feedlots often lack opportunities for grazing, which can reduce animal welfare and create environmental challenges related to manure management.

Automation and technological advancements also play a significant role in modern dairy farming. Automated rearing systems have been demonstrated to improve lactation yield while reducing labour demands. One of the most relevant examples of these technologies is probably the milking robot. It allows cows to be milked at their will, increasing the number of milking events per day, with a positive impact on productivity, which can reach up to 10% increase in dairy production. Milking robots are routinely linked to software management, where a large set of data from each cow is collected, helping farmers and veterinarians to take decisions. For instance, milking robots can assist in detecting when a cow has mastitis based on the somatic cell count regularly collected by the robot. Information on daily milk production may also help to decide the right moment for drying, or the need to optimise the diet. These innovations highlight the ongoing evolution of dairy production towards more technified systems with multiple goals, including the increasing of efficiency but at the same time improving the sustainability and animal welfare of dairy farms.

3 Effect of housing on productivity and emission intensity

A substantial body of research has explored housing management in dairy production, particularly concerning manure collection and storage, which significantly influence greenhouse gas (GHG) emissions (Adviento-Borbe et al., 2010). However, comparatively less attention has been given to the role of housing conditions of cows across different dairy production systems and their direct impact on productivity and GHG emissions intensity. Housing design, ventilation, bedding materials, and space allocation can affect animal health, feed efficiency, and overall milk yield, which in turn influence emissions per unit of milk produced. In this chapter, we will focus on this often-overlooked aspect, discussing how optimised housing strategies can enhance productivity while mitigating the carbon footprint of dairy cattle.

The way cows are housed has been acknowledged to be a key factor in improving the productivity and comfort of dairy cattle. In the past decades, research and innovation have focused on improving the housing of cows to make them comfortable, clean, and available to meet all their needs. The improvement in housing conditions has contributed, together with nutrition and genetics, to achieve unprecedented yields. For example, the average milk production of dairy cows in the 1940s was 2074 kg/y, compared with 9193 kg in 2007, which stands as a 443% increase (Capper et al., 2009). Improvements in housing conditions have focused on improving the welfare of cows because there is a close link between welfare and productivity. Among the strategies where housing strategies have shown more advancements include mitigation of thermal stress (mostly heat), improving the comfort of cows to facilitate resting, and promoting the natural behaviours of cows, mostly social interactions (Mainau et al., 2022).

In addition to improving animal welfare and productivity, the housing of livestock animals can contribute significantly to reducing waste and greenhouse gas (GHG) emissions through various measures. Intensification refers to the enhanced use of external inputs and services to improve system efficiency, which is often linked to reduced greenhouse gas emission intensity (GHG Ei) (Crosson et al., 2011). The emission intensity refers to the amount of GHG emissions, typically measured as carbon dioxide equivalents (CO_2eq), produced per unit of animal product, such as milk, meat, or eggs. It encompasses emissions from all different sources, including digestion (enteric fermentation), manure management, feed production, and energy use. Emission intensity varies by species, production system, and region, with ruminants like cattle having higher emission intensities due to methane production during digestion.

Reducing emission intensity is a critical strategy for mitigating livestock's environmental impact. Strategies include improving feed efficiency to reduce methane emissions, adopting precision farming techniques, and using feed additives or inhibitors to lower enteric fermentation. Additionally, improving animal nutrition and health and optimising breeding for higher efficiency can significantly reduce emissions per unit of product.

Intensification aims for the highest profitability of the system based on resource optimisation. This allows for improved system efficiency, achieving less waste per unit of product and reducing the intensity of GHG emissions. This has been reported by extensive literature, reviewed by Gerber et al. (2013) and Hyland et al. (2016) among others. Restricted access to pasture minimises the energy animals expend on maintenance and enhances diet quality, thus improving overall efficiency. Confinement also facilitates the capture and management of excreta, helping to lower nitrous oxide (N_2O) emissions by preventing the uncontrolled release of waste into the environment. Improved control of animals in a confined system enables better monitoring of their health and behaviour, further enhancing productivity and resource utilisation. Additionally, housing systems allow for the provision of high-quality, digestible diets – often based on concentrated feed rather than fibre-rich forage from grazing, which reduces methane emissions during digestion. Collectively, these measures reduce the carbon footprint of intensive livestock systems by minimising emissions and waste per unit of production. However, these measures may have some impact on the welfare of animals if they are not implemented considering the animals' needs.

4 Trade-offs between sustainability, welfare, and production efficiency

The housing of dairy cows is intrinsically linked to the One Welfare concept, which highlights the interconnectedness of animal welfare, human well-being, and environmental sustainability (Pinillos et al. 2016). Housing directly impacts the welfare of dairy cows by influencing their health, comfort, and ability to express natural behaviours.

Implementing housing strategies to reduce emission intensity can pose several risks to animal welfare. Intensive housing systems may lead to overcrowding, limiting the space available for each animal and hindering their ability to express natural behaviours, which inevitably increases the risk of higher social stress among animals (Llonch et al., 2017). This confinement can result in heightened competition for resources such as food and water, leading to aggression and stress-related issues.

Additionally, changes in housing conditions can compromise the physical health of animals. For instance, flooring materials like concrete can contribute to lameness, especially in dairy cattle, due to the concussive effects and contact with slurry (Fjeldaas et al., 2011). Such health issues not only affect the animals' well-being but can also reduce productivity, counteracting the intended benefits of emission reduction strategies. The use of tie stalls may facilitate manure collection, facilitating a better management to minimise pollution, including GHG emissions and slurry leaks. However, tie stalls restrict the capacity for movement of cows, reduce their comfort and can restrict some important behaviours such as feeding (Popescu et al., 2014). On the other hand, tie-stall housing may provide a better hygiene and cleanliness for cows. To achieve the highest mitigation potential without harming animal welfare, the design of tie-stall systems is crucial, focusing on cubicle size (which should fit with the dimensions of the cow) and the number of cubicles per cow, which should be higher than the number of cows present in the herd (Eerdenburg and Ruud, 2021).

Restricting access to pasture or outdoor environments have been mentioned as a possible strategy to mitigate GHG emissions. However, this restriction can thwart the natural grazing behaviour of ruminants, leading to frustration and behavioural problems (Charlton and Rutter, 2017). This lack of access to natural settings can also impact the animals' mental health, as they are deprived of opportunities to engage in instinctual activities. Therefore, while aiming to reduce emission intensity, it is crucial to carefully consider and mitigate the potential negative impacts on animal needs and preferences, which may ultimately affect their welfare. Access to pastures in semi-extensive systems aligns more closely with natural grazing behaviours, enhancing welfare but potentially exposing cows to weather extremes and parasites. So, further research should be looking for housing systems that offer an equilibrium between the capacity of the animal to express natural behaviour and reducing other animal welfare risks such as climate hazards or predators and mitigating the GHG emissions.

From a human perspective, housing design affects farm workers' safety and productivity. Well-designed systems reduce labour intensity and risks associated with handling cattle, benefiting farmers' physical and mental health. From an environmental point of view, housing systems play a role in sustainability. Efficient manure management in confined housing can mitigate greenhouse gas emissions, while overstocked pastures may lead to soil degradation and biodiversity loss. By aligning housing practices with the One Welfare framework, producers can achieve a balance that supports animal welfare, promotes sustainable farming, and enhances human well-being.

Animal welfare may not only be taken as a risk in GHG emission mitigation but also as an opportunity to help in the mitigation goal. Healthier animals tend to have better productivity, which means they can produce more milk or meat with lower emissions per unit of product. As reviewed by Džermeikaitė and colleagues (2024), the impact of dairy health and welfare on productivity and emission intensity is highly significant, where ill animals show a greater carbon footprint due to a decrease in efficiency. A more controlled environment protects animals from extreme weather conditions, which can improve their health and productivity. By maintaining optimal conditions, livestock can achieve better growth rates and reproductive performance, leading to increased productivity and reduced emissions per unit of product. A practical example is lameness and mastitis, which can be prevented by improving the housing environment. When present, these diseases can reduce milk output, leading to higher non-CO_2 GHG emissions per litre of milk produced. By enhancing animal welfare, farmers can extend the productive lifespan of their herds, thereby decreasing the replacement rate of animals and reducing the associated emissions from breeding and raising new livestock.

Moreover, better welfare practices often lead to improved feed efficiency, as animals in good conditions utilise feed more effectively, resulting in lower methane emissions from enteric fermentation. Strategies focused on improving animal welfare, such as providing better housing, nutrition, and healthcare, can mitigate stress and enhance overall productivity (Llonch et al., 2017). This creates a 'win-win' scenario where both animal welfare and environmental sustainability are prioritised, ultimately leading to a more efficient livestock production system that aligns with societal demands for ethical and sustainable practices. Thus, integrating animal welfare into GHG mitigation strategies is essential for achieving both environmental and ethical goals in livestock production.

5 Housing strategies to reduce emissions and waste: managing access to pasture

The management of cows' housing is a key factor in mitigating greenhouse gas (GHG) emissions from dairy cattle farming by influencing productivity, feed efficiency, and manure handling. Implementing strategic housing measures can lower methane and nitrous oxide emissions, improve resource efficiency, and enhance overall sustainability. These measures include restricted access to pasture, increased stocking density, improved animal surveillance, better diet quality, and enhanced manure management. This section explores each of these measures, describing how to achieve a reduction in the carbon footprint of dairy cattle farming.

One of the most visible measures of intensification of animal farming is the restriction of access to pasture. Restricting cows' access to pasture is an effective measure to reduce greenhouse gas (GHG) emissions in dairy farming. Grazing restriction has been shown to mitigate both N_2O and CH_4 emissions. DeRamus et al. (2003) demonstrated that limiting grazing time improved the efficiency of forage conversion into meat and milk, resulting in a 22% reduction in projected annual CH_4 emissions per animal. Similarly, De Klein et al. (2006) observed a 40% to 57% reduction in N_2O emissions from cattle when grazing was limited to 3 hours per day, compared to unrestricted access.

This measure is strictly linked to pasture availability and productivity and is based on resource management. Under poor pasture productivity, reducing grazing load may optimise the use of resources. Moreover, manure distribution while grazing might be linked to nutrient requirements from vegetation; otherwise, overfertilisation may lead to higher emission rates, mainly from N_2O. In addition, ration quality usually decreases when based on pasture if compared to concentrate, leading to higher CH_4 emissions from enteric fermentation. By contrast, the carbon footprint of concentrate feed may be higher than forages, usually linked to a high need of resources to produce them (fertilisation, irrigation, etc.) and the transport to bring the feed from the producing area to the farm. This equilibrium might be considered when designing grazing strategies to optimise the environmental impact of the process.

Restricted grazing systems may have benefits other than reducing the carbon footprint, such as providing the capacity to show natural behaviour, therefore improving animal welfare. High-performing dairy cows with substantial nutritional demands may achieve better health and body condition in intensive systems where feed is carefully managed. If grazing systems fail to meet these nutritional needs, animals may experience hunger and poor body condition, forcing them to compromise between essential behaviours like eating and resting (Rutter, 2010). Additional advantages of indoor housing include protection from adverse weather, predators, and parasites.

However, limiting access to pasture may have some adverse effects on animal welfare. Grazing animals, such as cattle, evolved as natural foragers and exhibit a strong preference for pasture access, provided their nutritional needs are met (Legrand et al., 2009). Denying pasture access can hinder natural behaviour, causing frustration and potentially diminishing welfare (Rutter, 2010). A balanced view of animal welfare suggests it should encompass not only harm prevention but also opportunities for positive experiences, which for grazing animals include access to pasture. Also, restricted grazing is necessarily linked to housing confinement, which increases the risk of lameness due to slurry exposure and the impact of walking on concrete surfaces (Haskell et al., 2006).

Mixed systems that combine periods of grazing with indoor housing seem to be a suitable strategy to reach the equilibrium between mitigation of the environmental impact and the welfare of cows. This can be achieved by combining giving access to pasture during some periods, including the rearing phase until the first service and the dry period, with grazing restriction during the milking period. Such systems allow animals to benefit from both controlled feeding environments and the ability to express natural grazing behaviours.

6 Housing strategies to reduce emissions and waste: increased stocking density

Intensive dairy systems are also characterised by reduced land area per animal (increased stocking rate). In dairy cattle, a 33% increase in stocking rate has been associated with a 38% increase in milk production per hectare, as indicated by the DairyMod model (Johnson et al., 2008). While higher stocking rates result in a 26% increase in CH_4 emissions per hectare, they concurrently reduce CO_2-equivalent emissions per litre of milk by 19%. For effective GHG mitigation, a high stocking density must be accompanied by an increase in feed supply, as increasing stocking density alone could lead to reduced production and higher GHG emission intensity per animal (Baudracco et al., 2011). Additionally, exceeding a certain stocking rate threshold in grazed systems may surpass the carbon sink capacity of pastures, depending on the pasture ecosystem type (Soussana et al., 2004). Further GHG reductions in intensive systems can be achieved through improved diet digestibility (e.g. grain-based versus forage diets), reduced energy allocation to maintenance in confined animals (Di Marco and Aello, 2001), and enhanced excreta management to mitigate N_2O emissions.

However, this strategy may not be effective in all production systems. For instance, Runting et al. (2024) highlight that maintaining high stocking rates in northern Australia's tropical savannas is environmentally unsustainable due to accelerated land degradation, which is exacerbated by climate change. Their integrated, spatiotemporal modelling approach demonstrates that lowering stocking density below the modelled carrying capacity reduces pressure on pasture resources, improving land conditions and decreasing methane (CH_4) and nitrous oxide (N_2O) emissions.

Additionally, a lower stocking rate results in decreased manure deposition per unit area, mitigating N_2O emissions associated with nitrogen cycling in soils. Runting et al. (2024) also find that reducing stocking density, in combination with fire management, can lower total GHG emissions by 23% while increasing profitability. This suggests that extensive livestock systems can enhance both economic and environmental outcomes by aligning stocking rates with ecological carrying capacities and adopting adaptive land management strategies.

Given that reducing GHG emissions in ruminants may require a 30% to 50% increase in stocking densities (Pinares-Patiño et al., 2007), significant adverse effects on animal health and overall welfare are likely. Elevating stocking rates can negatively affect animal welfare. Higher stocking densities often intensify competition for resources, leading to increased agonistic interactions and heightened social stress, particularly in indoor farming systems (Veissier et al., 2008). Greater population densities lead to escalated aggressive behaviour, causing social stress.

Another trade-off of increasing the stocking rate in ruminants is, at least in extensive systems, a greater risk of parasitic infections due to increased exposure to pathogens (Taylor, 2012). Cows suffering from diseases (including parasitic) may experience a reduction in production efficiency and the associated increase of GHG emission intensity. Thus, if stocking density is increased to improve environmental sustainability, careful attention should be given to the health status of cows, so the effects of poor health do not counteract the expected positive effects.

7 Housing strategies to reduce emissions and waste: improved animal surveillance

Livestock monitoring is essential not only to assure their welfare but also for improving efficiency and environmental sustainability. Continuous observation allows for the early detection of health issues and optimising reproductive performance and enhancing feed efficiency by adapting to the animal's needs.

Precision livestock farming technologies are expanding in modern livestock production (Berckmans, 2017). It is based on the use of sensors, data analytics, and automation to monitor and manage livestock in real time. PLF includes tools such as accelerometers, automated feeding systems, and environmental sensors to collect data on animal behaviour, health, and their environment, enabling precise decision-making.

PLF can play a significant role in mitigating GHG emissions by enhancing production efficiency and reducing waste, by optimising feeding, health, and breeding practices (Tullo et al., 2019). Enhanced feed management through individualised programmes minimises waste and reduces emissions linked to feed production. Health monitoring via sensors enables early disease detection, decreasing medication use and associated emissions while improving animal welfare. Additionally, PLF aids in optimising manure management, reducing emissions from manure decomposition, and improving farm environmental conditions. Overall, PLF presents a promising strategy for sustainable livestock farming, aligning with global environmental objectives with higher health and welfare.

Ferguson et al. (2024) evaluated the potential of PLF technologies to mitigate greenhouse gas (GHG) emissions in dairy farming by improving efficiency. They modelled the impact of accelerometer-based sensors on fertility, milk yield, and health across two Scottish dairy farm systems (8000 L and 10 000 L herds). The greatest total emissions reduction (−1.42%) occurred in the 10 000 L herd with fertility improvements and management interventions, while the largest per-unit emission reduction (−6.72% per kg fat and protein-corrected milk) was achieved through improved fertility and yield. These findings demonstrate that PLF technologies provide a viable strategy for reducing emissions in both low- and high-input dairy systems by optimising reproductive performance, increasing milk production efficiency, and improving herd health.

As reviewed by Stygar et al. (2021), PLF technologies enable continuous monitoring of key welfare indicators, such as milk yield fluctuations, movement patterns, and feeding behaviour, allowing early detection of health issues and stress. Data from automated milking systems, for instance, can be used to identify welfare concerns and even contribute to genetic evaluations. Additionally, production data have been explored for disease surveillance, such as detecting outbreaks of Bluetongue and Schmallenberg viruses. Beyond the farm level, PLF technologies facilitate welfare monitoring during transport and at slaughter, enhancing transparency in livestock production. Some certification schemes are using real-time animal behaviour data for certification and welfare labelling (Stygar et al., 2022). However, effective utilisation of PLF data requires improved methodologies for integration, processing, and interpretation. Ensuring data privacy, ownership rights, and computational capacity remains a challenge, but ongoing advancements in digital agriculture and high-performance computing are addressing these issues.

8 Housing strategies to reduce emissions and waste: improving diet quality

The housing of cows has a direct effect on the way they are fed. Cows housed in extensive or pasture-based systems have access to forage-based diets with high content in fibre. Instead, in indoor housing systems, feeding is normally based on high-energy diets, with a lower quantity of fibre. Digestion of fibre is associated with methane production, which is commonly named enteric methane. Methane emissions from enteric fermentation are a major output affecting the cattle's environmental impact. Ruminants produce CH_4 as a byproduct of carbohydrate degradation in the rumen. Making it simple, the lower digestibility of these carbohydrates will result in higher CH_4 emissions (Ellis et al., 2010). Improving diet quality by increasing digestibility is an effective way to reduce GHG emissions at farm level (Dida et al., 2024).

However, this measure presents some constraints in terms of environmental impact. To improve diet quality, including cereals or other 'high value' ingredients in diets is the most straightforward measure. Nevertheless, these ingredients tend to present a higher environmental footprint than other low-quality products, such as forages or some by-products. Models using Life Cycle Assessments include the carbon footprint of feed production and transport, which is not small (Rotz et al., 2010). Keeping a balance between diet quality and ingredients environmental footprint is key. Fortunately, there are available tools that might help the farmer to predict emissions from enteric fermentation as well as to determine the environmental footprint of the diet to optimise this process.

One of the significant risks associated with improving diet quality is dietary imbalance, which can have detrimental effects on animal welfare (Llonch et al., 2017). Dietary imbalance occurs when the nutritional composition of the feed does not adequately meet the specific needs of the animals. This normally happens due to an over-reliance on concentrated feeds, which are high in energy but may lack essential nutrients like fibre. For instance, feeding high levels of grains or protein supplements can lead to digestive disorders, particularly in ruminants. Conditions such as acidosis arise when there is an excessive build-up of lactic acid in the rumen due to the rapid fermentation of highly fermentable carbohydrates. This can cause significant discomfort, pain, and even severe health issues, including laminitis and liver abscesses, which compromise the overall welfare of the animal.

The introduction of new feed ingredients or supplements can lead to sudden dietary changes that animals may not adapt to quickly, which may provoke gastrointestinal disturbances. Animals may not adapt quickly to these changes, resulting in reduced feed intake, bloating, or diarrhoea. Such digestive issues can further exacerbate health problems and lead to a decline in overall productivity and welfare, ultimately increasing the emission intensity.

Additionally, impaired nutrition can thwart the immune function, making animals more susceptible to diseases and infections, leading to increased morbidity and mortality rates within herds (Ingvartsen and Moyes, 2013). This not only affects the individual animals but can also have economic implications for farmers due to increased veterinary costs and reduced productivity.

9 Housing strategies to reduce emissions and waste: manure management

When considering indoor housing, manure management is a key source of pollution. Manure nutrients follow a degradation process after excretion. Under adequate conditions on humidity and O_2 availability, nitrogen might be emitted as ammonia and nitrous oxide, depending on storage conditions. Moreover, manure fermentation might also lead to degrading carbohydrates into methane under anaerobic conditions (Chadwick et al., 2011).

Manure management practices are influenced by the housing system, mainly on the manure collection and storage phases (Owen and Silver, 2015). In tie stalls, manure is automatically collected and stored in pits or lagoons, which can be covered and/or treated (e.g. anaerobic digestion or acidification of the slurry) to reduce their emission potential. On the other hand, manure from loose housing is usually collected in a solid phase (farmyard manure), which can be composted. Both systems offer potential for GHG mitigation and the decision on which system to use may depend on the environmental conditions (notably temperature and humidity) (Petersen, 2018).

Despite the fact that different manure management systems and housing designs may imply a variety of measures to mitigate emissions, there are some general guidelines that might be adaptable to all situations. All measures rely on the same considerations: the less time the manure is stored, the lower the emissions; low humidity reduces emissions and aeration management might affect nutrient degradation. Based on these premises, several alternatives can be found.

One of the main strategies aimed at reducing emissions from manure is frequent removal. The lower exposition of manure to atmospheric conditions and continuous aeration will reduce mainly nitrogen-related emissions. This measure must be accompanied by a lower emission storage of manure hereafter. When dealing with solid manure (free-stall systems), reducing humidity is key to mitigating gaseous emissions. Several alternatives can be implemented to this aim based on increased air velocity at manure level, by adding more litter material, or through litter aeration. This last measure is controversial because it can reduce humidity, resulting in lower CH_4 and NH_3 emissions, but, at the same time, litter aeration seems to enhance N_2O emissions. Manure management strategies such as anaerobic digestion may reduce (almost completely) gaseous emissions, since the conditions on an anaerobic reactor impair ammonia and nitrous oxide emissions and methane is recovered for energy production. Composting may also mitigate gaseous emissions depending on management and manure composition (Liu et al., 2023).

10 Conclusion

Dairy farming encompasses diverse production systems adapted to specific environmental, economic, and market conditions, making establishing universally applicable GHG mitigation strategies challenging. However, improving efficiency remains a key approach, as it reduces emissions per unit of milk produced while optimising resource use. Housing strategies play a crucial role in this process, with well-managed indoor systems enhancing animal health, productivity, and feed efficiency, ultimately lowering GHG emissions. Nonetheless, limiting pasture access – despite its potential benefits in reducing enteric methane and soil-based nitrous oxide emissions – can negatively impact animal welfare by restricting natural grazing behaviours. Mixed systems that combine indoor housing with controlled pasture access provide a balanced approach to achieving environmental and welfare objectives.

Further, increasing stocking density can enhance land-use efficiency but requires careful management to prevent competition for resources and social stress. Precision livestock farming technologies offer a promising solution by enabling data-driven management of feeding, health, and reproduction, thereby improving efficiency and lowering emissions per unit of product. Dietary strategies based on increasing the quantity of energy in the diet through reduction of the fibre content also contribute significantly to emission reductions by enhancing digestibility and reducing enteric methane production. However, diet formulation must carefully balance environmental impact with animal health and welfare considerations. Additionally, manure management remains a critical factor in emission control, where frequent removal, moisture reduction, and improved storage conditions can substantially reduce methane, ammonia, and nitrous oxide emissions.

Overall, housing strategies should align with the health and welfare of animals to optimise production efficiency and, importantly, do not thwart other important societal demands such as improved welfare of cows. Future research and policy development should continue refining these strategies to optimise trade-offs between productivity, sustainability, and welfare in dairy farming.

11 Where to look for further information

Further information can be retrieved from the reference list below. Additionally, information on current trends on livestock GHG mitigation strategies can be found in the Food and Agriculture Organization website (FAO; www.fao.org). Also, details about methodologies to monitor GHG emission can be retrieved from the Internation Panel on Climate Change website (IPCC, www.ipcc.ch).

12 References

Adviento-Borbe, M. A. A., et al. (2010). Ammonia and greenhouse gas flux from manure in freestall barn with dairy cows on precision fed rations. *Transactions of the ASABE*, *53*, 1251–1266.

Baudracco, J., et al. (2011). Effects of stocking rate on pasture production, milk production and reproduction of supplemented crossbred Holstein–Jersey dairy cows grazing lucerne pasture. *Animal Feed Science and Technology*, *168*, 131–143.

Berckmans, D. (2017). General introduction to precision livestock farming. *Animal Frontiers*, *7*, 6–11.

Capper, J. L., et al. (2009). The environmental impact of dairy production: 1944 compared with 2007. *Journal of Animal Science*, *87*, 2160–2167.

Chadwick, D., et al. (2011). Manure management: Implications for greenhouse gas emissions. *Animal Feed Science and Technology*, *166–167*, 514–531.

Charlton, G. L. & Rutter, S. M. (2017). The behaviour of housed dairy cattle with and without pasture access: A review. *Applied Animal Behaviour Science*, *192*, 2–9.

Crosson, P., et al. (2011). A review of whole farm systems models of greenhouse gas emissions from beef and dairy cattle production systems. *Animal Feed Science and Technology*, *166*, 29–45.

De Klein, C. A. M., et al. (2006). Restricted autumn grazing to reduce nitrous oxide emissions from dairy pastures in Southland, New Zealand. *Agriculture, Ecosystems & Environment*, *112*, 192–199.

DeRamus, H. A., et al. (2003). Methane emissions of beef cattle on forages: efficiency of grazing management systems. *Journal of Environmental Quality*, *32*, 269–277.

Di Marco, O. N., & Aello, M. S. (2001). Energy expenditure due to forage intake and walking of grazing cattle. *Arquivo Brasileiro de Medicina Veterinária e Zootecnia*, *53*, 105–110.

Dida, M. F., et al. (2024). Dietary concentrate supplementation increases milk production and reduces predicted greenhouse gas emission intensity in pasture-based commercial dairy farms. *Journal of Dairy Science*, *107*, 5639–5652.

Džermeikaitė, K., et al. (2024). Relationship between dairy cow health and intensity of greenhouse gas emissions. *Animals*, *14*, 829.

Ellis, J. L., et al. (2010). Evaluation of enteric methane prediction equations for dairy cows used in whole farm models. *Global Change Biology*, *16*, 3246–3256.

Ferguson, H. J., et al. (2024). The impacts of precision livestock farming tools on the greenhouse gas emissions of an average Scottish dairy farm. *Frontiers in Sustainable Food Systems*, *8*, 1385672.

Fjeldaas, T., et al. (2011). Locomotion and claw disorders in Norwegian dairy cows housed in freestalls with slatted concrete, solid concrete, or solid rubber flooring in the alleys. *Journal of Dairy Science*, *94*(3), 1243–1255.

Gerber, P. J., et al. (2013). Technical options for the mitigation of direct methane and nitrous oxide emissions from livestock: a review. *Animal*, *7*(s2), 220–234.

Haskell, M. J., et al. (2006). Housing system, milk production, and zero-grazing effects on lameness and leg injury in dairy cows. *Journal of Dairy Science*, *89*, 4259–4266.

Hyland, J. J., et al. (2016). Improving livestock production efficiencies presents a major opportunity to reduce sectoral greenhouse gas emissions. *Agricultural Systems*, *147*, 123–131.

Ingvartsen, K. L. & Moyes, K. (2013). Nutrition, immune function and health of dairy cattle. *Animal*, *7*, 112–122.

Johnson, I. R., et al. (2008). DairyMod and EcoMod: biophysical pasture-simulation models for Australia and New Zealand. *Australian Journal of Experimental Agriculture*, *48*, 621–631.

Legrand, A. L., et al. (2009). Preference and usage of pasture versus free-stall housing by lactating dairy cattle. *Journal of dairy science*, *92*(8), 3651–3658.

Liu, Y., et al. (2023). A global meta-analysis of greenhouse gas emissions and carbon and nitrogen losses during livestock manure composting: influencing factors and mitigation strategies. *Science of the Total Environment*, *885*, 163900.

Llonch, P., et al. (2017). Current available strategies to mitigate greenhouse gas emissions in livestock systems: an animal welfare perspective. *Animal*, *11*, 274–284.

Mainau, E., et al. (2022). *Dairy cattle welfare in practice*. Ed. Manteca, X., 5m Publishing Ltd (Essex, The United Kingdom).

Owen, J. J. & Silver, W. L. (2015). Greenhouse gas emissions from dairy manure management: a review of field-based studies. *Global Change Biology*, *21*(2), 550–565.

Petersen, S. O. (2018). Greenhouse gas emissions from liquid dairy manure: Prediction and mitigation. *Journal of Dairy Science*, *101*(7), 6642–6654.

Pinares-Patiño, C. S., et al. (2007). Effects of stocking rate on methane and carbon dioxide emissions from grazing cattle. *Agriculture, Ecosystems & Environment*, *121*, 30–46.

Pinillos, R. G., et al. (2016). One Welfare–a platform for improving human and animal welfare. *Veterinary Record*, *179*(16), 412–413.

Popescu, S., et al. (2014). The effect of the housing system on the welfare quality of dairy cows. *Italian Journal of Animal Science*, *13*(1), 2940.

Rotz, C. A., et al. (2010). The carbon footprint of dairy production systems through partial life cycle assessment. *Journal of Dairy Science*, *93*(3), 1266–1282.

Runting, R. K., et al. (2024). Balancing livestock production and environmental outcomes in northern Australia's tropical savanna under global change. *Environmental Research Letters*, *19*, 104014.

Rutter, S. M. (2010). Grazing preferences in sheep and cattle: implications for production, the environment and animal welfare. *Canadian Journal of Animal Science*, *90*(3), 285–293.

Soussana, J. F., et al. (2004). Carbon cycling and sequestration opportunities in temperate grasslands. *Soil Use and Management*, *20*, 219–230.

Stygar, A. H., et al. (2021). A systematic review on commercially available and validated sensor technologies for welfare assessment of dairy cattle. *Frontiers in Veterinary Science*, *8*, 634338.

Stygar, A. H., et al. (2022). How far are we from data-driven and animal-based welfare assessment? A critical analysis of European quality schemes. *Frontiers in Animal Science*, *3*, 874260.

Taylor, M. A. (2012). Emerging parasitic diseases of sheep. *Veterinary Parasitology*, *189*, 2–7.

Tullo, E., et al. (2019). Environmental impact of livestock farming and precision livestock farming as a mitigation strategy. *Science of the Total Environment*, *650*, 2751–2760.
van Eerdenburg, F. J. & Ruud, L. E. (2021). Design of free stalls for dairy herds: a review. *Ruminants*, *1*(1), 1–22.
Veissier, I., et al. (2008). European approaches to ensure good animal welfare. *Applied Animal Behaviour Science*, *113*, 279–297.

Chapter 9

Development and use of methane-inhibiting feed supplements in dairy farming

John Newbold, Scotland's Rural College (SRUC), UK

1 Introduction

Despite decades of research, feed supplements that inhibit enteric methane production are not in widespread use. This chapter explores the range of challenges they face in achieving broad acceptance and application.

To access markets, such supplements must satisfy several criteria, which takes time and investment. They must obtain regulatory authorisation, with requirements differing between jurisdictions. They must be effective (with a sufficient body of evidence), and their owners must build supply chains from point of origin to point of consumption by the target animal. In general, methane inhibition does not translate into increased animal productivity, so does not increase revenue. Other incentives (financial or regulatory) are therefore needed to drive the use of methane-inhibiting feed supplements.

http://dx.doi.org/10.19103/AS.2025.0157.18

Adoption may be limited if methods are not available to monitor and verify their use and effectiveness, and rate of methane production is difficult to measure on farms. There is also the issue of securing wider user and societal acceptance. Continued effort is required to invent and develop supplements that meet these criteria and enter into widespread use.

2 Opportunities to inhibit methane through nutrition

Methane arises as a co-product of anaerobic fermentation in compartments of the digestive tract of many animals, notably the reticulo-rumen of ruminants such as domestic cattle, sheep and goats. Consortia of bacteria, protozoa and fungi degrade complex chemicals (e.g. cellulose and hemicellulose in plant cell walls) into simpler molecules (e.g. hexose sugars) that they ferment, generating energy to drive their growth, with volatile fatty acids (VFA), CO_2 and H_2 as by-products (Figure 1). Hydrogenotrophic archaea generate energy by using 2H to reduce CO_2 with the creation of CH_4, lost to the environment mainly by eructation or, after absorption, exhalation.[1] More detailed descriptions of these processes can be found in many textbooks and reviews, such as Ungerfeld et al. (2023) and Mackie et al. (2024).

There are perhaps four general strategies that can be used to inhibit methane production in the rumen: (1) reduce the production of hydrogen during fermentation, (2) provide alternatives to methanogenesis as pathways for hydrogen utilisation, (3) inhibit methanogenesis (or methanogenic archaea) directly or (4) metabolise methane before it leaves the rumen (Figure 1).

Increasing the dietary concentration of fat will generally reduce methane production by substituting for fermentable carbohydrates, thus reducing the amount of hydrogen produced. Increasing the dietary concentration of starch (generally at the expense of plant cell walls) favours the production of propionate, as a co-product, over acetate, which reduces the amount of hydrogen to be otherwise disposed of (e.g. through methanogenesis). For forage-based diets, substituting starch and/or fat for plant cell walls (especially forage plant cell walls), or increasing forage digestibility, is likely to increase the rate of feed intake, due to relief of the physical restriction of rumen fill (Allen et al., 1999). The rate of feed intake is the single best predictor of rate of methane production (Reynolds et al., 2011). Faster feed intake (kg/d) will mean faster methane production (g/d), but the associated increase in the rate of passage of both liquids and solids from the rumen (Krizsan et al., 2010) will flush out some of the relatively slow-growing methanogens (Pitta et al., 2022) and reduce methane yield (g/kg DM). Substituting starch and/or fat for plant

1 While hydrogenotrophic methanogens are the dominant producers of methane in the rumen (and thus the focus of this general description), methane is also produced from methanol and methylamines by methylotrophic methanogens (Poulsen et al., 2013), released from plant cell walls during their digestion.

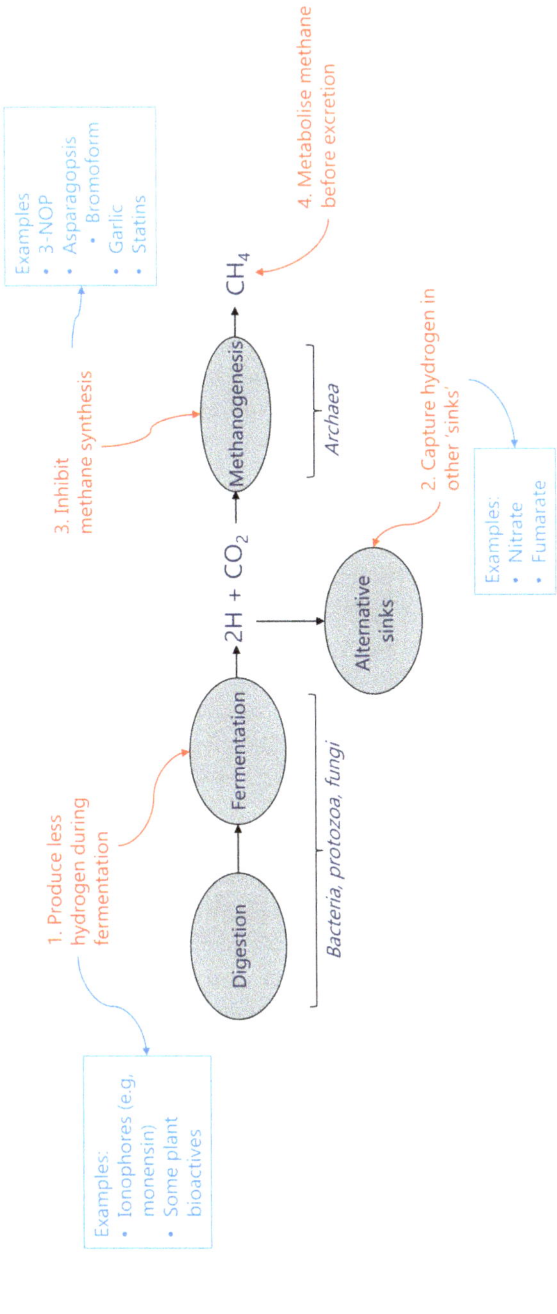

Figure 1 General strategies to reduce enteric methane emission.

cell walls in forage-based diets will increase dietary Metabolisable Energy concentration and intake, generally promoting animal productivity, and further reducing methane intensity (g/kg milk or growth).

These general nutritional effects are captured in many empirical equations that can be used to predict methane production rate, yield and intensity from diet composition (e.g. Niu et al., 2018, Benaouda et al., 2019) and are not reviewed further here.

3 Methane-inhibiting feed supplements

In the European Union and United Kingdom, the term 'Feed Additive' has a distinct regulatory meaning (see Tricarico et al., 2025, for discussion of the approaches taken by different legal jurisdictions to the authorisation and regulation of the use of methane-inhibiting feed supplements). In this chapter, 'feed supplement' is used as a generic term for feed products offered to ruminants for the specific purpose of slowing the rate of enteric methane production (expressed as g/d, litres/d or mol/d). In the EU and UK, some of these supplements are regulated as Feed Materials while others are regulated as Feed Additives.

Given the important contribution of enteric methane from ruminants to greenhouse gas (GHG) production (e.g. globally, enteric methane accounts for about 47% of ruminant GHG emissions, which in turn accounts for around 80% of total livestock GHG emissions, Opio et al., 2013), there is intense effort, around the world, to develop feed supplements to reduce enteric methane emissions from ruminants. These work through the strategies shown in Figure 1, although many have multiple touchpoints. For example, nitrate acts as an alternative sink for hydrogen, but has also been suggested to promote methane oxidation (Liu et al., 2017), while it's breakdown product nitrite may inhibit methanogens directly (Bozic et al., 2009). Antimicrobial essential oils may both slow fermentation, reducing the availability of hydrogen, and directly inhibit methanogens (Martin et al., 2021). Saponins are antiprotozoal, thereby reducing both the hydrogen generated by protozoa and the population of methanogens physically associated with protozoa (Guyader et al., 2017). Raising ruminal oxygen concentrations using chemical oxidising agents such as hydrogen peroxide may inhibit methanogenesis (as shown *in vitro* by O'Donnell et al., 2024) and slow fermentation.

Some studies have evaluated combinations of methane-mitigating feed supplements, hypothesising that effects of additives with different modes of action (Figure 1) should be at least additive. In one example, Guyader et al. (2015) found additive reductions in methane yield (g/kg DMI) in non-lactating dairy cows when nitrate and linseed oil were offered in combination. In a contrasting example, Maigaard et al. (2024) tested for interactions between a fat supplement (cracked rapeseed), nitrate and 3-nitrooxypropanol (3-NOP). In reducing methane rate (g/d), yield (g/kg DMI) and intensity (g/kg ECM), no combination of additives was more effective than the best of the three used individually: rather than having additive effects, the three supplements were antagonistic.

With the rapid generation of new research data, it is not surprising that reviews and meta-analyses describing effects of feed supplements on methane production appear with great regularity. For up-to-date technical assessments of methane-mitigating feed supplements currently in development, interested readers are referred to general reviews by Hegarty et al. (2021), Arndt et al. (2022), Tseten et al. (2024), Hristov et al. (2023) and Hristov (2024). More selective reviews are provided for plant-derived bioactive compounds by Martin et al. (2021) and for probiotics (direct-fed microbials) by Doyle et al. (2021).

It is a long road from the idea for a new methane-mitigating feed supplement to its widespread adoption in farming practice: arguably, it is a journey that no product has yet completed. Steps along this road include the initial invention or discovery, the creation and protection of intellectual property, the navigation of relevant regulatory pathways, the development of a supply chain (all the way to the rumen of the target animal), the presence of incentives to adopt methane-mitigating feed supplements, methods to verify that such supplements have been used and have been effective, and their acceptance by farmers, their customers in the supply chain, and wider society. Some of these steps are considered below, using as examples specific methane-inhibiting feed supplements currently in various stages of development.

4 Invention and discovery of methane mitigators

The first wave of rumen methane mitigation research from the 1960s onwards was driven by the potential to increase Metabolisable Energy supply (and thereby improve animal performance) by cutting the 2–12% of Gross Energy lost as methane (Johnson and Johnson, 1995). The ideas identified and evaluated in this context include various halogenated analogues of methane, such as chloroform (Bauchop, 1967), unsaturated fatty acids such as those in linseed oil (Czerkawski et al., 1966) and nitrate (Farra and Satter, 1971).

Some methane mitigators have been discovered by screening multiple candidates in short-term (e.g. 24h) rumen fermentation experiments conducted *in vitro*. Such studies are underpinned by hypotheses to varying degrees. For example, Durmic et al. (2013) selected a range of materials for testing based partly on prior knowledge of the likely effect of some of their constituents, such as long-chain polyunsaturated fatty acids or tannins. Although such screening experiments are quick and low cost, the return, in terms of promising candidates identified, can be low. For example, Bodas et al. (2008) screened 450 materials and found only 15 that reduced methane production, and only 6 that did so without generally suppressing rumen fermentation.

The use of *in vitro* fermentation techniques in methane research has been reviewed by Durmic et al. (2025). Challenges include the testing of materials at rates that would be infeasible *in vivo*, and the need to separate alteration of fermentation pathways from general suppression of fermentation (e.g. by expressing the ratio of CH_4 to VFA, or the ratio of CH_4 to CO_2). Short-term *in vitro* methods cannot, by definition, evaluate materials that might modulate the rumen microbiome over periods of several weeks. For example, Hart et al. (2019) evaluated a blend of essential oils *in vivo* in dairy cows, where a reduction in methane intensity (g/kg milk) only became evident after at least 21 days.

As an alternative to empirical screening, inhibitors can be designed based on an understanding of the biochemistry of methane synthesis or other aspects of the biology of microorganisms involved, directly or indirectly, in methanogenesis. The potent inhibitor 3-NOP, commercialised as a Feed Additive under the brand name Bovaer 10®, was discovered in a developmental programme designing specific small molecule inhibitors of methyl coenzyme-M reductase, the enzyme that catalyses the final step of methanogenesis (Duval and Kindermann, 2012). Other molecular targets have been discussed by Belanche et al. (2025).

5 Intellectual property and regulatory issues

5.1 Intellectual property

With the potential for regulatory or commercial incentives to use methane-inhibiting feed supplements, it is not surprising that inventors and developers of such products have sought to protect their intellectual property through patenting (e.g. 3-NOP, nitrate) or by keeping product formulations as closely held secrets (e.g. some blends of phytochemicals, for which the ingredients may be disclosed, but not the quantitative formulation).

However, some inhibitors described in the published literature do not appear to have patent protection or have patents that have expired. One such is the class of compounds known as statins, developed to lower cholesterol in humans by inhibiting the enzyme HMG-CoA-reductase. This enzyme is also involved in the synthesis of isoprenoid alcohols, found in the cell membranes of archaea but not in other rumen microorganisms. The use of statins to inhibit methane was patented in 1999, shortly before its description in the literature by the inventors (Miller and Wolin, 2001). A review by Abrego-Gacia et al. (2019) found an average reduction in methane yield (g/kg DMI, 6 studies) of 19%. To this author's knowledge, further steps to commercialise the use of statins to reduce methane production have not been taken. Could one reason be that the limited options for future intellectual property protection disincentivise investment in further product development? Are there other 'orphan' technologies which could be developed using governmental, or philanthropic, rather than commercial finance?

5.2 Authorisation and regulation

Tricarico et al. (2025) have compared regulatory pathways for methane-mitigating feed supplements across several jurisdictions. Comments here are limited to the EU and UK (which has largely retained the procedures in place at the time of exit from the EU).

Methane-mitigating feed supplements could be regulated as Feed Additives (specifically, for methane mitigators, as zootechnical Feed Additives 'used to favourably affect the environment' (EC Regulation 1831/2003)) or as Feed Materials (EC Regulation 767/2009). The distinction is important, in that the burden of proof of efficacy is higher for zootechnical Feed Additives than for Feed Materials, while the 10 years of brand specific authorisation afforded to zootechnical Feed Additives provides some incentive to pursue that route. Direct competitors would have to invest in the development work necessary to build their own dossier on efficacy and safety sufficient to secure their own authorisation as a zootechnical Feed Additive.

Monensin is a methane-mitigating agent Feed Additive whose use (in the EU and UK) is prevented by regulation. This molecule, an ionophore antibiotic not used in human or veterinary medicine, selectively inhibits Gram-negative bacteria, increasing the proportion of propionate in rumen VFA (reviewed by McGuffey et al., 2001). As fermentation of glucose via succinate to propionate is a net user of reducing power, this shift reduces the amount of hydrogen to be disposed of (Figure 1). Monensin also inhibits protozoa (further reducing hydrogen supply), although there is evidence for protozoal adaptation over time (Sylvester et al., 2009). Appuhamy et al. (2013) explored the consequences of these effects for methane production and yield, using experiments in both dairy and beef cattle: overall, monensin reduced methane yield (expressed as a percentage of gross energy intake) by 9%.

The authorisation for monensin in the EU was withdrawn from 1st January 2006, but it continues in use in many other jurisdictions. It is suggested that essential oils can mimic the mode of action of monensin (Calsamiglia et al., 2007), although a recent meta-analysis of beef cattle experiments reported lower propionate (as a percentage of rumen VFA) in beef cattle fed essential oils instead of monensin (Torres et al., 2021). In this study, methane yield tended to be higher with essential oils than with monensin. To date, no satisfactory alternative to monensin (as a low-inclusion Feed Additive reducing methane by altering fermentation pathways to favour propionate) has been developed.

To date,[2] only one product, Bovaer 10®, has been authorised as a methane-mitigating Feed Additive in the EU (including, for regulatory purposes, Northern Ireland) and Great Britain. The authorisation is for *'dairy cows and cows for reproduction'*, with an application to extend this to other ruminants currently under regulatory review. This is telling and suggests that the costs of pursuing authorisation may be prohibitive for smaller companies who may have promising products but limited financial and personnel resources.

In addition to demonstrating safety (for animals, consumers of foods from those animals, workers in the supply chain, and the environment), applicants seeking EU authorisation of zootechnical additives must demonstrate efficacy in three independent long-term experiments. For dairy cows, for example, experiments must be conducted with cows less than 16 weeks postpartum when enrolled, yielding more than 30 kg milk/d, and must last a minimum of 12 weeks (Bampidis et al., 2024).

Whether this level of proof is sufficient to justify authorisation is a societal and therefore political decision. Clearly, not all possible interactions between the technology (the potential Feed Additive) and the environment in which it will be used (different breeds of animal, different basal diets, different feeding systems, the presence of other Feed Additives, etc.) can be explored in just three experiments: there is always a need for further research to widen the inference space. As time goes by and data accumulate, technologies progress up a pyramid of evidence, securing authorisation at some point along the way (Figure 2).

The conundrum is to balance the risk of Type 1 error (believing something to be true – in this case, that a potential Feed Additive is effective – when it is not) with the risk of Type 2 error (believing something to be ineffective, when in fact it works). In the case of methane-mitigating Feed Additives, the consequence of a Type 1 error could be 'greenwashing', wasting time and money, while the consequence of a Type 2 error could be that society misses out on an effective tool to combat climate change.

Article 13 of EC Regulation 767/2009 permits suppliers of Feed Materials to *'draw particular attention to the presence or the absence of a substance in the feed…or to a specific function'*, provided that the supplier can provide *'scientific substantiation of the claim, either by reference to publicly available scientific evidence or through documented company research'*. Examples of substances used first as Feed Materials that have subsequently been found to have methane-mitigating effects are calcium nitrate double salt (originally a source of calcium and nitrogen) and one particular genus of seaweed.

2 Search of EU Feed Additives Register, 8 Feb 2025 https://ec.europa.eu/food/food-feed-portal/screen/feed-additives/search.

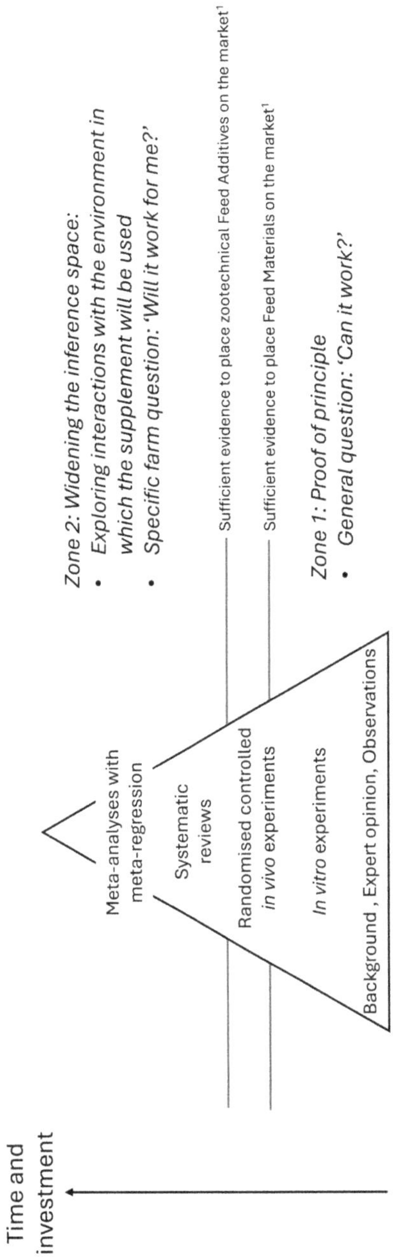

Figure 2 Building the pyramid of evidence for methane-inhibiting feed supplements.

The development of products from seaweed perhaps illustrates a regulatory grey area. Two species of red seaweed in the genus *Asparagopsis* have been shown to potently reduce methane production and yield (reviewed by Camer-Pesci et al., 2023). This is due mainly to their content of bromoform ($CHBr_3$), which directly inhibits methanogenesis (see Glasson et al. (2022) for detailed mode of action). From available literature, Eason and Fennessy (2025) established a linear reduction in methane yield (g/kg DMI) as the dietary concentration of bromoform increased.

Seaweed has nutritional value for ruminants, providing digestible carbohydrate and protein (see Bikker et al. (2020) for review), and has a long history of use as animal feed. Therefore, seaweed was included in the original EU Catalogue of Feed Materials (EC Regulation 68/2013). The most recent version of the Catalogue (EC Regulation 1104/2022) also includes *'Algae meal from Asparagopsis'* as a specific entry, *'obtained by drying and crushing macroalgae of the genus Asparagopsis'*.

Kelly et al. (2025) have provided the first *in vivo* evaluation, in feedlot beef cattle, of synthetic bromoform. Presented in oil form to provide 32 mg $CHBr_3$/kg diet DM, synthetic bromoform reduced methane production rate (g/d), methane yield (g/kg DMI) and methane intensity (mg/kg ADG) by around 95%. The regulatory path for this product in the EU and UK would be as a zootechnical Feed Additive, requiring generation and acceptance of a technical dossier of a similar scale and depth (and cost) as that submitted for 3-NOP.

Various concerns have been expressed about the safety of bromoform, including potential carcinogenicity and its transfer into milk and meat (reviewed by Eason and Fennessy, 2025). It seems inconsistent that the regulatory path for bromoform presented as *Asparagopsis* (a Feed Material) should be less onerous than for bromoform presented in synthetic form (a candidate Feed Additive).

Finally, it is worth noting that suppliers of both Feed Materials and Feed Additives have an overriding legal responsibility to ensure that their products are safe (EC Regulation 178/2002, carried over into UK law).

6 Accounting for secondary effects

Feed supplements that reduce methane could have positive or negative effects on animal productivity, behaviour and welfare. For Feed Additives, the authorisation process provides some assurance that products are both effective and safe for the animal (as well as for workers in the supply chain, the environment and consumers of animal-sourced foods). However, as noted earlier, the evidence required for authorisation of a Feed Additive will provide limited understanding of interactions with farming systems, basal diets or other technologies, while less evidence may be available before a supplier decides to place a Feed Material on the market. Refining estimates of efficacy and defining secondary effects on productivity, behaviour and welfare will require continuing research.

Two long-established dogmas are relevant to the expectation of how inhibition of methanogenesis will affect animal performance. Firstly, reductions in the Gross Energy intake lost as methane should increase the supply of Metabolisable Energy and thus increase animal performance. Czerkawski and Breckenridge (1975) argued that performance would be increased if hydrogen not used for methanogenesis was used in microbial protein or lipid synthesis, the production of propionate rather than acetate, or in the saturation of unsaturated fatty acids. However, the use of feed supplements to inhibit methane has not been found to generally improve animal performance. Feed supplements that inhibit methane production (especially low-inclusion Feed Additives that have a minor effect on diet composition) have limited effects on feed intake, milk yield, milk composition and growth rate. Ungerfeld (2018) concluded, from a meta-analysis of experiments testing chemical methane inhibitors, that '*inhibiting rumen methanogenesis has not consistently translated into greater animal productivity*'. He argued that the extent of methane inhibition achieved in many of these experiments may have been too small to result in statistically detectable productivity gains.

The second dogma is that direct inhibition of methane will increase the partial pressure of hydrogen in the rumen, preventing the oxidation of nicotinamide adenine dinucleotide (NADH to NAD^+ and H_2) and inhibiting the growth of bacteria responsible for plant cell wall digestion (Wolin, 1975; Janssen, 2010). This leads to the expectation that direct inhibition of methane will compromise digestibility and reduce animal performance.

These two dogmas point in opposite directions when considering secondary responses to direct inhibition of methanogenesis. To explore the net effect, Ungerfeld et al. (2022) conducted a further meta-analysis using studies where greater than 50% reduction in methane production rate had been demonstrated. These profound reductions were associated with lower DMI, and reductions in DMI were greater when methanogenesis was inhibited with halogenated compounds, such as bromoform, than with 3-NOP. Some of the energy 'saved' when methane is inhibited is lost through increased emissions of H_2, accounting for 11% of the energy saved when methane was inhibited by 50% (Ungerfeld et al., 2022), but these authors found no evidence for reduced digestibility when methane was inhibited, against theoretical expectations. Elevated hydrogen resulting from marked inhibition of methanogenesis will shift rumen fermentation away from acetate towards propionate, which may partly explain reduced DMI and which may favour increased efficiency of energy utilisation.

In an update to this meta-analysis, using data where methane was inhibited by more than 30%, Morgavi et al. (2023) found no relationship between rates of production of methane and energy-corrected milk, and only a numerical (P = 0.11) improvement in the growth rate of beef cattle.

Substantial research effort has been devoted to the methane inhibitor 3-NOP since the publication of its patent by Duval and Kindermann (2012) making it the most comprehensively studied methane-mitigating feed supplement. There are sufficient data to permit meta-analyses of both its efficacy (Kebreab et al., 2023) and its secondary effects (Martins et al., 2025) in dairy cows. Use of 3-NOP resulted in reduced dry matter intake and milk yield (0.98 kg/d), but increased concentrations of milk fat and milk protein. Martins et al. (2025), assuming zero cost of 3-NOP and using US feed costs and milk prices, calculated that Income Over Feed Cost was increased by just US$0.14/cow/d, and concluded that other incentives would be needed to drive adoption of 3-NOP on commercial farms.

The availability of low-inclusion additives with precise, specific and significant effects on methanogenesis (e.g. 3-NOP and bromoform) should create new opportunities to conduct long-term studies in which methane is inhibited to a large degree (>50%) and the potentially negative (reduced fibre digestion) and potentially positive (increased supply of propionate) consequences are measured. However, achieving very high methane inhibition is challenging, as illustrated by the lack of additivity of effects of individual supplements reported by Maigaard et al. (2024).

Kjeldsen et al. (2022) took reduced feed intake as a sign of sickness or malaise, perhaps associated with the increase in ruminal H_2 gas pressure that follows direct inhibition of methanogenesis. Maigaard et al. (2024) reported altered feeding behaviour in cows offered 3-NOP or nitrate, who made more attempts to access feed (not containing 3-NOP or nitrate) in bins to which they were not assigned. Kjeldsen et al. (2022) concluded that further research is needed to understand the magnitude, causes and welfare consequences of altered feeding behaviour when methane-mitigating feed supplements are used.

In conclusion, current evidence suggests that secondary benefits of improved animal performance are unlikely to cover the cost of at least some methane mitigators, still less provide an additional margin. Direct incentives (regulation or financial reward) will have to be provided to drive uptake of methane-mitigating feed supplements.

7 Developing feed supplement supply chains

The challenge of developing practical supply chains of a scale able to achieve real impacts on methane varies markedly between different feed supplements.

At one extreme, there are well-established supply chains for various fat supplements, used as energy sources in ruminant diets. Rumen-inert fats reduce methane yield and intensity simply by displacing fermentable carbohydrates. Other fat supplements may also have direct effects in the rumen through a combination of mechanisms 1, 2 and 3 (Figure 1). In a meta-analysis of dairy cow experiments, Niu et al. (2018) found that an increase in dietary fat of 10 g/kg DM reduced methane production rate, methane yield and methane intensity by 3%, 2% and 3.5%, respectively. These effects may be small (e.g. in relation to the *circa* 30% reductions achieved with 3-NOP) but in many markets the supply chain is in place to allow them to be achieved.

Nitrate reduces methane by providing an alternative sink for hydrogen (Figure 1), and a meta-analysis found reductions in methane production rate and methane yield, in dairy cows, of 20% and 15.5%, respectively (Feng et al., 2020). For reasons of safety and availability, most published work with nitrate has used a calcium ammonium double salt ($5Ca(NO_3)2 \cdot NH_4NO_3 \cdot 10H_2O$). This is widely used as a nitrogenous fertiliser, so an established supply chain exists.

In the case of 3-NOP – a novel product with no pre-existing use – a groundbreaking ceremony for a new factory in Scotland took place in late 2022, with commercial production taking several years (and significant investment) to come on stream.

According to Nilsson and Martin (2022), '*the largest hurdle in implementing a wider use of the seaweed feed supplement* [i.e. *Asparagopsis*] *is to scale up the production.*' Vijn et al. (2020) have reviewed the challenges of wild harvesting, sea-based seaweed farming and on-land cultivation. This supply chain challenge is one driver for the development of synthetic bromoform as an alternative to products derived from *Asparagopsis*.

For every methane-inhibiting feed supplement, Life Cycle Assessment (cradle-to-farm-gate or cradle-to-grave) must account for GHG emissions generated during product manufacture and transport. For low-inclusion Feed Additives in particular, such emissions are likely to be small in relation to savings in methane emissions resulting from their use.

8 Incentivising feed supplement use

If methane-mitigating feed supplements do not benefit animal performance (resulting in increased revenue or reducing costs), other means must be established to incentivise their use. These could take the forms of sticks (taxation, regulation, government mandate) or carrots (e.g. government subsidies, direct payment by milk processors and/or retailers, or the generation and sale of carbon credits).

In New Zealand, where more than half of GHG emissions are due to agriculture, the current government has postponed an earlier (2022) plan to tax farm GHG emissions (as reported by Lyubomirova, 2024). The world's first tax on enteric methane is now expected in Denmark, where a 'Green Tripartite

Agreement' between the government and environmental and agricultural organisations was agreed in June 2024 and is now being converted to proposed legislation (Anonymous, 2024a). For enteric methane, the agreement foresees both a 'stick', in the form of a tax on CO_2e emissions, but also a 'carrot' in the form of tax deductions for farmers adopting GHG mitigation measures.

There are two types of market for carbon credits: compliance and voluntary. Since its establishment in 2005, the EU Emissions Trading Scheme (ETS) has reduced CO_2e emissions from electricity, industry and heat generation by more than a third (European Commission, 2022). However, neither the EU ETS nor the UK equivalent encompass agriculture, so reductions in enteric methane do not generate high-value credits saleable on compliance markets. The European Commission has published a review of options for rewarding emissions reductions in the agri-food sector, including the introduction of an 'AgETS' (European Commission, 2023). In the UK, a recent report from the upper level of parliament recommended that including methane in the UK ETS *'would be a sensible step'*, and that, if this were done, methane should be considered separately from other GHG such as CO_2 (House of Lords Environment and Climate Change Committee, 2024).

From 2022, government subsidy is available in the Flanders region of Belgium allowing cattle farmers to choose from five methane-mitigating feed practices, including 3-NOP, nitrate, or fat supplements such as extruded linseed (Anonymous, 2022). Ungerfeld et al. (2022) referenced a scheme of Payment for Environmental Services in France to stimulate the use of certain methane-mitigating Feed Materials, such as linseed. At the time of writing, no direct government subsidies for the use of methane-mitigating feed supplements are available in the UK.

The Science Based Targets initiative (SBTi), funded primarily from philanthropic sources, introduced its Forest, Land and Agriculture (FLAG) Science Based Target Setting Guidance in 2022. Large businesses are motivated to set SBTi targets – including Scope 3 targets covering emissions in their supply chains – to improve the confidence of their investors, reduce regulatory risk and encourage innovation to enable those targets to be met. Large dairy processors operating in the UK have now set (e.g. Arla, Lactalis) or are committed to set (e.g. Muller, Leprino, Saputo) SBTi approved targets to make substantial reductions in their emissions by 2030. SBTi targets set by individual companies (which can be searched at https://sciencebasedtargets.org/target-dashboard) will encourage the incentivisation of dairy farms to reduce the GHG intensity of the milk they produce.

Some dairy supply chains already incentivise improved environmental sustainability via the milk price. For example, the 'FarmAhead' programme of Arla UK includes milk price increases for steps such as improved feed efficiency and lower use of inorganic fertiliser, both of which will reduce greenhouse gas intensity, but does not currently directly incentivise the use of methane-mitigating feed supplements (Arla, 2022). Dutch dairy company Royal Friesland Campina has announced successful results of pilot use of 3-NOP across 20 000 cows in 158 farms (Royal Friesland Campina, 2023) and in April 2024 UK food retailer Marks & Spencer announced that it would invest £1M to support the use of 3-NOP by its 40 'Select Dairy Farms' (Anonymous, 2024b).

At least two platforms for trading voluntary carbon credits include methodologies for using feed supplements to reduce enteric methane production from ruminants. Gold Standard is *'a best practice standard for climate and sustainable development interventions'* established by a group of NGOs including WWF. The Gold Standard methodology is compliant with guidelines of the FAO Livestock Environmental Assessment and Performance (LEAP) Partnership (FAO, 2020). These guidelines are a general, universal framework allowing the contribution of any Feed Additive to emissions reduction to be quantified. In December 2018, supported by DSM, Gold Standard launched a methodology to quantify effects of feed supplements on methane emissions (TREES Consulting, 2018). In this methodology, the efficacy of the feed supplement can either be measured directly on farm or inferred from 'peer reviewed data'. Chocolate manufacturer Barry Callebaut has undertaken a Gold Standard certified pilot project to use a Feed Additive (Agolin Ruminant, a blend of plant-derived bioactives including eugenol and linolool) on its supplier farms in the Netherlands and the USA (Barry Callebaut, 2021).

Washington DC-based Verra offers a 'verified carbon standard' to allow certified projects to turn GHG emission reductions and removals into tradable carbon credits. The 'UK CowCredit Project' was a Verra-certified project of UK/Swiss company Mootral, generating saleable carbon credits from a small number of UK farms using their garlic-based feed supplement (Verra, 2020). Another Verra-certified project is using Agolin Ruminant to reduce methane emissions from ruminant farms in Australia and New Zealand (Verra, 2024). Both projects use Verra methodology VM0041, which offers two options to calculate the emissions saved. Option 2 allows for direct on-farm measurement of methane emission, while Option 1 allows calculation of an emission reduction *from 'meta-analysis of at least three peer-reviewed publications'*. If these publications were *in vitro* studies, they would constitute a lower burden of proof of efficacy than is required for authorisation of zootechnical Feed Additives in the EU or UK.

Nitrate (specifically calcium nitrate) was certified (in 2014) as a method to reduce the methane emissions of beef cattle in the Australian Government Carbon Credit Union Scheme. However, the method was withdrawn in December 2024 after a review concluded that it was 'uncertain' that the method was additional, measurable and verifiable (DCCEEW, 2024).

In summary, the emergence of direct incentives for the use of methane-mitigating feed supplements is taking time, with some false starts. There is significant ongoing activity in this space by governments and businesses, so this section of this chapter will rapidly become outdated.

9 Measurement, reporting and verification

If adoption of methane-mitigating feed supplements must be specifically incentivised, there will be a requirement to measure and report the responses, and to have those responses verified.

Verifying that a supplement has been delivered to farm is straightforward in developed markets. For example, SilvAir™ (the brand name for nitrate placed on the market by patent-holder Cargill Animal Nutrition) is mainly available as an ingredient of pelleted compound feed, giving some assurance that it will be delivered to target farms and is likely to be consumed by target animals. Verifying that consumed products have been effective is more challenging.

The House of Lords Environment and Climate Change Committee (2024), in their chapter on agriculture, stated that '*enhanced monitoring and verification of emissions at smaller scales not only improves the quality and accuracy of data for the UK's National Inventory Report but also makes it possible to reward and encourage good practice in mitigating emissions by individual businesses or individuals.*' It is not clear whether 'smaller scales' refers to individual farms. The report calls for *'work to improve the accuracy and reliability of agricultural emissions calculators for use on individual farms'* but gives no other suggestions for how the use of methane-mitigating feed supplements might be monitored, reported and verified.

Measuring the rate of methane emission on farm is technically difficult and/or prohibitively expensive (reviewed by Zhao et al., 2020; Coates et al., 2021). The concentration of methane in exhalations and eructations can be measured using 'sniffer' systems that spot-sample air when animals are confined during milking or feeding (Garnsworthy et al., 2012), or by using laser methane detector 'guns' at a known distance in front of individual animals at a feed bunk (reviewed by Sorg, 2022). The estimation of methane rate (g/d) from spot samples of methane concentration is the subject of continuing research (e.g. Suzuki et al., 2021) and debate (Huhtanen et al., 2020). There is also continuing work to develop proxies for methane emission, such as prediction from the milk mid-infrared spectrum (e.g. Shadpour et al., 2022).

As with the development of incentives to encourage their use, the development of Measurement, Reporting and Verification (MRV) tools relevant to methane-mitigating feed supplements is a topic of ongoing research.

10 Securing societal acceptance

In late 2024, dairy cooperative Arla issued a press release describing farm trials being conducted by some of their UK member farmers with Bovaer 10®. This initiative follows earlier, larger scale trialling on farms in Denmark, Sweden and Germany. Arla's UK press release triggered a wave of misinformation on social media, with reports of consumers discarding milk in protest (for example, see https://www.bbc.co.uk/news/articles/c8rjdgre3vpo).

This episode serves as a reminder that social acceptance of new technologies is not fixed and permanent, but vulnerable to challenge during both development and implementation.

This recent challenge to 3-NOP may have been driven partly by scepticism over the role of enteric methane in climate change. The recognition that enteric methane is of biogenic origin, cycling relatively quickly between atmospheric CO_2 and C captured by plants used as feed, creates an opportunity for farmers to make a rapid contribution by cutting methane emissions, buying time for society to deal with CO_2 of fossil fuel origin.

Using feed supplements to reduce methane emissions may be seen as fighting nature, especially when based on synthetic compounds, rather than material of natural origin. Future challenges to the acceptability of methane-mitigating feed supplements, even when authorised and effective, are to be expected, but may be difficult to predict.

11 Conclusions

No feed supplements aimed specifically at methane mitigation from ruminants have yet been fully developed and widely adopted (although methane mitigation is a side benefit of some existing supplements, developed for other purposes, such as the ionophore antibiotic monensin). Multiple criteria must be satisfied, and hurdles overcome, to achieve widespread adoption, and the limiting factor varies between supplements and over time. For example, current limitations to the wider adoption of 3-NOP (as Bovaer 10®) include lack of incentive (who pays for its use?), some societal resistance and, in some markets (such as the EU and UK), limited breadth of regulatory authorisation across different classes of ruminant. Widespread use of *Asparagopsis* seaweed is limited by lack of a large-scale supply chain. A somewhat personal view of the status of selected products and technologies against these criteria is illustrated in Figure 3.

There is probably no single best route to navigate the pathway from idea to adoption, no single best order in which to jump the hurdles. Costs of developing methane-mitigating feed supplements (e.g. funding necessary research and pursuing regulatory authorisation) are high, prohibitively so for

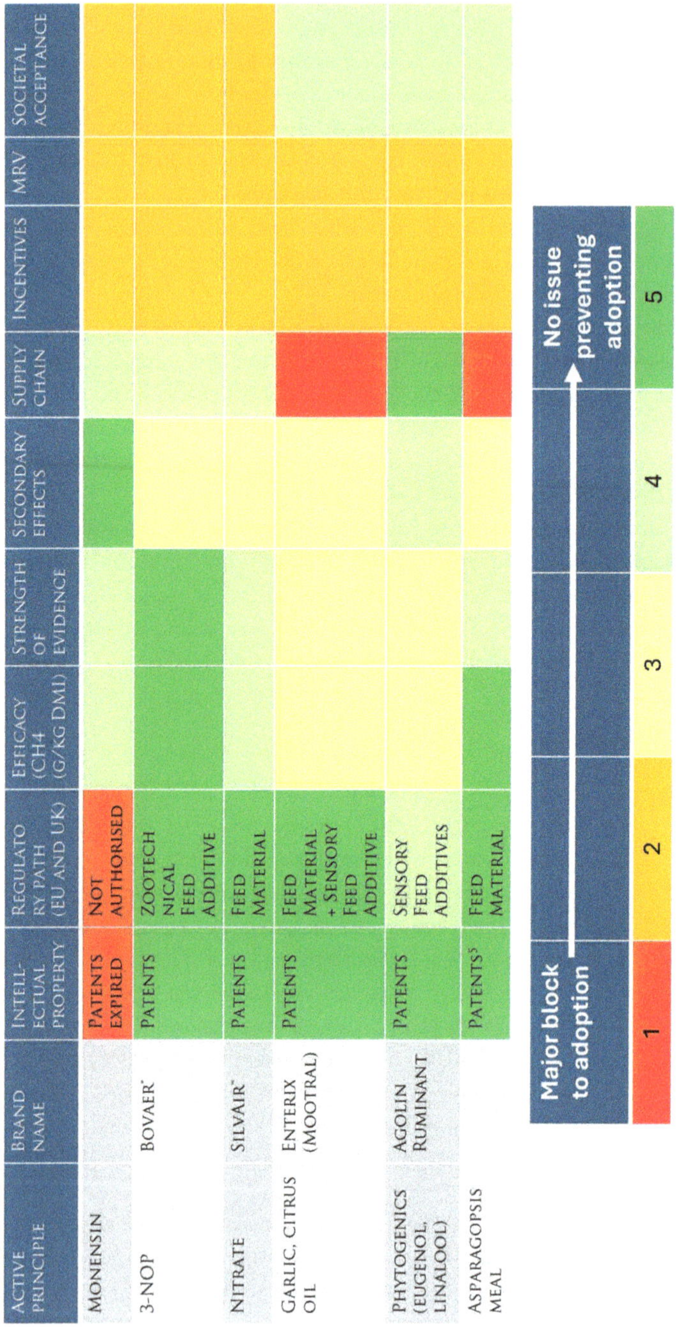

Figure 3 Status of selected feed supplements against steps towards the market.

many businesses. There is therefore an important role for governments, through matching grants, and philanthropic initiatives (such as the Global Methane Hub[3]) to facilitate and support new product development, including some existing ideas for which opportunities for protection of intellectual property may be limited.

Future food systems will need to be increasingly circular, characterised by replacement of finite resources (e.g. fossil fuels) by regenerative resources (e.g. renewable energy), minimal leakage of natural resources (including C as methane) and reuse/recycling (Van Zanten et al., 2019). Ruminants play important roles in circular food systems, using feed resources (forages, food processing co-products and some former foodstuffs) not consumed by humans, that are often grown on land not suitable for arable cropping. For example, Hennessy et al. (2021) found that a dairy/beef system contributed positively to the Human Edible Protein Efficiency and optimal use of land (metrics of circularity) of the food system in Ireland.

One way to reduce methane emissions from ruminants (and methane intensity) is to supplement diets with sources of starch or fat, but many of these are human edible. There can therefore be a trade-off between methane and circularity (Newbold, 2024). This reinforces the importance of developing feed supplements that are effective, over the long term (e.g. whole lactations of dairy cows), in high forage diets. This is true for both temperate regions with highly productive ruminant livestock systems, such as the UK, and tropical and semi-tropical systems where productivity is low, and methane intensity very high. There is a real need for further invention to fill the funnel of candidate supplements making their long and difficult journey towards implementation, at scale, on farm.

12 Where to look for further information

Various aspects of this topic are discussed in a special edition of the Journal of Dairy Science (volume 108, issue 1. January 2025). This collection of six articles is the first outcome of a flagship project by the Feed and Nutrition Network of the Global Research Alliance on Agricultural Greenhouse Gases. Written by a team of 60 leading researchers from 46 institutions across 23 countries, articles cover, in turn, the discovery and initial testing of methane-mitigating feed supplements, evaluation of efficacy and safety, the modelling and prediction of synergies and trade-offs, modes of action, regulatory pathways and recommendations for quantification of their effects

3 The Global Methane Hub convenes partners and funds work toward the common goal of reducing methane emissions https://www.globalmethanehub.org/.

13 References

Abrego-Gacia, A., et al. (2019) Lovastatin as a supplement to mitigate rumen methanogenesis: an overview. *Journal of Animal Science and Biotechnology* 12: 123.

Allen, M.S., et al. (1999) Equation to predict feed intake response by lactating cows to factors related to the filling effect of rations. *Journal of Dairy Science* 102: 7961–7969.

Anonymous (2022) Belgium: Flemish cattle farmers receive a subsidy for using feed that inhibits methane emissions. https://www.tridge.com/news/flemish-farmers-receive-a-subsidy-for-using-feed-t.

Anonymous (2024a) Agreement on a Green Denmark (in Danish) https://oem.dk/media/ul2jcmou/aftale-om-et-groent-danmark-24-juni-2024-a.pdf.

Anonymous (2024b) UK retail giant to cut its dairy cow emissions. https://www.dairyglobal.net/health-and-nutrition/nutrition/uk-retail-giant-marks-and-spencer-to-cut-its-emissions/.

Appuhamy, J.A.D.R.N., et al. (2013) Anti-methanogenic effects of monensin in dairy and beef cattle: a meta-analysis. *Journal of Dairy Science* 96: 5161–5173.

Arla (2022) Arla foods climate check report 2022. https://cdb.arla.com/api/assets/arla-uk/b42024a3-cd61-4948-a8ae-c678cfa93b6a/climate-check-report-2022-final300922pdf-1-.pdf/.

Arndt, C. et al. (2022) Full adoption of the most effective strategies to mitigate methane emissions by ruminants can help meet the 1.5°C target by 2030 but not 2050. *Proceedings of the National Academy of Sciences* 119: e2111294119.

Bampidis, V., et al. (2024) Guidance on the assessment of the efficacy of feed additives. *EFSA Journal* 22: e8856. https://doi.org/10.2903/j.efsa.2024.8856.

Barry Callebaut, A.G. (2021) Setting the standard for CO_2 reduction in dairy production. https://www.barry-callebaut.com/en/about-us/media/news-stories/setting-standard-co2-reduction-dairy-production.

Bauchop, T. (1967) Inhibition of rumen methanogenesis by methane analogues. *Journal of Bacteriology* 94: 171–175.

Belanche, A., et al. (2025) Feed additives for methane mitigation: a guideline to uncover the mode of action of antimethanogenic feed additives for ruminants. *Journal of Dairy Science* 108: 375–394.

Benaouda, M., et al. (2019) Evaluation of the performance of existing mathematical models predicting enteric methane emissions from ruminants: animal categories and dietary mitigation strategies. *Animal Feed Science and Technology* 255: 114207.

Bikker, P., et al. (2020) Evaluation of seaweeds from marine waters in Northwestern Europe for application in animal nutrition. *Animal Feed Science and Technology* 263: 114460.

Bodas, R., et al. (2008) In vitro screening of the potential of numerous plant species as antimethanogenic feed additives for ruminants. *Animal Feed Science and Technology* 145: 245–258.

Bozic, A.K., et al. (2009) Effects of the methane-inhibitors nitrate, nitroethane, lauric acid, Lauricidin and the Hawaiian marine algae Chaetoceros on ruminal fermentation in vitro. *Bioresource Technology* 100: 4017–4025.

Calsamiglia, S., et al. (2007) Invited review: essential oils as modifiers of rumen microbial fermentation. *Journal of Dairy Science* 90: 2580–2595.

Camer-Pesci, B., et al. (2023) Opportunities of Asparagopsis sp. cultivation to reduce methanogenesis in ruminants: a critical review. *Algal Research* 76: 103308.

Coates, T., et al. (2021) Measuring methane emissions from livestock. In Baines, R. (ed.), *Reducing greenhouse gas emissions from livestock production*, Burleigh Dodds Science Publishing, Cambridge, UK. www.bdspublishing.com.

Czerkawski, J.W. and Breckenridge, G. (1975) New inhibitors of methane production by rumen micro-organisms. Development and testing of inhibitors in vitro. *British Journal of Nutrition* 34: 429–446.

Czerkawski, J.W., et al. (1966) The metabolism of oleic, linoleic and linolenic acids by sheep with reference to their effects on methane production. *British Journal of Nutrition* 20: 349–362.

DCCEEW (2024) Beef cattle herd management method - Periodic review, Department of Climate Change, Energy, the Environment and Water, Canberra, December 2024.

Doyle, N., et al. (2021) The use of feed supplements to reduce livestock greenhouse gas emissions: direct-fed microbials. In Baines, R. (ed.), *Reducing greenhouse gas emissions from livestock production*, Burleigh Dodds Science Publishing, Cambridge, UK.

Durmic, Z., et al. (2013) In vitro screening of selected feed additives, plant essential oils and plant extracts for rumen methane mitigation. *Journal of the Science of Food and Agriculture* 94: 1191–1196.

Durmic, Z., et al. (2025) Feed additives for methane mitigation: Recommendations for identification and selection of bioactive compounds to develop antimethanogenic feed additives. *Journal of Dairy Science* 108: 302–321.

Duval, S. and Kindermann, M. (2012) Use of nitrooxy organic molecules in feed for reducing enteric methane emissions in ruminants, and/or to improve ruminant performance. World Intellectual Property Organization. International Patent Application WO 2012/084629 A1.

Eason, C.T. and Fennessy, P.F. (2025) Methane reduction, health and regulatory considerations regarding Asparagopsis and bromoform for ruminants. *New Zealand Journal of Agricultural Research* 68: 1–30.

European Commission (2022). Report from the Commission to the European Parliament and the Council on the Functioning of the European carbon market in 2021 pursuant to Articles. https://eur-lex.europa.eu/legal-content/EN/TXT/?uri=COM:2022:516:FIN (vol. 10)(5) and 21(2) of Directive 2003/87. European Commission (as amended by Directive 2009/29. EC and Directive (EU) 2018/410).

European Commission (2023) Pricing agricultural emissions and rewarding climate action in the agri-food value chain. European Commission Directorate for Climate Action, Brussels. https://doi.org/10.2834/200.

FAO (2020) Environmental performance of feed additives in livestock supply chains – Guidelines for assessment – Version 1. Livestock Environmental Assessment and Performance Partnership (FAO LEAP), Rome. https://doi.org/10.4060/ca9744en.

Farra, P.A. and Satter, L.D. (1971) Manipulation of the ruminal fermentation. III. Effect of nitrate on ruminal volatile fatty acid production and milk composition. *Journal of Dairy Science* 54: 1018–1024.

Feng, X.Y., et al. (2020) Antimethanogenic effects of nitrate supplementation in cattle: a meta-analysis. *Journal of Dairy Science* 103: 11375–11385.

Garnsworthy, P.C., et al. (2012) On-farm methane measurements during milking correlate with total methane production by individual dairy cows. *Journal of Dairy Science* 95: 3166–3180.

Glasson, C.R.K., et al. (2022) Benefits and risks of including the bromoform containing seaweed Asparagopsis in feed for the reduction of methane production from ruminants. *Algal Research* 64: 102673.

Guyader, J., et al. (2015) Additive methane-mitigating effect between linseed oil and nitrate fed to cattle. *Journal of Animal Science* 93: 3564–3577.

Guyader, J., et al. (2017). Tea saponin reduced methanogenesis in vitro but increased methane yield in lactating dairy cows. *Journal of Dairy Science* 100: 1845–1855.

Hart, K.J., et al. (2019) An essential oil blend decreases methane emissions and increases milk yield in dairy cows. *Open Journal of Animal Sciences* 9: 259–267.

Hegarty, R.S., et al. (2021) An evaluation of emerging feed additives to reduce methane emissions from livestock. Edition 1. A report coordinated by Climate Change, Agriculture and Food Security (CCAFS) and the New Zealand Agricultural Greenhouse Gas Research Centre (NZAGRC) initiative of the Global Research Alliance (GRA).

Hennessy, D.P., et al. (2021) The net contribution of livestock to the supply of human edible protein: the case of Ireland. *Journal of Agricultural Science, Cambridge* 159: 463–471.

House of Lords Environment and Climate Change Committee (2024) Methane: keep up the momentum. House of Lords, London.

Hristov, A.N. (2024) Invited review: advances in nutrition and feed additives to mitigate enteric methane emissions. *Journal of Dairy Science* 107: 4129–4146.

Hristov, A.N., et al. (2023) The use of supplements to mitigate enteric methane emission in dairy cattle. In Hristov, A. N. (ed.), *Advances in sustainable dairy cattle nutrition*, pp. 321–346, Burleigh Dodds Science Publishing, Cambridge, UK.

Huhtanen, P., et al. (2020) Short communication: Variation in feed efficiency hampers use of carbon dioxide as a tracer gas in measuring methane emissions in on-farm conditions. *Journal of Dairy Science* 103: 9090–9095.

Janssen, P.H. (2010) Influence of hydrogen on rumen methane formation and fermentation balances through microbial growth kinetics and fermentation thermodynamics. *Animal Feed Science and Technology* 160: 1–22.

Johnson, K.A. and Johnson, D.E. (1995) Methane emissions from cattle. *Journal of Animal Science* 73: 2483–2492.

Kebreab, E., et al. (2023) A meta-analysis of effects of 3-nitrooxypropanol on methane production, yield, and intensity in dairy cattle. *Journal of Dairy Science* 106: 927–936.

Kelly, L., et al. (2025) The effect of Rumin8 Investigational Veterinary Product – a bromoform based feed additive – on enteric methane emissions, animal production parameters, and the rumen environment in feedlot cattle. *Translational Animal Science* 9: txaf028.

Kjeldsen, M.H., et al. (2022) Potent methane reducing feed additives in a Danish context, and their reduction potential, additive effects, risks related to animal welfare and carry-over to milk, and potential trade-offs. Advisory report from DCA – Danish Centre for Food and Agriculture, Aarhus University.

Krizsan, S.J., et al. (2010) A meta-analysis of passage rate estimated by rumen evacuation with cattle and evaluation of passage rate prediction models. *Journal of Dairy Science* 93: 5890–5901.

Liu, L., et al. (2017) Nitrate decreases methane production also by increasing methane oxidation through stimulating NC10 population in ruminal culture. *AMB Express* 7: 76.

Lyubomirova, T. (2024) New Zealand abandons plan to price farm emissions – for now. Dairy Reporter, 13 June 2024. https://www.dairyreporter.com/Article/2024/06/13/new-zealand-shelves-farm-emissions-levy-plans/#:~:text=The%20National%20party%2Dled%20government,pricing%20system%20can%20be%20effective.%E2%80%9D.

Mackie, R.I., et al. (2024) Hydrogen production and hydrogen utilization in the rumen: key to mitigating enteric methane production. *Animal Bioscience* 37: 323–336.

Maigaard, M., et al. (2024) Effects of dietary fat, nitrate, and 3-nitrooxypropanol and their combinations on methane emission, feed intake, and milk production in dairy cows. *Journal of Dairy Science* 107: 220–241.

Martin, C., et al. (2021) The use of plant bioactive compounds to reduce greenhouse gas emissions from farmed ruminants. In Baines, R. (ed.), *Reducing greenhouse gas emissions from livestock production*, Burleigh Dodds Science Publishing, Cambridge, UK.

Martins, L.F., et al. (2025) Lactational performance effects of 3-nitrooxypropanol supplementation to dairy cows: a meta-regression. *Journal of Dairy Science* 108: 1538–1553.

McGuffey, R.K., et al. (2001) Ionophores for dairy cattle: Current status and future outlook. *Journal of Dairy Science* 84(E. Suppl.): E194–E203.

Miller, T.L. and Wolin, M.J. (2001) Inhibition of growth of methane-producing bacteria of the ruminant forestomach by hydroxymethylglutaryl~SCoA reductase inhibitors. *Journal of Dairy Science* 84: 1445–1448.

Morgavi, D.P., et al. (2023) Reducing enteric methane emissions improves energy metabolism in livestock: is the tenet right? *Animal* 17: 100830.

Newbold, J.R. (2024) Methane in the context of circular dairy farming. In: M. Eastridge (ed.), *Proceedings of the Tri-State Dairy Nutrition Conference*, Fort Wayne, Indiana, USA.

Nilsson, J. and Martin, M. (2022) Exploratory environmental assessment of large-scale cultivation of seaweed use to reduce enteric methane emissions. *Sustainable Production and Consumption* 30: 413–423.

Niu, M., et al. (2018) Prediction of enteric methane production, yield, and intensity in dairy cattle using an intercontinental database. *Global Change Biology* 64: 3368–3389.

O'Donnell, C., et al. (2024) Novel oxidising feed additives reduce in vitro methane emissions using the rumen simulation technique. *Science of the Total Environment* 926: 171808.

Opio, C., et al. (2013) Greenhouse gas emissions from ruminant supply chains – A global life cycle assessment. Food and Agriculture Organization of the United Nations (FAO), Rome.

Pitta, D., et al. (2022) Understanding the role of the rumen microbiome in enteric methane mitigation and productivity in dairy cows. *Journal of Dairy Science* 105: 8569–8585.

Poulsen, M., et al. (2013) Methylotrophic methanogenic *Thermoplasmata* implicated in reduced methane emissions from bovine rumen. *Nature Communications* 4: 1428.

Reynolds, C.K., et al. (2011). Improving the efficiency of energy utilisation in cattle. *Animal Production Science* 51: 6–12.

Royal Friesland Campina (2023) Large scale pilot proves methane-reducing feed additive Bovaer is ready for roll-out in the Netherlands. https://www.frieslandcampina.com/news/large-scale-pilot-proves-methane-reducing-feed-additive-bovaer-is-ready-for-roll-out-in-the-netherlands/.

Shadpour, S., et al. (2022) Predicting methane emission in Canadian Holstein dairy cattle using milk mid-infrared reflectance spectroscopy and other commonly available predictors via artificial neural networks. *Journal of Dairy Science* 105: 8272–8285.

Sorg, D. (2022) Measuring livestock CH_4 emissions with the laser methane detector: a review. *Methane* 1: 38–57.

Suzuki, T., et al. (2021) Prediction of enteric methane emissions from lactating cows using methane to carbon dioxide ratio in the breath. *Animal Science Journal* 92: e13637.

Sylvester, J.T., et al. (2009) Rumen ciliated protozoa decrease generation time and adjust 18S ribosomal DNA copies to adapt to decreased transfer interval, starvation, and monensin. *Journal of Dairy Science* 92: 256–269.

Torres, R.N.S., et al. (2021) Meta-analysis of the effects of essential oil as an alternative to monensin in diets for beef cattle. *The Veterinary Journal* 272: 105659.

TREES Consulting LLC (2018) Reducing methane emissions from enteric fermentation in dairy cows through application of feed supplements. https://globalgoals.goldstandard.org/standards/404_V0.9_LUF_AGR_Livestock_Enteric-fermentation-in-dairy-cows-through-application-of-feed-supplements.pdf.

Tricarico, J.M., et al. (2025) Regulatory frameworks and scientific evidence requirements for the authorization of feed additives to mitigate ruminant methane emissions. *Journal of Dairy Science* 108: 395–410.

Tseten, T., et al. (2024) Strategies to mitigate enteric methane emissions from ruminant animals. *Journal of Microbiology and Biotechnology* 32: 269–277.

Ungerfeld, E.M. (2018) Inhibition of rumen methanogenesis and ruminant productivity: a meta-analysis. *Frontiers in Veterinary Science* 5: 113.

Ungerfeld, E.M., et al. (2022) Current perspectives on achieving pronounced enteric methane mitigation from ruminant production. *Frontiers in Animal Science* 2: 795200.

Ungerfeld, E.M., et al. (2023) Fermentation in the rumen. In Hurst, C.J. (ed.), *Microbial fermentations in nature and as designed processes*, Wiley, Hoboken, USA.

Van Zanten, H.H.E., et al. (2019) The role of farm animals in a circular food system. *Global Food Security* 21: 18–22.

Verra (2020) UK CowCredit project: A UK dairy initiative to reduce methane from enteric fermentation. Verified Carbon Standard Project 2072. https://registry.verra.org/app/projectDetail/VCS/2072.

Verra (2024) Feedworks Agolin® enteric methane abatement in ruminants, Australia & New Zealand. Verified Carbon Standard Project 4880. https://registry.verra.org/app/projectDetail/VCS/4880.

Vijn, S., et al. (2020) Key considerations for the use of seaweed to reduce enteric methane emissions from cattle. *Frontiers in Veterinary Science* 7: 597430.

Wolin, M. (1975) Interactions between the bacterial species of the rumen. In McDonald, I.W. and Warner, A.C.I. (ed.), *Digestion & metabolism in the ruminant*, pp. 134–148, University of New England, Armidale.

Zhao, Y., et al. (2020) A review of enteric methane emission measurement techniques in ruminants. *Animals*, 10: 1004.

Part 3

Strategies to achieve net zero: system-level improvements

Chapter 10

Manure management and processing

Alice Rocha, University of California Davis, USA

1 Introduction

Despite its environmental impact, animal agriculture is essential to healthy human diets. Animal-source protein (ASP) provides essential macronutrients like protein, as well as micronutrients like vitamins, minerals, and essential fatty acids [1, 2]. Research shows that without animal agriculture, U.S. diets will be higher in calories and severely deficient in essential nutrients [1]. Even so, the environmental impact of the sector cannot be ignored and therefore presents a wicked challenge for consumers, producers, researchers, and policymakers alike, as populations grow and the global demand for ASP increases.

To quantify and continue advancing the sustainability of the livestock industry and protect these vital sources of high-quality protein, the entire production system must be scrutinized. It is well known that almost 75% of greenhouse gas (GHG) emissions associated with ASP production occur on farm, before any product leaves for additional processing [3, 4]. These on-farm emissions are split between major hotspots: enteric fermentation and manure management, which equal about 25% and 24% of the carbon footprint, respectively [3, 4]. This makes the agricultural sector a significant contributor to carbon dioxide (CO_2), methane (CH_4), and nitrous oxide (N_2O) emissions, and as of 2021, agriculture was responsible for about 9.4% of all U.S. GHG emissions [5].

http://dx.doi.org/10.19103/AS.2025.0157.21

1.1 Manure management in the United States

Manure management is the fourth largest source of anthropogenic sources of CH_4 and N_2O and responsible for 13.8% of U.S. agricultural GHG emissions [6]. Manure management emissions have increased by 55.6% from 1990 to 2022, predominantly attributed to growth in the dairy and swine industry [6]. This highlights the critical importance of reducing emissions from manure management systems. However, it is easier said than done, as these waste management systems are complex and dependent on the resources and capital available to farmers at the local and regional level [7]. Therefore, the remainder of this chapter will focus exclusively on manure management in the dairy industry.

The most common manure management strategies are composed of steps meant to aid in the collecting, handling, storage, and eventual application of dairy manure at the farm level while minimizing and mitigating pollution risks to air, water, and land [7, 8]. There are three core interconnected steps: (1) manure handling, defined as the means through which a farmer collects/removes manure from animal housing and/or grazing areas; (2) manure storage, which includes the infrastructure needed to handle manure and urine in large volumes until it can be used; and (3) manure application, which is how a farmer applies or utilizes stored manure [7, 9].

Throughout the three steps, there is a wide variety of options dairy farmers can adopt. Handling/collection ranges from minimal effort, like little to no handling of manure deposited directly on pasture, to the more complicated infrastructure of daily flushing/scraping. Scraping is a dry method of manure collection, relying on either wide shovels pushing through barn alleys around freestalls/tie stalls or other designated rest areas that collect manure [10–13]. Machinery, like vacuum trucks, can also be used to scrape away manure. Another type of manure collection is called flushing. This is when recycled water, usually previously used to clean milking equipment, flushes through a barn to remove manure [13, 14]. Flushing consumes significant amounts of water compared to scraping, despite using recycled water, and leads to manure being stored anaerobically. The consumption of water and production of GHGs from anaerobic storage is of environmental concern for liquid manure systems.

Manure storage is also complex, since storage, as well as application and handling, are interdependent and are often influenced by the economic capital available to a dairy farmer, geographical limitations, government policy, farm size, land availability, and investment costs [7, 15]. There are multiple different manure storage types, still categorized as dry or liquid. The most popular dry system is called pasture/range deposition, with over 23% of dairy farms in the U.S. utilizing this system [6]. Other dry options include solid storage (e.g. dry lot), daily spread, and composting. The commonality between all these dry forms of storage is the use of large piles of manure that are kept away from animals in dry, open areas [7, 11]. Dry manure, after composting, is frequently utilized as cheap animal bedding [10, 11]. By storing dry manure, farmers can often avoid the anaerobic environment necessary for gas formation. However, depending on the size of the dairy, space may become a limiting factor.

Liquid storage is frequently linked to flushing systems and often seen in larger dairy operations, where the manure of thousands of animals needs to be managed daily [7, 11, 15]. The large basins used to hold the manure, sometimes for six months or longer, are called lagoons. These basins are left uncovered and receive a daily influx of fresh manure, often multiple times per day. Lagoons are cost effective, allow for longer storage periods, help stabilize volatile compounds, and can handle large volumes of feces and urine [16]. The dominant issue with this system is linked to the GHG production potential associated with liquid manure storage [17]. Lagoons are the perfect environment for gas production especially because of long retention times and large surface areas, which result in high emission rates [16, 18, 19]. The volatile solids from manure deposited into lagoons pass through a series of chemical reactions to form CH_4, as well as other gases [20].

With policy pressure to reduce on-farm emissions, improve water quality, reduce malodors, and more [21, 22], the dairy industry has begun to shift away from lagoons. In the last decade, the popularity of certain manure management systems has changed. Liquid-based storage, like slurry/liquid and anaerobic lagoons, have decreased in prominence in the U.S. by 17% and 5%, respectively [6]. However, alternative manure management systems with proven abilities to reduce GHG emissions, like anaerobic digesters, have increased in popularity. As of 2022, about 4.9% of U.S. dairy farms employ anaerobic digesters [6].

There are some storage options that constitute hybrid solid-liquid systems. Solid separators and settling basins, for example, are meant to remove the volatile organic solids from manure to prevent GHG formation [10]. This results in two types of manure needing to be stored for later use: the liquid fraction, which goes to a smaller lagoon or basin, and the solid fraction, which is usually placed in a dry lot. These options will be further discussed in Section 3.2.

Application is the end-of-life usage of manure and traditionally results in manure being applied as fertilizer to a field. Again, there are a variety of ways that manure can be applied to a field. One of the cheapest ways is broadcasting [23]. This technique can be done with solid and liquid manure and can be incorporated or unincorporated (i.e. mixed into the soil or left on top of the soil undisturbed) [23]. Regardless of the manure state, unincorporated broadcasting has been associated with higher rates of nitrogen emissions, especially NH_3 volatilization [23, 24].

The next major type of manure application is injection, where manure is injected directly into the soil, which is used for liquid manures [23]. Placing manure below the soil surface helps prevent NH_3 volatilization and odor. Injection has been shown to reduce most GHG emissions, except for N_2O, by about 10% [25]. Another common end use of manure is composting. Composting helps further break down organic matter, reduce pathogenic load, decrease odor, and lower CH_4 emissions by about 45% [26, 27]. Compost can be used as a soil amendment, providing a rich source of nutrients, and improves the fertility, water holding capacity, and overall biological properties of soil [26, 28]. It is also commonly used as cheap bedding for animals.

Ultimately, not all manure management systems employ all three steps. For example, pasture-based systems do not employ any of the above-mentioned handling/collection as manure is deposited directly onto land and is otherwise largely unmanaged [7, 15]. Other systems can utilize a combination of techniques, designed to best suit a farmer's needs, which are variable and depend on factors like farm size, animal type and breed, climate, finances, and more [7]. Before delving into the potential mitigative strategies of manure management (Section 3), it is prudent to understand the emissions resulting from manure management.

2 Greenhouse gas emissions

Three main agricultural GHGs, CO_2, CH_4, and N_2O are emitted by various parts of the dairy sector. The agricultural GHGs of concern in manure management are CH_4 and N_2O. For CH_4, the global warming potential (GWP) is categorized as biogenic or anthropogenic, with GWPs of 27.2 and 29.8, respectively [29]. Anthropogenic emissions are those originating from human-controlled activities like the burning of fossil fuels, whereas biogenic emissions are those originating from biomass carbon that are part of the natural carbon cycle [30]. Manure management is categorized as anthropogenic CH_4, like landfills, burning biomass, coal mining, and ruminant fermentation. Methane is synthesized from the breakdown and fermentation of carbohydrates [31–33]. Once in the atmosphere, CH_4 can be oxidized by hydroxyl radicals back to CO_2 and other atmospheric gases, with a half-life of about 10–12 years [34–36]. As the carbon in CH_4 becomes CO_2 again, it reenters the biogenic carbon cycle and either becomes plant matter once more or soil carbon [37]. In U.S., livestock agriculture emits 36% of total anthropogenic CH_4 emissions [9]. Manure management was responsible for 7.2% of total U.S. anthropogenic CH_4 emissions [9].

On the other hand, the other GHG from agriculture and livestock production, N_2O has a GWP value of 273 [29]. And unlike CH_4, N_2O is a long-lived climate pollutant, capable of remaining in the atmosphere for over 100 years [5, 38]. However, it is emitted in relatively low quantities compared to the other GHGs (i.e. CH_4 and CO_2), with about 5% of total

U.S. N_2O emissions coming from manure management practices and 74% coming from land management practices, due to the application of artificial and organic fertilizers to soil [9]. Nitrogen (N) is an essential nutrient for the growth and development of both animals and plants [39–41]. The formation of N_2O is complicated, requiring both aerobic and anaerobic environments and ammonium (NH_4^+) oxidizing and denitrifying microbes, prior to emission into the atmosphere [42, 43]. Currently, many interventions designed to curb GHG emissions focus on CH_4 and CO_2, whereas only in manure management are nitrogenous emissions of greater importance. Research on acidification and urease/nitrification inhibitors has helped reduce N_2O emissions. These interventions interrupt the N cycle and prevent the formation of nitrogenous gases by stopping key compounds like nitrites and nitrates from forming [39, 44].

Another key nitrogenous pollutant is NH_3. It forms from a chemical reaction between organic N in the form of urea in urine and the enzyme urease in feces, either in animal housing facilities or in manure storage areas [45, 46]. However, NH_3 is a vital part of the N cycle, as organic N, urea and protein are transformed into NH_4^+, nitrates (NO_3^-), then nitrites (NO_2^-), prior to becoming NH_3 or other nitrogenous gases [40, 43].

While NH_3 is not a GHG, it is no less problematic when it comes to its impact on the environment and animal/human health, considered hazardous at high concentrations [45, 47]. Volatilization can consume up to 70% of excreted N and be deposited into water and land-based ecosystems, leading to eutrophication, acidification, and toxicity [46, 48]. However, the majority of inorganic N present in slurry and fresh manure is in the form of NH_4^+, which exists at equilibrium with NH_3 in aqueous solution [45].

In addition to threats to natural ecosystems, NH_3 is also a precursor to the formation of small aerosols called particulate matter 2.5 ($PM_{2.5}$), capable of surviving in the atmosphere for over two weeks and carrying pathogens into deep lung tissue if inhaled [13]. Furthermore, NH_3 is a precursor to the formation of N_2O [28, 43, 46]. Ammonia's ability to negatively impact a broad range of ecosystems and human health makes it a particularly potent pollutant that must be aggressively managed.

3 Existing manure management emissions interventions

The dairy industry has grown, steadily increasing production over the last few decades and leading to a greater quantity of manure and urine being managed [49, 50]. Because digestion is an incomplete process, there is still organic matter present in manure after defecation. The decomposition of volatile solids results in the production of GHGs [16, 51]. The volatile solid content of manure is an important part of models and equations developed to estimate gas production potential from manure [29, 52].

Due to the role manure management plays in GHG production, the U.S. dairy industry has put forth aggressive emission reduction goals, aiming to be carbon neutral by 2050 [53]. Some identified interventions that reduce the negative environmental impact of manure management include nutrient management planning, anaerobic digesters, composting, solid separators, crusting, daily spread, and reducing reliance on liquid manure storage [27, 53]. Current interventions aim to change how manure is managed in the hopes of decreasing associated emissions. Such options include solid separation and dry storage, aiming to eliminate anaerobic environments leading to CH_4 formation [10]. The following sections will delve into popular manure intervention options in further detail to better elucidate their mitigative potential.

3.1 Anaerobic digesters

A potent option for GHG emission prevention is an anaerobic digester, which captures biogas as a source of energy. Cattle manure has a very high volatile solids and energy content, which makes it an excellent substrate for gasification [54]. Digesters on dairy farms can be implemented in two primary ways: (1) covering a pre-existing lagoon, also known as cover and flare or fixed film [55, 56], (2) or constructing a new standalone digester [56]. The digester maintains the anaerobic, carbon-rich environment necessary for the production of biogas–a mixture of about 60% CH_4 and 40% CO_2–following the fermentation of volatile solids [57].

The fermentation process that generates CH_4 in anaerobic digesters, as well as uncovered lagoons, mimics the biochemical process in the rumen. Like rumen fermentation, organic matter undergoes hydrolysis, i.e. breaking down large molecules into their respective monomers like amino acids or monosaccharides. The next phase is called acidogenic fermentation, with the digestive enzymes of acidogens [58, 59]. Monomers from hydrolysis are then fermented into VFAs (i.e. butyrate, acetate, propionate) and other compounds like alcohols, H_2, and CO_2 [58, 60]. Acetogens, a small group of homo-acetic bacteria, then utilize H_2 and CO_2 resulting from previous fermentation steps to generate more acetate. Lastly, strictly anaerobic methanogens consume the byproducts of fermentation like acetate, formate, CO_2, H_2, and use methanogenesis to synthesize CH_4 [58, 60]. Similar to processes in the rumen, the hydrogenotrophic pathway dominates methanogenesis in digesters and lagoons [61].

Seasonal temperature fluctuations impact CH_4 emissions when manure is stored outdoors. Longer storage periods have been shown to produce more CH_4, especially at temperatures greater than 15°C [61]. The relationship between CH_4 production and temperature is linked to the efficacy of anaerobic digesters within certain climatic regions. Because the production of biogas is dependent on warmer ambient temperatures, this is a risk that anaerobic digesters built in cooler regions will not be sufficiently productive to justify construction costs [15]. However, the warm climate in California makes it the ideal region for anaerobic digesters [62]. Research has long investigated co-digestion and the use of accelerants to overcome climatic limitations and enhance biogas production [63–65].

Anaerobic digesters combat emissions by preventing biogas from entering the atmosphere and have become increasingly popular in regions like California [60, 62]. Since 2014, anaerobic digesters have offset 21.02 MMT CO_2e in California following the installation of 114 digesters [66]. A study based in New York found that from 1992 to 2022, the implementation of 662 cover and flare digesters reduced 62% of GHG emissions coming from manure storage [67]. And according to the U.S. EPA, as of 2022, states like Arizona, Indiana, Massachusetts, and Oregon have about 20–30% of dairy cow manure managed via anaerobic digester [6].

Additionally, the generated biogas can be used as renewable natural gas fuel in machinery or burned for electricity, helping to reduce dependency on fossil fuels [60]. However, anaerobic digesters are not often able to be implemented on medium-to-small sized farms (<1000 head) due to the associated start-up and maintenance costs [7, 10, 15, 68]. The EPA reports only 221 operational dairy anaerobic digesters as of 2023 across the entire U.S. [69]. While there are financial aid programs available, sponsored by local government like the California Department of Food and Agriculture [66], adoption remains limited as not all farms that apply will qualify for aid. However, AgSTAR through the EPA has developed project planning and financing guidelines for dairies in the hopes of highlighting the benefits associated with anaerobic digesters and key considerations and aid for those considering this system [70, 71].

3.2 Solids separators

Another alternative system supported for its ability to reduce GHG emissions from manure is solids separation. This technique reduces the amount of solids deposited into lagoons, thereby reducing the organic load managed in the system. It is estimated that about 81% of dairy farms in California's San Joaquin Valley use some form of solid separation [18]. There are a diverse collection of separation technologies available, including but not limited to roller drums, screw press, stationary/vibrating screens, and centrifuge [72]. The diversity in available technologies is without a doubt of benefit to farmers, lending additional flexibility to those adopting these systems.

Gas formation is directly linked to volatile solid content of manure, so when solids are separated, this reduces hydrolysis, which is considered the rate limiting step in CH_4 formation in lagoons and anaerobic digestion, by limiting the availability of organic substrates [57, 58, 73]. Additionally, the solids could be composted to be used as bedding, which would help offset some on-farm costs associated with purchased bedding [15]. Like anaerobic digestion, solids separators are still not the perfect system when it comes to mitigating the environmental impact of MMS. First of all, it depends on the construction of a separator and additional space on which to store the solids [10]. While solid separation remains a more cost-effective option compared to anaerobic digestion, it is likely not a suitable option for farms that do not utilize some type of liquid manure storage.

3.3 Converting to scraping collection

Flushing is commonly used for manure collection, using water in flushing barns to remove manure. However, the addition of water to excreta creates an anaerobic environment [10]. Converting from flushing to scraping is one of the alternative systems being funded by the CDFA's Alternative Manure Management Program (AMMP), as handling dry manure has been shown to reduce GHG emissions [10, 66]. However, scraping rather than flushing results in severe pollution swapping between CH_4 and NH_3. A study comparing flushing versus scraping manure at different daily frequencies showed that while CH_4 emissions significantly decrease with scraping, NH_3 emissions increased by 152–175% [13]. Without additional interventions to mitigate increased NH_3 volatilization, any benefits of reduced CH_4 from scraping are lost due to pollutant swapping.

There are plenty of other alternative MMS available that are ultimately designed to reduce air, land, and water pollution. However, not all the options are equally available, effective, or feasible for farmers given the diversity in MMS that exist within the United States. Furthermore, many of these mechanical interventions do little when it comes to addressing N pollution. To holistically address the environmental risks associated with MMS, all potential sources of pollution need to be simultaneously addressed. Recent research has shifted to focus on changing the manure entering the systems, rather than the system itself.

4 Conclusion and future trends

While there are limitations associated with changing the infrastructure of a manure management system, research shows that changes made directly to the excreta lead to significant reductions in GHG emissions. Much of this research has focused on altering the chemical composition of manure, either by disrupting the natural environment or directly affecting the microbial population and thereby inhibiting gas formation [57, 61, 74].

4.1 Acidification and inhibitors

One potential alternative intervention is acidification. A review by Kupper, Häni, Neftel, Kincaid, Bühler, Amon, and VanderZaag [57] found that acidification resulted in 70% and 96% reductions in NH_3 and CH_4, respectively, during the storage period. Another study treated slurry with sulfuric acid and saw up to 99% long-term reductions of CH_4 emissions across a range of ambient temperatures [44]. Acidification disrupts the neutral environment preferred by microbes that make CH_4 and NH_3 [57, 75]. Lowering the pH destroys the negative redox environment needed for CH_4 formation and shifts the equilibrium between NH_4^+ and NH_3 toward NH_4^+ formation [28, 47, 75]. The only gas seemingly unaffected by acidification was N_2O.

Another option is using nitrification and urease inhibitors. These compounds are meant to interrupt the N cycle and slow the formation of nitrates and NH_3 [39, 76]. A study comparing urease and nitrification inhibitors found that plots of grass treated with urease or nitrification inhibitors, compared to an untreated control, had better plant regrowth, N recovery, and lower NH_3 losses by 26% and N_2O emissions by 50.2% [39]. Another study found that nitrification inhibitor 3,4-dimethylpyrazole phosphate reduced N_2O emissions by 81% when paired with injecting manure into soil [77]. Further research is needed to determine if the use of acidification or inhibitors is feasible at the commercial scale.

4.2 Manure additives

While infrastructural changes are proven in their ability to reduce the environmental impact of manure management, especially by reducing GHG emissions and other air quality pollutants, research has started looking into alternative options that would benefit dairy farms of all sizes. Like feed additives, manure additives have been developed to reduce emissions from forming at the source. The main area of concern is during manure storage, when excreta is left to sit until application, use, or disposal, and depending on the storage type, this can lead to higher emissions rates [7, 9, 15].

SOP Lagoon is a commercially available manure additive, meant to be added to slurry or lagoon water to reduce GHG and NH_3 emissions and odor [78, 79]. SOP Lagoon is a calcium sulfate dihydrate compound formulated via proprietary technology [78]. Initially tested in laboratory settings, a dose of 61.6 g SOP Lagoon/m^3 lagoon water significantly reduced CO_2, CH_4, NH_3, and N_2O emissions and also reduced odor intensity [78]. A follow up commercial on-farm trial in Italy found 80% reductions in CH_4 and 75% reductions in CO_2 after two months of treatment compared to untreated liquid manure [79]. SOP Lagoon contains gypsum, which has been linked to acidification and reduced GHG emissions [57, 78, 79]. However, the mode of action of SOP Lagoon is under investigation, with research needed to explore microbial interactions [79].

Another option is Eminex®, a calcium cyanamide ($CaCN_2$) compound traditionally used in pesticides and artificial fertilizers, which has recently entered the European market [80]. Following a 26-week incubation within a closed anaerobic system, a one-time application of Eminex® significantly reduced GHG emissions in both dairy and swine slurry [80]. Total CH_4 and CO_2 were reduced by up to 99% at a treatment rate of 2.07 g Eminex®/kg slurry in dairy cattle slurry and fattening swine slurry, and total N_2O was decreased by up to 80% and 60% for swine and dairy cattle slurry, respectively, when compared to a control receiving no Eminex® [80]. However, NH_3, across both slurry types, did not significantly decrease with treatment [80]. Eminex® was recently tested in liquid storage systems and was found to be similarly effective in reducing GHG emissions from dairy lagoon water. Rocha et al. [81] applied Eminex® at the manufacturer-recommended rate (1.0 kg/m^3 manure) to dairy lagoon wastewater and measured reductions in CO_2, CH_4, and N_2O of 12%, 85.1%, and 82.7%, respectively. The same methodology was applied to fresh dairy slurry, and measured reductions in CO_2, CH_4, and NH_3 of 49.3%, 30.4%, and 34.9%, respectively [81]. The mode of action of this product is also still being explored, but it has been linked to its dicyandiamide content and acidification potential [80], which directly inhibit nitrification and disrupt the microbial environment needed for gas formation [39, 57, 76].

Manure additives are newer mitigative tools, when compared to anaerobic digesters and solid separators, but by no means less important. While infrastructural changes benefit larger farmers, the use and ease of application of manure additives will be vital to helping smaller dairy farms address their own manure emissions while minimizing cost and labor demands.

Liquid-based manure handling and storage has led to increased absolute emissions and emission intensity over time, predominantly due to CH_4, as N_2O emissions remained stable around 0.03 MMT/year [15, 82]. Depending on the methodology, manure-based GHG emissions are a greater contributor to global warming compared to enteric fermentation [82]. When comparing the standard GWP_{100} metric to the newer GWP*, increasing manure CH_4 emissions contributed significantly more warming equivalents compared to enteric CH_4, at 90.8 MMT CO_2-warming equivalents (CO_2-we) versus 89.2 MMT CO_2-we, respectively. Conversely, when compared to manure CH_4, GWP_{100} presented enteric CH_4 as the greater contributor by about 206% [82]. It is important to note that this does not mean that manure CH_4 is more damaging than enteric CH_4, but rather that enteric emissions have stabilized over time while manure CH_4

emissions increased. The overall contribution of the two different sources of CH_4 calls into question metric use and when certain calculation methodologies are appropriate over others. However, additional speculations fall outside the scope of this chapter.

5 Where to look for further information

Additional information on manure management interventions and mitigative efforts is outlined in the following literature:

- Niles et al. (2019) Tradeoffs in US dairy manure GHG emissions, productivity, climate, and manure management strategies. *Environ Res Common*. 1:1–15. https://doi.org/10.1088/2515-7620/ab2dec.
- Varma et al. (2021) Dairy and swine manure management – Challenges and perspectives for sustainable treatment technology. *Sci Total Environ*. 778:1–18. https://doi.org/10.1016/j.scitotenv.2021.146319.
- Niles et al. (2022) Manure management strategies are interconnected with complexity across U.S. dairy farms. *PLoS One*. 17:1–19. https://doi.org/10.1371/journal.pone.0267731.
- El Mashad et al. (2023) Anaerobic Digestion and Alternative Manure Management Technologies for Methane Emissions Mitigation on Californian Dairies. *Atmosphere*. 14:1–18. https://doi.org/10.3390/atmos14010120.

Key research in this area can also be found at the following organizations:

- California Department of Food and Agriculture Alternative Manure Management Program, https://www.cdfa.ca.gov/oefi/AMMP/.
- California Department of Food and Agriculture Dairy Digester Research & Development Program, https://www.cdfa.ca.gov/oefi/ddrdp/.
- US Environmental Protection Agency AgSTAR, https://www.epa.gov/agstar.

6 References

1. White, R.R. and Hall, M.B. Nutritional and greenhouse gas impacts of removing animals from US agriculture. *Proc. Natl. Acad. Sci. U. S. A.* 2017, *114*, E10301–E10308. https://doi.org/10.1073/pnas.1707322114.
2. Leroy, F., et al. The role of meat in the human diet: evolutionary aspects and nutritional value. *Anim. Front.* 2023, *13*, 11–18. https://doi.org/10.1093/af/vfac093.
3. Wattiaux, M.A., et al. Invited Review: Emission and mitigation of greenhouse gases from dairy farms: The cow, the manure, and the field. *Appl. Anim. Sci.* 2019, *35*, 238–254. https://doi.org/10.15232/aas.2018-01803.

4. Thoma, G., et al. Greenhouse gas emissions from milk production and consumption in the United States: A cradle-to-grave life cycle assessment circa 2008. *Int. Dairy J.* 2013, *31*, S3–S14. http://dx.doi.org/10.1016/j.idairyj.2012.08.013.
5. EPA. Inventory of U.S. Greenhouse Gas Emissions and Sinks: 1990–2021. 2023, 1–881.
6. EPA. *Inventory of U.S. Greenhouse Gas Emissions and Sinks: 1990–2022*; EPA 430-R-24-004; U.S. Environmental Protection Agency: 2024.
7. Niles, M.T., et al. Manure management strategies are interconnected with complexity across U.S. dairy farms. *PLoS One* 2022, *17*, 1–19. https://doi.org/10.1371/journal.pone.0267731.
8. Lim, T., et al. *Increasing the Value of Animal Manure for Farmers*; AP-109; U.S. Department of Agriculture Economic Research Service: 2023.
9. EPA. *Inventory of U.S. Greenhouse Gas Emissions and Sinks: 1990–2020*; Environmental Protection Agency: 2022.
10. El Mashad, H.M., et al. Anaerobic digestion and alternative manure management technologies for methane emissions mitigation on californian dairies. *Atmosphere* 2023, *14*, 1–18. https://doi.org/10.3390/atmos14010120.
11. Kaffka, S., et al. *Evaluation of Dairy Manure Management Practices for Greenhouse Gas Emissions Mitigation in California*; UC Davis: 2016; pp. 1–127.
12. Meyer, D., et al. Survey of dairy housing and manure management practices in California. *J. Dairy Sci.* 2011, *94*, 4744–4750. https://doi.org/10.3168/jds.2010-3761.
13. Ross, E.G., et al. Manure Flushing vs. Scraping in Dairy Freestall Lane Reduces Gaseous Emissions. *Sustainability* 2021, *13*, 1–12. https://doi.org/10.3390/su13105363.
14. Elder, J.D. and Petty, R. Flush Systems and Sand. In Proceedings of the Fifth International Dairy Housing Proceedings, Fort Worht, TX, USA, 2003; pp. 386–390.
15. Niles, M.T. and Wiltshire, S. Tradeoffs in US dairy manure greenhouse gas emissions, productivity, climate, and manure management strategies. *Environ. Res. Commun.* 2019, *1*, 1–15. https://doi.org/10.1088/2515-7620/ab2dec.
16. Petersen, S.O. Symposium review: Greenhouse gas emissions from liquid dairy manure: Prediction and mitigation. *J. Dairy Sci.* 2018, *101*, 6642–6654. https://doi.org/10.3168/jds.2017-13301.
17. Owen, J.J. and Silver, W.L. Greenhouse gas emissions from dairy manure management: a review of field-based studies. *Glob Chang Biol.* 2015, *21*, 550–565. https://doi.org/10.1111/gcb.12687.
18. Meyer, D., et al. *Characterize Physical and Chemical Properties of Manure in California Dairy Systems to Improve Greenhouse Gas Emission Estimates*; University of California, Davis: 2019; pp. 1–70.
19. Hilgert, J.E., et al. Methane Emissions from Livestock Slurry: Effects of Storage Temperature and Changes in Chemical Composition. *Sustainability* 2022, *14*, 1–19. https://doi.org/10.3390/su14169934.
20. Hanafiah, M.M., et al. Emissions of carbon dioxide and methane from dairy cattle manure. In Proceedings of the IOP Conference Series: Earth and Environmental Science, 2021.
21. Lara, S., et al. Senate Bill No. 1383. 2016.

22. EPA. Compendium of State Approaches for Manure Management. Available online: https://www.epa.gov/npdes/compendium-state-approaches-manure-management (accessed on July 26).
23. UMNE. Manure application methods and nitrogen losses. Available online: https://extension.umn.edu/manure-management/manure-application-methods-and-nitrogen-losses (accessed on July 26).
24. Miller, D.J., et al. Cropland soil nitrogen oxide emissions vary with dairy manureincorporation methods. *AGE* 2023, *7*, 1–13.
25. Harmon, J.D., et al. *Land Application–Application Methods, Application: used for during feld application of manure to reduce odors and emissions*; Iowa State University: 2014; pp. 1–3.
26. Keena, M.A. *Composting Animal Manures: A guide to the process and management of animal manure compost*; NDSU Extension: 2022; pp. 1–8.
27. EPA. Practices to Reduce Methane Emissions from Livestock Manure Management. Available online: https://www.epa.gov/agstar/practices-reduce-methane-emissions-livestock-manure-management#two.
28. ES, F. Opportunities to Reduce GHG Emissions through Manure Management. In *Environmental Stewardship Continuous Improvement Reference Manual 2017*; Farm ES: 2017; pp. 58–71.
29. IPCC. *Sixth Assessment Report: 2019 Refinement to the 2006 IPCC Guidelines for National Greenhouse Gas Inventories*; IPCC: 2021.
30. Tian, H., et al. The terrestrial biosphere as a net source of greenhouse gases to the atmosphere. *Nature* 2016, *531*, 225–228. https://doi.org/10.1038/nature16946.
31. McSweeney, C.S. and Mackie, R.I. *Improving Rumen Function*; McSweeney, C.S.M., R.I., Ed.; Burleigh Dodds Science Publishing: 2020.
32. Huws, S.A., et al. Addressing Global Ruminant Agricultural Challenges Through Understanding the Rumen Microbiome: Past, Present, and Future. *Front. Microbiol.* 2018, *9*, 2161. https://doi.org/10.3389/fmicb.2018.02161.
33. Russell, J.P. Rumen. *Encyclopedia of Microbiology* 2009, 163–174. https://doi.org/10.1016/B978-012373944-5.00061-4.
34. Cain, M., et al. Improved calculation of warming-equivalent emissions for short-lived climate pollutants. *Nature* 2019, *2*.
35. Lynch, J., et al. Agriculture's Contribution to Climate Change and Role in Mitigation Is Distinct From Predominantly Fossil CO2-Emitting Sectors. *Front. Sustain. Food Syst.* 2021, *4*, 1–9. https://doi.org/10.3389/fsufs.2020.518039.
36. Hill, J., et al. Measuring Methane Production from Ruminants. *Trends Biotechnol.* 2016, *34*, 26–35. https://doi.org/10.1016/j.tibtech.2015.10.004.
37. Johnston, C.A., et al. Carbon cycling in soil. *Front. Ecol. Environ.* 2004, *2*, 522–528. https://doi.org/10.1890/1540-9295(2004)002[0522:CCIS]2.0.CO;2.
38. Prather, M.J., et al. Measuring and modeling the lifetime of nitrous oxide including its variability. *J. Geophys. Res. Atmos.* 2015, *120*, 5693–5705. https://doi.org/10.1002/2015JD023267.
39. Park, S.H., et al. Urease and nitrification inhibitors with pig slurry effects on ammonia and nitrous oxide emissions, nitrate leaching, and nitrogen use efficiency in perennial ryegrass sward. *Anim. Biosci.* 2021, *34*, 2023–2033. https://doi.org/10.5713/ab.21.0046
40. Fowler, D., et al. The global nitrogen cycle in the twenty-first century. *Philos. Trans. R. Soc. B* 2013, 1–13.

41. NASEM. *Nutrient Requirements of Dairy Cattle: Eighth Revised Edition*, 8 ed.; The National Academies Press: 2021.
42. Sabba, F., et al. Mechanisms of nitrous oxide (N2O) formation and reduction in denitrifying biofilms. *Biotechnol. Bioeng.* 2017, *112*. https://doi.org/10.1002/bit.26399.
43. Khairunisa, B.H., et al. Nitrogen transformation processes catalyzed by manure microbiomes in earthen pit and concrete storages on commercial dairy farms. *Environ. Microbiome* 2023, *18*, 32. https://doi.org/10.1186/s40793-023-00483-z.
44. Sokolov, V.K., et al. Dairy manure acidification reduces CH4 emissions over short and long-term. *Environ. Technol.* 2021, *42*, 2797–2804. https://doi.org/10.1080/09593330.2020.1714744.
45. Van Horn, H.H., et al. Components of Dairy Manure Management Systems. *J. Dairy Sci.* 1994, *77*, 2008–2030.
46. Aguirre-Villegas, H.A. and Larson, R.A. Evaluating greenhouse gas emissions from dairy manure management practices using survey data and lifecycle tools. *J. Cleaner Prod.* 2017, *143*, 169–179. https://doi.org/http://dx.doi.org/10.1016/j.jclepro.2016.12.133.
47. Hristov, A.N., et al. Review: Ammonia emissions from dairy farms and beef feedlots. *Can. J. Anim. Sci.* 2011, *91*, 1–35. https://doi.org/10.4141/CJAS10034.
48. Sutton, M.A., et al., Eds.; *Atmospheric Ammonia*; 2009: Springer Dordrecht; p. 464.
49. Barth, C., et al. Agricultural Waste Characteristics. In *Agricultural Waste Management Field Handbook*; USDA NRCS: 2008.
50. Varma, V.S., et al. Dairy and swine manure management - Challenges and perspectives for sustainable treatment technology. *Sci. Total Environ.* 2021, *778*, 1–18. https://doi.org/10.1016/j.scitotenv.2021.146319.
51. Jiang, Y., et al. Volatile organic compounds conversion pathways and odor gas emission characteristics in chicken manure composting process. *Front. Ecol. Evol.* 2023, *11*, 1–12. https://doi.org/10.3389/fevo.2023.1192132.
52. Leytem, A.B., et al. Methane emissions from dairy lagoons in the western United States. *J. Dairy Sci.* 2017, *100*, 6785–6803. https://doi.org/10.3168/jds.2017-12777.
53. Innovation, U.C.f.D. 2020 Dairy Sustainability Report. Available online: https://2020report.usdairy.com/images/IC_US_Dairy-SustainabilityReport2020.pdf.
54. Font-Palma, C. Methods for the treatment of cattle manure–a review. *C* 2019, *5*, 1–20. https://doi.org/https:doi.org/10.3390/c5020027.
55. Oliver, J. and Ray, L. Manure storage impermeable cover and flare systems – Potential climate benefits and considerations. *ProDairy* 2023, pp. 1–2.
56. Wilkie, A.C. Anaerobic Digestion of Dairy Manure: Design and Process Considerations. In Proceedings of the Dairy Manure Management Conference, Ithaca, NY, 2005; pp. 301–312.
57. Kupper, T., et al. Ammonia and greenhouse gas emissions from slurry storage -- A review. *Agric. Ecosyst. Environ.* 2020, *300*, 1–18. https://doi.org/10.1016/j.agee.2020.106963.
58. Li, Y., et al. Enhancement of methane production in anaerobic digestion process: a review. *Appl. Energy* 2019, *24*, 120–137. https://doi.org/10.1016/j.apenergy.2019.01.243.
59. Reis, M., et al. 6.51 – Mixed culture processes for polyhydroxyalkanoate production from agro-industrial surplus/wastes as feedstocks. In *Comprehensive Biotechnology*; Academic Press: 2011; Volume 6, pp. 669–683.

60. D'Silva, T.C., et al. Enhancing methane production in anaerobic digestion through hydrogen assisted pathways – A state-of-the-art review. *Renew. Sust. Energ. Rev.* 2021, *151*. https://doi.org/10.1016/j.rser.2021.111536.
61. Cárdenas, A., et al. Methane emissions from the storage of liquid dairy manure: Influences of season, temperature and storage duration. *Waste Manag.* 2021, *121*, 392–402. https://doi.org/10.1016/j.wasman.2020.12.026.
62. Marklein, A.R., et al. Facility-scale inventory of dairy methane emissions in California: implications for mitigation. *Earth Syst. Sci. Data* 2021, *13*, 1151–1166. https://doi.org/10.5194/essd-13-1151-2021.
63. Kassem, N., et al. Combining anaerobic digestion and hydrothermal liquefaction in the conversion of dairy waste into energy: A techno economic model for New York State. *Waste Manag* 2020, *103*, 228–239. https://doi.org/10.1016/j.wasman.2019.12.029.
64. Zhang, C., et al. Low-cost composited accelerants for anaerobic digestion of dairy manure: Focusing on methane yield, digestate utilization, and energy evaluation. *Biosource Techno* 2018, *263*, 517–524. https://doi.org/10.1016/j.biortech.2018.05.042.
65. Bella, K. and Venkateswara Rao, P. Anaerobic digestion of dairy wastewater: effect of different parameters and co-digestion options–a review. *Biomass Conv Bioref* 2021, *13*, 2527–2552. https://doi.org/10/1007/s13399-020-01247-2.
66. CDFA. Alternative Manure Management Program. Available online: https://www.cdfa.ca.gov/oefi/AMMP/.
67. Wightman, J.L. and Woodbury, P.B. New York Dairy Manure Management Greenhouse Gas Emissions and Mitigation Costs (1992–2022). *J. Environ. Qual.* 2016, *41*, 266–275. https://doi.org/10.2134/jeq2014,06.0269.
68. Cowley, C. and Brorsen, B.W. Anaerobic Digester Production and Cost Functions. *Ecol. Econ.* 2018, *152*, 347–357. https://doi.org/10.1016/j.ecolecon.2018.06.013.
69. EPA. Anaerobic Digestion on Dairy Farms. Available online: https://www.epa.gov/agstar/anaerobic-digestion-dairy-farms.
70. AgSTAR. Funding On-Farm Anaerobic Digestion. Available online: https://epa.gov/sites/default/files/2014-12/documents/funding_digestion.pdf.
71. AgSTAR. Project Planning and Financing. Available online: https://www.epa.gov/agstar/project-planning-and-financing (accessed on May 5 2024).
72. CARB. *Compliance Offset Protocol Livestock Projects: Capturing and Destroying Methane from Manure Management Systems*; California Air Resources Board: 2014; pp. 1–60.
73. Liao, P.H., et al. Effect of liquid-solid separation on biogas production from dairy manure. *Ener. Agric.* 1984, *3*, 61–69. https://doi.org/10.1016/0167-5826(84)90005-6.
74. Chadwick, D., et al. Manure management: Implications for greenhouse gas emissions. *Anim. Feed Sci. Technol.* 2011, *166*, 514–531. https://doi.org/10.1016/j.anifeedsci.2011.04.036.
75. Qu, Q. and Zhang, K. Effects of pH, total solids, temperature and storage duration on gas emissions from slurry storage: a systematic review. *Atmosphere* 2021, *12*, 1–16. https://doi.org/10.3390/atmos12091156.
76. Ning, J., et al. Dicyandiamide has more inhibitory activities on nitrification than thiosulfate. *PLoS One* 2018, *13*. https://doi.org/10.1371%2Fjournal.pone.0200598.

77. Lin, S., et al. Timing of Manure Injection and Nitrification Inhibitors Impacts on Nitrous Oxide Emissions and Nitrogen Transformations in a Barley Crop. *Soil Sci. Soc. Am. J.* 2018, *81*, 1595–1605. https://doi.org/10.2136/sssaj2017.03.0093.
78. Peterson, C.B., et al. Effects of SOP lagoon additive on gaseous emissions from stored liquid dairy manure. *Sustainability* 2020, *12*, 1–17.
79. Chiodini, M.E., et al. Real-scale study on methane and carbon dioxide emission reduction from dairy liquid manure with the commerical additive SOP LAGOON. *Sustainability* 2023, *15*, 1–13. https://doi.org/10.3390/su15031803.
80. Holtkamp, F., et al. Calcium cyanamide reduces methane and other trace gases during long-term storage of dairy cattle and fattening pig slurry. *Waste Manag.* 2023, *161*, 61–71. https://doi.org/10.1016/j.wasman.2023.02.018.
81. Rocha, A.S., et al. Effect of Eminex® on greenhouse gas and ammonia emissions from dairy slurry and lagoon wastewater. *Sustainability* 2024, *16*, 5778. https://doi.org/10.3390/su16135778.
82. Beck, M.R., et al. U.S. manure methane emissions represent a greater contributor to implied climate warming than enteric methane emissions using the global warming potential* methodology. *Front. Sustain. Food Syst.* 2023, *7*, Article 1209541. https://doi.org/10.3389/fsufs.2023.1209541.

Chapter 11

Optimizing manure and slurry application from intensive dairy farming operations

Shabtai Bittman and Derek Hunt, Agassiz Research and Development Centre - Agriculture and Agrifood Canada, Canada; Kirsten Hannam and Barbara Cade-Menun, Summerland Research and Development Centre - Agriculture and Agrifood Canada, Canada

1 Introduction

An important benefit of manure, long known and exploited, is the concentration of nutrients by confined animals that can be collected and spread onto cropland to improve crop production. Integrated farming of livestock and crops was widely practiced long before the benefits of transferring manure nutrients to crops were scientifically understood. The lack of adequate manure for expanding agriculture led in the late 1800s to the collection and importation of large quantities of bird guano from vast deposits in South America, which ultimately motivated the invention of lower-cost, more easily accessible manufactured nitrogen (N) fertilizers using the Haber-Bosch process, and the discovery of mined phosphorus (P), potassium (K), and other nutrients for use as fertilizer. As fertilizer availability increased, the logistical

http://dx.doi.org/10.19103/AS.2025.0157.23

disadvantages of manure management became more widely recognized than its value as a soil amendment, and manure was relegated, on many farms, as an unavoidable waste to be expediently discarded. There has often been insufficient regard to the possible benefits of manure as a soil amendment or to manure management practices that could mitigate widespread contamination of surface (esp. P, pathogens) and ground water (esp. nitrate (NO_3)) and of soil nutrient overloading. The impacts of manure on air quality, via atmospheric deposition of nutrients into oligotrophic ecosystems, and of fine particulate formation on human health are also often overlooked.

The growing interest in strategic manure use has stemmed less from an increasing appreciation of its intrinsic value and economic benefits than from growing concern by the public, and mounting scientific evidence of environmental degradation on farmland and surrounding ecosystems. These have, in turn, led to regulations such as application limits under the nitrate vulnerable zones (NVZs) and ammonia (NH_3) emission limits (National Emission Ceilings) directives in the EU, and the P indices imposed in several US jurisdictions. There is an inclination to overapply manure because it is free and more unwieldy to transport and apply than purchased fertilizer. The current coincidence of anxiety by farmers about loss of soil organic matter (SOM) and the need for preservation of soil health and by the public about our changing climate has positioned increased soil organic carbon (SOC) and the use of manure carbon (C) as possible redeemers of society's collective bad behaviour. From small to large, many farmers now see increased SOM as a tool to elevate crop productivity and improve land value and to serve as a possible revenue stream under remunerative C sequestration schemes. Counterposed are the inevitable emissions of nitrous oxide (N_2O) from applied manure, both directly after application and indirectly as secondary emissions from deposited gaseous NH_3, and particulate ammonium (NH_4) or leached NO_3, plus nutrient loss in run-off resulting in water quality degradation. The loss of methane (CH_4) from stored slurry manure, which contributes both to global warming and to tropospheric ozone formation, must also be considered (Sampedro et al. 2023).

Emission rates of NH_3 are determined by conditions at the manure–atmosphere interface, which are difficult to measure and are rarely done, especially in *in situ* studies. Emitted NH_3-N is mixed in the atmosphere and transported by air (Asman et al. 1998). Some gaseous NH_3 is deposited on nearby soil or vegetation depending on concentration gradients (referred to as dry deposition), but the majority is transported for one to many kilometers, mostly as secondary fine particulates and aerosols formed in the atmosphere by reaction of NH_3 with acid gases, such as NOx, SOx, Cl, and organic acids (Bittman et al. 2014b) (Fig. 1). These N-rich fine (<2.5 μm) particulates are deleterious to the health of humans (and animals) because they are capable of penetrating deep into lungs and causing cardio-pulmonary illnesses (Wyer

Figure 1 Air pollution caused by fine particulates containing ammonium from agricultural sources in south-coastal BC, Canada. Panel on the left depicts high-visibility wintertime conditions due to low ammonia emissions. The panel on the right shows the same view in mid-summer, where ammonia-based fine particulates reduced visibility (Barthelmie and Pryor 1998). Photos by S. Bittman, Agriculture and Agri-Food Canada.

et al. 2022). When these particles are deposited (wet deposition), usually at a distance from the source, they can acidify soils and harm oligotrophic ecosystems such as alpine meadows, lakes, and bogs (Clark et al. 2021).

Agriculture is the primary source of atmospheric NH_3 in industrial countries with large agricultural sectors and, unlike other air pollutants, NH_3 has been trending higher in recent decades due to the expansion of livestock production and increased application of NH_3-based fertilizers (Wyer et al. 2022). Emissions of NH_3 must be reported by all UNECE (United Nations Economic Commission for Europe) parties in 'ground up' emission inventories under the Convention for LongRange Transport of Air Pollution (CLRTAP). Methods for mitigation of NH_3 losses from agricultural sources, including land spreading of manure, in consideration of all reactive N species, are suggested for the UNECE in an official Guidance Document (Sutton et al. 2022), and global guidance is being developed under the International Nitrogen Management System (INMS; https://www.inms.international/).

Since manure sources have concentrated and expanded in past decades, often in ever larger animal facilities and livestock-intensive regions, it has become increasingly important and perhaps daunting to manage manure strategically (Kleinman et al. 2022). Whereas there are aspirations now towards processing manure so it can be transported and returned to distant cropland and remote cropping regions, and towards production of commercial soil enhancers such as compost, it remains the simplest approach for many farmers to use manure on their own operations to replace fertilizer and improve/maintain soil productivity and health. But using manure to replace fertilizer has proven to be challenging, all the more so in crowded landscapes with growing populations, especially around cities, and where there is competition for land, space, and water and ever-increasing concentration of nutrients. In many countries, the need for intensive agricultural production, including dairy, in such regions is forcing incentive and invention (Bittman and Hunt 2024).

Published by Burleigh Dodds Science Publishing Limited, 2026.

The objective of this chapter is to describe progress towards the use of manure for replacing commercial nutrients and enhancing SOM, in the face of longstanding and modern obstacles that include logistics and unpredictability, and avoiding environmental harm. The intention is also to consider novel approaches that may be needed for further advances. For this chapter, manure consists primarily of excreta (faeces, urine) and bedding materials, which are handled either as a liquid (called slurry) or solid (referred to as farmyard manure or FYM).

2 Factors affecting dairy slurry composition

Animal manures and slurries contain multiple elements in a variety of chemical forms. Common elements in dairy manure and slurry include N, P, C, calcium, K, sulphur, and zinc; however, this chapter will focus on N, P, and C. The forms and concentrations of these elements can be altered by a variety of factors, including animal diets and manure handling, processing, and storage (Ouatahar et al. 2021). For example, the C content of manure from 22 conventional and organic dairy farms in Texas ranged from 12% to 45% C (dry matter basis), depending on the type of forage provided (hay, pasture, haylage, maize silage), the use of bedding materials, storage type (e.g. stacking pad, pit, roofed storage, lagoon, windrow), and physical composition (He and Waldrip 2015). Slurry from dairy farms in coastal British Columbia (BC), Canada, averaged 2.5% C on a wet basis, with a C:N ratio of ~9.8.

Dairy manure C is typically dominated by aromatic (e.g. from lignin and tannins) and aliphatic (long chain, C-rich) biopolymers (e.g. waxes, suberins, and cutins) as shown by C-13 nuclear magnetic resonance (^{13}C NMR) spectroscopy (He and Waldrip 2015). This is because the more soluble C compounds (amino acids, simple sugars, and low-molecular weight organic acids) contained in feed are preferentially consumed during cattle digestion, leaving behind more complex, less easily degraded C compounds (He et al. 2015, He et al. 2009). Composting of manure leads to further microbial consumption of degradable carbohydrates and to accumulation of less- degradable proteins and phenolic and carboxylic forms of C in the compost (Liang et al. 1996).

On a fresh basis, dairy slurry usually varies from about 5–7% dry matter with about 0.25% (2.5 g kg^{-1}) total N; 0.13% TAN (total ammoniacal N = NH_4-N plus NH_3-N); and 0.05% (0.49 g kg^{-1}) total P. Typically, the TAN to total N ratio is 60–70% (higher with high protein diets), and the total N:total P ratio is 5:1 to 6:1 (e.g. Bittman et al. 2011). This means that a 10 000-litre tank of slurry contains just 25 kg of total N (of which 15-18 kg is readily available TAN) and about 5 kg of P (only a portion of which may be plant available), illustrating a key logistical challenge of using manure nutrients to replace chemical fertilizers. With a pH of around 6.8–7.2, there is a significant tendency for the NH_4-NH_3

Published by Burleigh Dodds Science Publishing Limited, 2026.

equilibrium to shift towards the production of gaseous NH_3; this promotes NH_3 losses from manure exposed to the atmosphere, especially under hot, windy conditions (Sommer et al. 1991). Thus, the C:N ratio of fresh dairy slurry is lower than farmyard manure (FYM) due to loss of volatile NH_3 in storage.

Nitrogen is excreted in urine as urea, which is rapidly hydrolysed to NH_3 and NH_4 (TAN) by the ubiquitous urease enzyme excreted in faeces. Separating urine from faeces at the time of excretion will preserve urea and slow the formation of TAN. This has been attempted, but is difficult to accomplish in housing, although this separation occurs naturally on pastures where NH_3 emissions are typically much lower than from housing. The organic N in slurry is mainly from excreted faeces, consisting largely of undigested protein bound to undigestible fibre and lignin and from other organic materials such as discarded or soiled feed and (low N) bedding materials. Farmers may also occasionally recycle spoiled feed into manure storages.

Ideally, manure application rates are informed by the content of TAN, N, and P. Unlike fertilizer, however, manure does not come with a ready analysis and because sampling stored manure is often very difficult (Ni and Lim 2022), testing is typically done infrequently. Although sampling from a manure pit after agitation or from the manure tanker itself is simpler, awaiting lab results would inevitably delay spreading. Furthermore, inaccuracies can arise due to inter-lab variability in analytical methods and the problem of subsampling heterogeneous material. Instead, application rates can be determined using *in situ* tests for TAN; test results can be validated with laboratory testing post-spreading, although this is not commonly done. A long established *in situ* tester for TAN concentration in slurry is the Nova/Agros meter (AGROS NOVA meter for liquid manure (novanna.co.uk)), which converts NH_4 from a manure sample placed in a sealed container to N_2, which is then measured using a pressure gauge attached to the container (Van Kessel and Reeves 2000). Concentrations of total N and P can then be estimated from previous laboratory-based manure analyses, because ratios of TAN:Total-N and Total N:P in manure slurry are relatively stable on farms under consistent feeding and manure management practices. Real-time analyses using selective ion sensors and near infra-red spectrometry are under development but not yet widely done (Reeves 2007; Feng et al. 2022; Piepel and Olfs 2023). These tools are used for liquid manures (slurry) but not solid manures, which contain less TAN and which are also especially difficult to sample before spreading.

Importantly, while total P in slurry can be estimated from ratios of total N:P, total P is not always a good predictor of plant-available phosphate. A range of inorganic and organic (bonded to C groups) P compounds can be found in manures and slurries. Inorganic P compounds include phosphate (HPO_4^{2-} or $H_2PO_4^-$ at the pH range of most samples), polyphosphates (chains of three or more phosphate groups linked by anhydrous bonds), and pyrophosphate (polyphosphates with two phosphate groups). Phosphates in dairy manure and slurry can originate directly from dietary mineral additives or from the hydrolysis of organic P compounds in feed during digestion. Pyrophosphate and polyphosphate in manures, including slurry, may come from undigested dietary plant material or from excreted rumen microbes, for which they are P storage compounds (Toor et al. 2005). Organic P compounds are grouped by bonding with C groups into orthophosphate monoesters [one C group per phosphate, linked by oxygen (O); P-O-C], orthophosphate diesters (two C-groups per phosphate, C-O-P-O-C), and phosphonates (direct C-P bond, without O). Orthophosphate monoesters include phytate (*myo*-inositol hexaphosphate), a P storage compound in grain and plant material (e.g. maize), and simple sugar phosphate such as glucose-6-phosphate. Orthophosphate diesters include so-called 'biological' P compounds, such as RNA and DNA, and phospholipids and lipoteichoic acids in cell walls of plant materials from feed and in rumen microbes. There is a range of phosphonate compounds, but the compound most detected in slurry and manure is 2-aminoethylphosphonic acid (2-AEP, also called ciliatine), from the cilia of rumen protozoa (Toor et al. 2005).

3 Dairy manure production in Canada

Dairy production in Canada is year-round (aseasonal) by animals that are totally or largely confined because of prolonged winters and consistent year-round milk demand. This is in marked contrast to seasonal beef cattle production in Canada, where calving occurs mainly in spring. Beef cattle are grazed in summer; in winter, they are housed (winter dry lots) or sometimes grazed with bale-feeding (Kelln et al. 2012; Chen et al. 2017). Eventually, cattle are moved to large feedlots for finishing for market. In countries where dairy animals are kept largely on pastures, such as Ireland and New Zealand, milk production is more seasonal than in Canada, with most milk produced on fresh grass in spring and much less in winter. From the time of excretion to land application, emission of NH_3 from manure is a critical loss pathway for N from dairy farms. Based on unit food protein produced, emissions from dairy production are similar to poultry (meat) but much lower than beef or pig production (Bittman et al. 2017). In Canada, 85% of large dairy farms with an average of 195 lactating cows handle manure as slurry compared to only 64% of farms averaging 40 lactating cows. Dairy farms in Canada are often family-owned operations with most milk cows kept on farms with less than 75 milking cows (VanderZaag et al. 2023). There has been a trend towards increasing farm size and milk production (Sheppard et al. 2011a), while total cow numbers in Canada have decreased over the past 20 years, thanks to increasing milk production per lactation, achieved by improved genetics and feeding practices (Bach et al. 2020).

Published by Burleigh Dodds Science Publishing Limited, 2026.

Precise feeding is managed by frequent feed testing and adjusting rations for different stages of growth and periods of lactation, with particular attention to digestible fibre and protein attributes and to P concentration, in order to maximize milk production at low feeding costs. Currently, N use efficiency (NUE) (N exported as milk and meat/N consumed by cattle) for dairy cows is about 25–30% (Pogue et al. 2025); the remainder of the N is excreted in faeces (40%) and urine (60%). As mentioned above, the animal diet will directly affect both the forms and concentrations of P compounds in manure. Dietary P is essential for milk synthesis, and cows with inadequate dietary P may have reduced milk production or adverse health effects, e.g. due to mobilization of P from bone (Wu et al. 2000). The P concentration of forages and grains can vary widely, so commercial P supplements are often added to the diets of lactating cows to ensure sufficient dietary P (Toor et al. 2005; Duplessis et al. 2021). This, in turn, will affect manure P concentration, with higher total P concentrations in manure from animals with higher dietary P (Toor et al. 2005). Dietary P will also alter manure P forms; e.g. manure from animals with higher dietary P from supplementation contains more phosphate (Toor et al. 2005; McDowell et al. 2008). However, diets high in phytate from grains do not result in manure with higher phytate concentrations likely due to hydrolysis of phytate to phosphate during digestion in the rumen (McDowell et al. 2008).

In Canada and elsewhere, there has been a shift away from tie-stall barns with solid manure to free-stall and loose housing barns where manure is handled as slurry that is collected several times per day by scraping or washing. In 2006, 51% of dairy farms were handling manure as slurry, and this proportion has since increased with farm size (Sheppard et al. 2011a). About 21% of TAN is emitted as NH_3 from dairy housing and 5.5% from storages, which is a significant loss of available N that effectively decreases the N:P ratio of applied slurry (Sheppard et al. 2011b). Many measures are used to minimize N losses (especially NH_3) from animal facilities. The slurry manure is stored over winter for as long as 6–8 months in concrete or steel tanks or large (usually lined) lagoons. Unlike in European countries where there is concern about limiting NH_3 emissions into the atmosphere, few storage facilities in Canada have covers, although some storages are under roofs to keep out precipitation and ensure sufficient capacity, which is particularly important in rainy years. Despite lack of covers, NH_3 emission from dairy storages are relatively small because the slurry in storage often forms a surface crust and because of cold winter-time temperatures (Sheppard et al. 2011b). Emission of NH_3 from manure, from the time of excretion to land application, is a critical loss pathway for dairy farm N. Of the excreted N by dairy animals, 27% is lost by land spreading manure, which is about 51% of all NH_3 losses on dairy operations.

The bedding used in dairy barns is mainly straw in grain-growing regions and wood chips or sawdust in regions where those products are abundant.

On some farms, sand bedding is used to optimize animal health (Singh et al. 2020). Much of the sand is recovered from manure by filtration prior to re-use, but sand is hard on manure handling equipment, which often requires more maintenance. Some farmers also recover organic solids from slurry for use as bedding after solid/liquid separation, followed by processing in a tumbling drip composter (Ackerman et al. 2018). Devices such as the BeddingMaster (https://mavasol.com/en/solutions/manure-management/products/beddingmaster) can be used to facilitate this process. Surplus recovered bedding is supplied to nearby dairy farms or to horticultural operations for use as mulch (Franzluebbers et al. 2021) or as a soil amendment; with its high C:N ratio (31.2), recovered bedding can be preferable to raw manures, such as poultry litter (C:N of 6.3) or even cured compost (C:N of 22.9), because of lower N leaching risk (Zhang et al. 2019). The choice of bedding materials can also substantially alter manure P forms and concentrations, as can subsequent manure treatment, such as composting. For example, straw and wood products can add phosphate and phytate (Noack et al. 2012; Cade-Menun et al. 2015), and separated solids used as bedding material can have much higher total P concentrations and lower N:P than the original dairy manures used to produce the materials (Ackerman et al. 2018). Composting can alter manure P composition compared to faeces, increasing both phosphate concentration and organic P species associated with the microbes involved in decomposition (Toor et al. 2005), although Hansen et al. (2004) suggested that the P forms in solid manure and lagoon manure were similar.

4 Using dairy manure and slurry for crop fertilization

4.1 Applying manure/slurry to fields

A primary goal of applying manure on cropland is to replace and perhaps obviate the need for chemical fertilizers. The logistics of spreading both slurry and FYM are complicated because of their low nutrient concentrations and the large quantities of material that must be handled and transported (Harrigan 2010). The drier and lighter FYM is often broadcast onto soils, with incorporation via tillage helping to increase the availability of nutrients to crops and minimize runoff losses (Kumaragamage and Akinremi 2018). However, tillage cannot be conducted on perennial field crops or under no-till systems. Manure that is not fully incorporated into the soil leaves high concentrations of phosphate and phytate at the soil surface (Cade-Menun et al. 2015), rendering it prone to loss from the field and contamination of surface waters, e.g., due to overland flow and even wind transport. There are some injection tools for soil application of poultry litter, but these methods are not widely used and are unsuitable for clumpier cattle FYM (Kulesza et al. 2016; Landry et al. 2011).

Slurry applications require hauling massive quantities onto fields over a short time window. All manure tanks are heavy, particularly rapid, self-loading vacuum tanks. As a consequence, soil compaction can be a common problem caused by slurry manure applications, especially on wet, poorly drained and/or fine-textured soils. Tramlines (travel on set tracks) used for applying fertilizer and pesticides on arable crops are rarely used for manure spreading for logistical reasons (see Mederle and Bernhardt 2017). Flotation tires can reduce the risk of soil compaction, but they are expensive and not suitable if extensive road travel is required, although some tractors can extend tire wear by adjusting tire pressure between road transport and field travel. Other measures to facilitate the spreading of slurry while preventing soil compaction include umbilical (dragline) systems, where long flexible hoses are used to transfer manure from stores to fields, obviating the need for heavy tanks. Only very large farms can afford this expensive equipment; smaller operations may engage contractors, where available. Farms with large volumes of manure (e.g. large operations with shallow, uncovered lagoons in high rainfall areas) may install underground pipes to rapidly move manure from stores to fields. Others collect manure in temporary flexible storage placed near crop fields to hasten spreading during critical application windows. The ability to spread manure rapidly is important when labour is limited and/or expensive and the application time is constrained by short seasons or by inclement weather.

The most rapid and least trouble-prone method of spreading slurry is by broadcasting (e.g. splash-plate applicator), especially on small family farms with limited resources. About 80% of dairy farms in Canada spread manure by broadcasting (in 2005). However, slurry applied via broadcasting is prone to substantial N losses by NH_3 volatilization and associated odours, which may restrict applications near residential areas. Furthermore, broadcast manure tends to be applied unevenly, particularly in windy conditions, and is more likely to contaminate protected areas or waterways due to drift. Some European countries (The Netherlands, Denmark) have banned slurry broadcasting to reduce NH_3 emissions into the atmosphere.

4.2 Determining manure application rates

Like fertilizer, manure must be applied at crop-appropriate rates, which is aided by *in situ* testing methods (see above). However, the availability of manure nutrients, especially N, is less predictable and less consistent than fertilizer, because of variable mineralization rates of the organic N fraction and, especially, the propensity for substantial NH_3 loss by volatilization in the hours and days after application. Manure and slurry application rates are most often based on crop N requirements, which will oversupply P because the ratio of N:P in manures and slurries does not match crop nutrient requirements

(Kumaragamage and Akinremi 2018). This ratio can be improved by using a separated dairy slurry rather than a whole dairy slurry, which contains a lower P:N relative to whole slurry (Table 1). Separation efficiency varies greatly with method and manure type (Möller et al. 2002).

Loss of NH_3 via volatilization is highly variable and can depend on many factors, including weather (temperature, rain, wind, relative humidity, sunshine), soil properties (porosity, water content, cation exchange capacity, pH), and manure composition (NH_3 and dry matter concentration, pH, viscosity) (Sommer et al. 2003). As well, crop residue or a living crop canopy may, paradoxically, affect emissions in opposite directions – on the one hand, emissions decrease when living plants or crop residues at the soil surface reduce airflow, incident solar radiation and soil surface temperatures, and facilitate absorption of NH_3 into leaves (Sutton et al. 1993); on the other hand, emissions increase when living plants or crop residue prevent manure from reaching the soil surface where NH_4 ions can be adsorbed.

Various low-emission applicators, some with a span >10 m for rapid application, have been developed to conserve manure N and reduce deleterious emissions. Two applicator types are particularly rapid and require little additional power: trailing hoses, which hang loosely at or near the ground surface, and trailing shoes, which float directly on the soil surface. Applicators that penetrate the soil and require more power include shallow injection with open slots (<10 cm deep), and deep- or closed-slot injection (>15 cm deep), where no manure is left exposed to air. In each case, the manure stream from the tank is divided with a chopper-distributor into multiple smaller streams that feed into individual emitters spaced across a draw bar. The chopper is designed to divide the flow evenly (ideally CV <10%) and prevent blockages in the smaller hoses, which is critical for efficient operation of low-emission applicators. A less refined tool with no active divider, consisting of a simple manifold assembled with PVC pipes, can be home built; pipes are kept rather large (approx. internal diameter of 7–10 cm) to avoid blockages and, hence,

Table 1 Properties of whole dairy slurry and liquid fraction separated by settling in a two-stage lagoon system, averaged over 6 years, in coastal BC, Canada

	Whole slurry	Liquid fraction
Dry matter, %	5.9	1.6
TAN, g kg^{-1}	1.31	0.95
Total-N, g kg^{-1}	2.54	1.45
TAN:Total-N	0.52	0.66
Total P, g kg^{-1}	0.49	0.15
Total N:P	5.18	9.67
pH	7.0	7.6

are spaced further apart. Even less ambitious, the pipes may be designed to terminate above the canopy and used as low-trajectory broadcasting devices.

4.3 Can soil-applied dairy manure replace commercial fertilizer?

To begin addressing this question, we compared the effect on grass yield and N uptake of conventionally applied (broadcast via splash plate) slurry, slurry applied by trailing shoe (also called a drag-shoe), and commercial fertilizer (Bittman et al. 1999). Drag-shoe application places surface bands of manure on the soil surface. The study was conducted over 3 years in a Mediterranean-type maritime climate on a stand of perennial tall fescue (*Festuca arundinacea* Scheb.). Figure 2 shows N uptake of grass after application of commercial fertilizer (curved lines) and slurry at equivalent rates of mineral N in spring, summer, and autumn. Nitrogen uptake response to broadcast slurry was variable, especially in the summer, but the response to drag-shoe applied slurry was consistently similar to the commercial fertilizer. Over the nine campaigns, total N recovery relative to the quantity of applied mineral N averaged 58% for fertilizer, 53% for drag-shoe applied slurry and 39% for broadcast applied slurry; thus, 14% of applied mineral N in slurry was conserved with low emission application relative to broadcast application. Of the total manure N applied, 27% and 20% was recovered after drag-shoe vs broadcast slurry application, respectively; the remainder was either lost to the environment or remained in soil. Low-emission application was superior but perhaps not economically attractive based on the value of N, especially if the farm has an abundance of manure. However, the added benefits of uniform application, less soil acidification than NH_4-based fertilizers, potentially less P loading, climate and NH_3 mitigation (less fertilizer consumption, less secondary N_2O), reduced odour (Lau et al. 2003), and better aesthetics (see Fig. 3) may encourage farmers to consider adoption. Although not measured here, NH_3 emission reductions of up to 60% are attributed, on average, to trailing shoe applicators (Webb et al. 2010; Misselbrook et al. 2022) due to reduced exposed area of the banded manure and to less slurry deposited on crop residue or stubble. Delayed application of fertilizer and slurry, which often occurs on farms due to weather or other factors, affected N uptake only in autumn when growing conditions decline.

The banded slurry gradually soaks into the soil, where the dissolved TAN is adsorbed. To facilitate infiltration of slurry into the soil, reduce NH_3 loss, and improve N recovery, we developed an applicator that bands slurry with trailing shoes directly over vertical pockets formed by rolling tines (20 cm deep at 19 cm × 47 cm spacing). These pockets can accommodate up to 26 t ha^{-1} of slurry, which is similar to that of slurry application via slot injection. While this is lower than typical slurry application rates in Canada (40–60 t ha^{-1}), the slight twist of

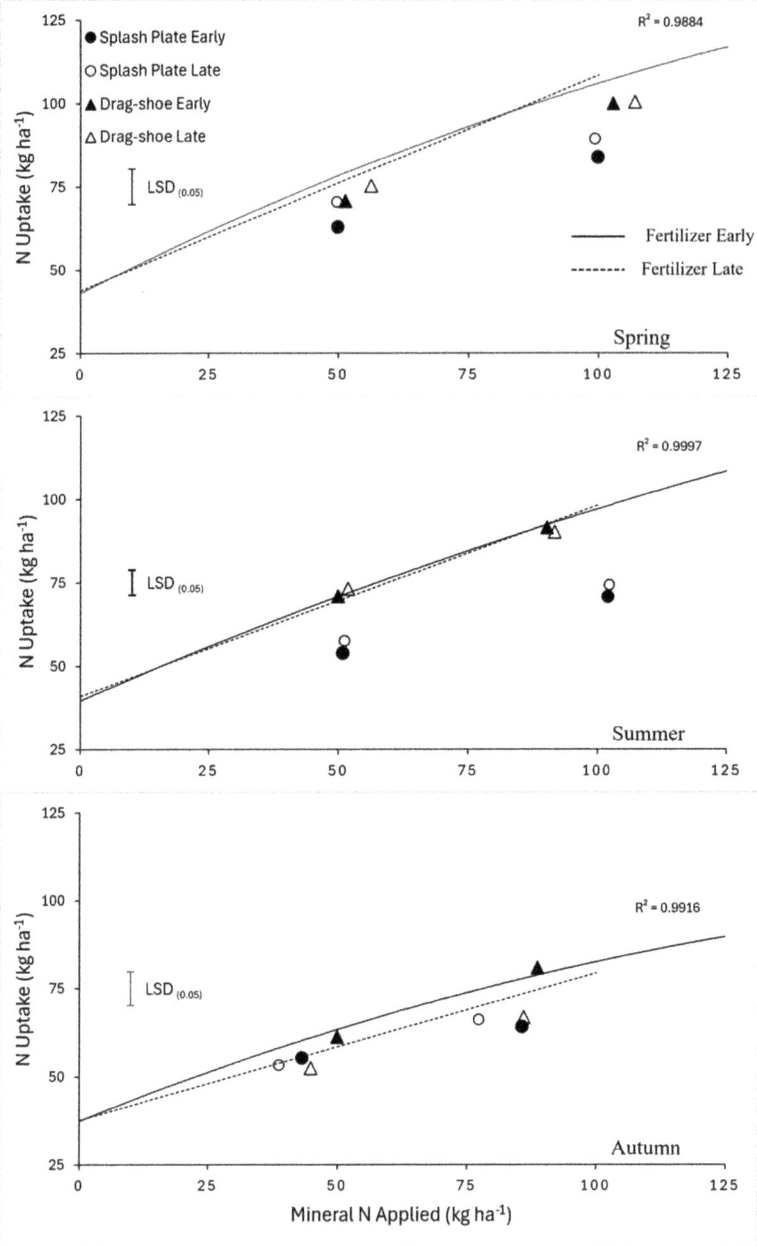

Figure 2 Nitrogen uptake (herbage yield x N concentration) by tall fescue as affected by NH_4-NO_3 fertilizer (lines) and dairy slurry spread with splash plate (circles) and drag-shoe applicators (triangles). Applications are either early (at beginning of growth) or late (7–10 days later) in spring, summer, and autumn. Data are means across 3 years (1994–1996). Source: Adapted from: Bittman et al. (1999).

Figure 3 Slurry applicator with surface banding over rolling tines in south-western BC, Canada. Photos by S. Bittman and D.E. Hunt. Agriculture and Agri-Food Canada.

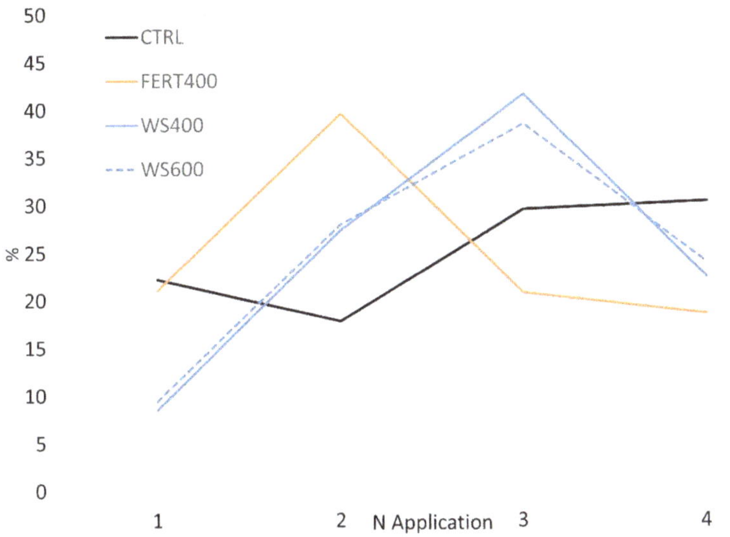

Figure 4 Relative loss of N_2O after applications of dairy slurry at low (WS400) and high (WS600) rates and NH_4-NO_3 fertilizer (FERT400) and Control (CTRL) on grass in April (1), June (2), August (3), and Sept. (4). Emissions are percent of annual, averaged over 5 years. Lowest emissions for slurry followed the first application due to cool soils. Emissions from chemical fertilizer occurred earlier than slurry due to the presence of NO_3 in the fertilizer but not in the slurry. Source: Adapted from: Hunt et al. (2019).

the tines is also designed to loosen the soil and further enhance infiltration of thinner slurry fractions with dissolved NH_4 (Fig. 3).

Field measurements using integrated horizontal flux or semi-closed chambers showed 46–48% less NH_3 emissions with the rolling tine applicator compared to broadcasting, although N uptake by grass was increased by only 14% (Bittman et al. 2005b). Dell et al. (2012), using the same commercial implement, reported NH_3 reductions of 30–70% over 3 years but no agronomic benefit over broadcasting. Similarly, Sherman et al. (2021) reported a 52% reduction in NH_3 emissions by application with rolling tines compared to broadcasting, which was less than reduction by shallow injection but greater than with banding; treatment effects on yield were inconsistent. Sherman et al. (2021) did report more N_2O emissions using rolling tines than broadcast manure but less than shallow injection, cautioning about environmental tradeoffs. Inconsistent results of the rolling tine relative to broadcasting may be related to higher manure dry matter or wet soil conditions that reduce the capacity for enhanced slurry infiltration.

Of note, the rolling tine can be used on stony and sloped land where it can reduce surface run-off of dissolved P and suspended solids after manure application relative to broadcasting (Table 2), especially in uncompacted and unsaturated soil (van Vliet et al. 2006; Johnson et al. 2011). In contrast to the rolling tine, shallow, open furrows by shallow injection may facilitate manure run-off on sloped fields (farmer observation). The rolling tines may also reduce preferential flow of slurry in soils with substantial macropores, thereby mitigating the risk of microbial and nutrient contamination of water bodies (Lapen et al. 2008).

Manure application timing can reduce emissions of NH_3 (and N_2O^- see below), but care is required to avoid environmental trade-offs. Strategic reduction in emissions by timing has been called Application Timing Management System and is recognized by the UNECE as a mitigation method for NH_3 with some caveats (Bittman et al. 2014b). For example, applying slurry in cool, moist weather (e.g. spring vs summer) can significantly reduce NH_3 emissions and improve N-uptake (Lalor et al. 2011; Fig. 1), but might also be associated with a higher risk of leaching and run-off. Emissions of N_2O could also be reduced by spreading manure in early spring (when soils are cool) and avoiding late spring and early summer applications when soils are warm and

Table 2 Effect of soil aeration with rolling tines in uncompacted soil (Holland Equipment Ltd., Norwich, ON) on runoff volumes, and loadings of suspended solids and dissolved reactive P after autumn application of dairy slurry in south coastal BC, Canada

	1998–1999		2001–2002	
	(high rain)		(low rain)	
Parameter	Control	Aeration	Control	Aeration
Runoff amount (L)	10 460	1990	3550	1890
Suspended solids load ($kg\ ha^{-1}$)	130	40	400	210
Dissolved reactive P load ($kg\ ha^{-1}$)	0.71	0.03	0.65	0.11

Data from the driest and wettest of four winters are shown in the table.
Source: Adapted from van Vliet et al. (2006).

moist (Fig. 4; Hunt et al. 2019). This measure for mitigating N_2O is consistent with those described above for increased plant N uptake and reduced loss of NH_3.

Researchers in Washington State, USA, developed a scheduling tool called Application Risk Management (ARM) to mitigate the risk of leaching and run-off (Embertson 2016; https://agri-nmp-msa.apps.silver.devops.gov.bc.ca/). The tool can reduce the risk associated with spreading manure on grass in cool winter weather by taking account of both field conditions and near-term weather forecasts (72 h). On-farm studies showed that successful mitigation of leaching losses depends on field conditions and that, in general, ARM management reduced concentrations of soil NO_3 (Cox et al. 2018).

This concept was tested in a controlled field study in nearby coastal BC. Lower soil NO_3 concentrations were observed for dairy slurry applied in winter (February) than in autumn (November) and overall lower concentrations were observed for dairy slurry than for NH_4-NO_3 fertilizer applied at the same timing (Fig. 5, below; Hunt et al, unpublished data). These findings suggest that dairy manure can be safely applied on (unfrozen) soil in late winter to early spring to mitigate NH_3 loss while also reducing NO_3 potential leaching compared

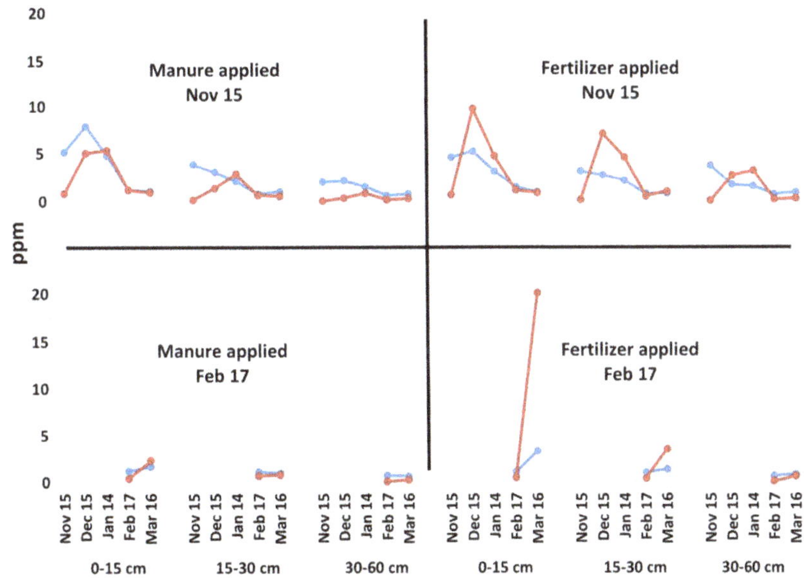

Figure 5 Soil NO_3-N (red) and NH_4-N (blue) concentration after application of dairy slurry and NH_4-NO_3 fertilizer in November (top panels) and February (bottom panels) (2014-2015) at three soils depths (0–15 cm; 15–30 cm, and 30–60 cm) under high winter rainfall conditions in south-coastal BC, Canada. The data suggests lower risk of NO_3 leaching after manure application in winter (mid-February) compared to Fall (mid-November) (unpublished data).

to autumn applications, provided that tools such as ARM are employed to reduce runoff risk. Importantly, manure and slurry application onto wet soils or in cool, rainy weather can increase the loss of P in run-off or preferential flow (Kumaragamage and Akinremi 2018). As such, the relative risks of run-off losses of N and P vs emissions of NH_3 and N_2O (see below) must be balanced when considering the timing of manure applications to soil.

The agronomic benefits of early slurry application on grass include matching the slow rate of crop N uptake in spring to the gradual microbial nitrification of NH_4 in the applied slurry. This concept is also referenced under the 'TSUM' degree-day indicator used for timing application of N fertilizer on grass in early spring (Kowalenko et al. 1989: see www.farmwest.com), but now used by farmers for timing of both fertilizer and manure application. TSUM units are accumulated growing degree days (0°C base) from January 1 until a total of 200 is reached, which in coastal BC occurs in mid-February when spreading slurry on grass is deemed to be agronomically efficient and safe on flat fields with low run-off as seen above. Early application of slurry is important to farmers with limited storage capacity. Also, frequent emptying of slurry stores will reduce CH_4 emissions (VanderZaag et al. 2013).

Indeed, applying manure during the cool non-growing season is essential for farms with limited manure storage capacity. To reduce the risk of leaching, however, regulations in many regions require that there be a well-established crop to capture nutrients from manure applied between autumn and spring. Of course, this is not possible on land used to grow maize silage unless a robust cover crop is successfully established after maize harvest, which may be difficult in years with a late maize harvest or when autumn weather is too wet, dry, or cold for successful establishment of cover crops. An alternative to autumn planting is interplanting the cover crop into the growing maize so that it is well established in the maize understory and can grow rapidly through the shoulder season, soon after the maize is harvested. One crop well suited for intercropping with maize is biennial Italian ryegrass (*Lolium multiflorum* L.), a high-protein, highly palatable cover crop that has proven over years of research and on-farm experience to tolerate shade, drought, and disease loading in the maize understory and to recover well from traffic caused by maize-harvesting equipment (Fig. 6). Seeding the Italian ryegrass when maize is at the 5–6 leaf stage maximizes establishment while minimizing competition with maize by the grass.

A well-established, actively growing cover crop can recover 40–70 kg ha^{-1} of residual soil N after maize harvest in autumn; residual soil N is almost unavoidable in manure-treated fields because maize ceases to take up N in late summer, while organic N from manure and soil continues to mineralize. The cover crop will also respond well to applications of 20–30 m^3 of slurry (100–150 kg total-N ha^{-1}) in late February and March, thereby allowing the

Figure 6 Italian ryegrass inter-seeded between rows of maize at the six-leaf stage, shown in late summer (left) and in early next spring (right). Photos by D.E. Hunt, Agriculture and Agri-Food Canada.

Table 3 Fall and spring yield and N uptake of Italian ryegrass inter-seeded at 6-leaf stage of corn, Agassiz 2018-2019

Sampling stage	Sampling date	Biomass (t DM ha^{-1})	N Uptake (kg N ha^{-1})
Under maize canopy at the time of silage harvest	2-Oct-18	0.5	17
Autumn growth after maize harvest but before winter freezing conditions	19-Nov-18	1.3	39
Spring growth harvested as a grass silage crop	29-Apr-19	3.7	79

Source: Adapted from: Hunt and Bittman (unpublished data).

farmer to clear manure storage space while also promoting growth of a forage crop that can be used to complement summer maize grown on the same land. Table 3 shows some typical Italian ryegrass yield and N uptake for Agassiz, BC.

Another approach to facilitating infiltration of slurry TAN into the soil in order to reduce NH_3 emissions and improve crop N uptake involves separating slurry into liquid and solid fractions. The thinner and less viscous liquid fraction contains most of the manure TAN, which will be easily adsorbed in the soil (Table 1).

Currently, slurry separation is done on farms by mechanical filtration, such as the screwpress separator (Möller et al. 2000), to remove solids for re-use directly or after composting as a high C:N soil amendment on farms with low or declining soil OM (Zhang et al. 2019; Franzluebbers et al. 2021). It is also now used for bedding and mulch for certain high-value horticultural crops such as small fruits. Thus, separated manure solids can help remove some nutrients off-farm. The liquid fraction retained after filtration will have a lower C:N ratio but a similar P:N ratio to the original slurry because the particles containing P will pass through the filters (Möller et al. 2002). Instead, large batch-type centrifuges are required to effectively balance the N:P ratio of slurry to match crop needs; these may be particularly appropriate for farms needing to remove manure off-site and export large amounts of excessive P. In an on-farm study in coastal BC, manure separation via mobile centrifuges significantly improved the N:P ratio of the supernatant but at a significant financial and energy cost (https://www.agproud.com/articles/45270-centrifuges-an-option-for-dairy-manure-nutrient-management). Given the low commercial demand for the high-P sediment produced, there is currently little interest in this technology in Canada.

In slurry with <7% solids, natural settling of solids in tanks or ponds is a low-cost alternative to centrifugation. Table 1 shows the properties of the whole slurry compared to the supernatant (liquid fraction) after several months of storage in a two-stage lagoon joined by a weeping wall. The percent dry matter of the slurry was decreased from 5.9% (a typical value for dairy slurry) to 1.6% (a typical value for pig slurry) while the ratio of TAN:total N increased from 0.52% to 0.66% and the ratio of N:P increased from 5.2 to 9.7. The triple benefits of this liquid fraction, from a nutrient-use perspective, are (i) faster potential soil infiltration, (ii) a higher proportion of rapidly available N, and (iii) relatively less P loading. That being said, the higher pH (7.6 vs 7.0) of the liquid fraction is a potential driver for greater NH_3 emissions if soil infiltration is delayed, e.g. in soils that are saturated, hydrophobic, compacted, or crusted.

A 6-year trial using a low-emission trailing shoe to apply whole slurry or the liquid fraction (obtained from a two-stage settling lagoon) showed that plant N uptake from the slurry liquid fraction was significantly higher than from whole slurry applied at equivalent rates of total-N (Fig. 7; Bittman et al. 2011). Apparent plant N recovery was 41.6% of total-N from the liquid fraction compared to 25.5% for the whole slurry which averaged over the 6-year trial, likely included some uptake of N released from legacy organic N from previous years, especially for the organic N-rich whole slurry treatment. The higher rate of N recovery from the liquid fraction can be attributed largely to reduced NH_3 loss after application (Pedersen et al. 2021) although the effect of separation on NH_3 emissions is not always consistent (Bhandral et al. 2009) probably due to poorly understood complications at the manure–air or manure–soil interface, like surface crusting or hydrophobic soils in hot weather. Percent recovery of applied total-N was greater from mineral fertilizer (54.6%) than from the separated slurry (41.6%), but similar based on the mineral N fraction (Fig. 7).

In the same trial, much lower total P was applied in the liquid fraction than in the whole slurry treatment: at an application rate of 400 kg total-N ha^{-1} (nominal), the liquid fraction supplied about 47 kg P ha^{-1} while the whole slurry supplied about 78 kg P ha^{-1} (Bittman et al. 2011). Using data from Zhang et al. (2021), it was determined that the annual P surplus (in excess of crop uptake)

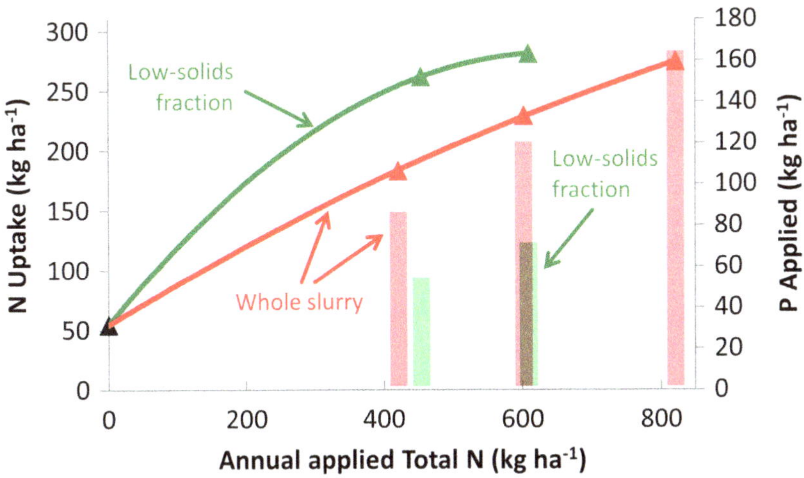

Figure 7 N uptake by tall fescue (6-year mean) receiving either whole dairy slurry (red) or the liquid fraction (green) after settling in a two-stage lagoon system. The vertical bars indicate the P loading at two application rates of whole slurry and liquid fraction. The manure properties are shown in Table 1 based on Bittman et al. (2011).

was 15.1 kg P ha^{-1} for the separated liquid fraction compared to 36.8 kg P ha^{-1} for the whole slurry treatment. After 11 years of applying separated dairy liquid, Zhang et al. (2021) reported that there was no evidence of excess P loading in soils in terms of extractable P (Kelowna extract) or total P although some P loading was evident at high rates of whole slurry. There was also little effect of legacy P (or N) from application of the liquid fraction on maize planted after the grass was terminated. With more effective settling techniques to remove greater volumes of solids, or a small addition of N fertilizer, the liquid fraction could be used for optimum yield and N uptake of grass at low cost and without the risk of applying surplus P; such an approach would represent a significant advance in manure management on intensive dairy farms. A companion study (Hunt et al. 2019) showed that the N_2O emission factor was similar for applications of whole manure and the separated liquid fraction (~0.63) and that emissions from whole manure were similar to or higher than from separated liquid manure, based on crop yield, but lower than from mineral fertilizer. This shows that manure separation did not cause N_2O pollution swapping despite the higher moisture content and likely lower NH_3 loss for the liquid fraction than whole slurry.

The strategy of separating the liquid manure fraction in order to reduce emissions and conserve N has been questioned because of limited options for effectively using the remaining solid fraction, which can give off NH_3 during storage and land application (Bittman et al. 2014a). We assessed the sludge in the bottom third of the manure tank, after decanting the thinner supernatant, as a replacement for commercial mineral P fertilizer for maize (Bittman et al. 2012; Hunt et al. 2016; Hunt and Bittman 2021). Extensive trials on dairy farms in coastal BC have shown that P fertilizer (mono- or diammonium phosphate) applied using a mechanical seeder in bands placed 5 cm below and to the side of the maize seed furrow, benefited early maize growth and advanced maturity by a few days even on high P soils (https://farmwest.com/resources/books/advanced-silage-maize-management-2004/chapter-3/). Two trials were conducted in which the sludge was applied at ~30 kg P ha^{-1} by injection (15 cm deep) at maize row spacing (75 cm). After 2–3 days, the maize was planted 0–15 cm off the injection furrow. The first study showed that the sludge did not impede root colonization by arbuscular mycorrhizae and that at close placement (5 cm) all yield parameters were similar for maize receiving commercial fertilizer and for maize receiving the separated thick slurry fraction (Bittman et al. 2006, 2012). In the second study, the silage maize received three sludge rates which were supplemented with commercial fertilizer to provide equivalent rates of total-N, i.e. fertilizer plus precision-placed dairy sludge (Hunt and Bittman 2021; Fig. 8). This trial showed that crop P recovery from sludge was similar to that from mineral P fertilizer at equivalent rates of total N. However, crop yields were slightly reduced in the lower N sludge treatments because these received a higher proportion of organic N from manure than from chemical fertilizer. In the moderate sludge treatment (32.5 kg P ha^{-1} and 160 kg N ha^{-1}), yield was 1.3 t ha^{-1} lower for the sludge treatment, but 22% of the total N applied was not-readily-available organic N. Apparent P recovery rate from slurry P was affected by N application and could not be fully teased out. Note that at the lower sludge and high N rate, apparent P recovery was close to 90% but yield was about 1 t ha^{-1} lower than the high sludge rate at equivalent total N (more mineral N). The injected slurry at 32.5 kg P ha^{-1} fully supplied the early (not shown) and season-long P requirements of maize, obviating the need for mineral fertilizer. Successful use of precision-placed (whole) dairy slurry for starter P has been reported in Germany, where this practice is now used commercially (Federolf et al. 2015), in Denmark (Pedersen et al. 2020) and The Netherlands (Schröeder et al. 2015), Note that injection of sludge substantially increased N_2O emissions compared to broadcasting, a tradeoff that was substantially addressed by mixing the nitrification inhibitor dicyandiamide into the sludge before application (Hunt and Bittman unpublished).

When assessing plant uptake of manure N, it is important to consider the influence of the crop species. We compared uptake of N from slurry and commercial fertilizer by three perennial grasses: orchardgrass (*Dactylis glomerata*), tall fescue, and perennial ryegrass (*Lolium perenne*) from 2004 to 2010 on soils pre-treated with manure since 1996 (Table 4) (Bittman et al. 2014b). Percent N uptake across grass species averaged over the 7 years was greater for fertilizer than for slurry N (72.3% vs 65.1%, respectively) at an N

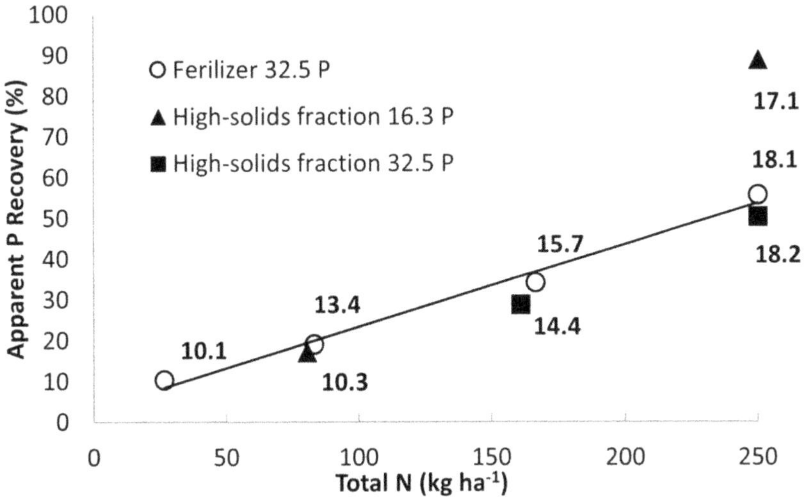

Figure 8 Apparent P recovery (%) by silage corn (recovery relative to applied P adjusted for control) treated with precision-injected (<10 cm from corn row) separated dairy slurry (high solids fraction); the slurry fraction was applied at 16.3 (triangles) and 32.5 (squares) kg P ha^{-1}. Total N applied in the high-solids fraction was 80.5 kg ha^{-1} and 161 kg ha^{-1} at the low and high P application rates, the remaining N coming from commercial fertilizer (total applied N on X-axis). Response to mineral N fertilizer rates with 32.5 kg P ha^{-1} is depicted by the open circles and the regression line. P recovery is shown at equivalent total N rates. The data show that manure and fertilizer P recovery are similar at equivalent total N rates. The values associated with each dot represent yield (t DM ha^{-1}) which is somewhat lower with slurry than mineral fertilizer due to less available N in slurry, but improved with increasing fertilizer N; at high addition of mineral N (250 kg N ha^{-1}), corn yield is equivalent from manure and fertilizer (~18.1 t DM ha^{-11}) showing similar efficacy of manure and fertilizer P. Note, at low manure and high total N rates, P recovery >90% but yield is slightly depressed likely by insufficient manure P. (Data is mean of 3 years; based on Hunt and Bittman 2021).

application rate of 400 kg total-N ha^{-1} (nominal). The average N uptake, across all N treatments, was consistently lowest for perennial ryegrass, due to lower total yield potential and, possibly, shallower roots, less drought tolerance in summer, and less winter hardiness. In general, orchardgrass responded better to commercial fertilizer while tall fescue responded better to (whole) slurry, especially during vigorous spring growth. We attribute this to more consistent season-long growth for tall fescue than orchardgrass, which better matches nutrient availability from slurry than from mineral fertilizer. In the first-cut (taken in mid-May), tall fescue took up more N (16–37 kg ha^{-1}) from slurry than orchardgrass, perhaps due to more vigorous spring growth or relative preference for NH_4, but annual differences in N uptake were somewhat smaller. Crops appear to differ in their preference for NH_4 and NO_3 but this requires more study. The uptake of slurry N by tall fescue also generally increased over

Table 4 Annual N uptake by three forage grasses (2004-2010) treated with dairy slurry, fertilizer, or both dairy slurry + fertilizer at contrasting total N rates (shown)

Treatment		N Rate	Orchardgrass		Tall fescue		Perennial ryegrass		Mean	NUE[b]
		kg ha^{-1} year^{-1}								
Control		0	100	J[a]	94	J	77	K	90	
Fertilizer	Low	200	239	G	210	H	178	I	209	104.5
Fertilizer	**High**	**400**	**310**	**D**	**289**	**E**	**268**	**F**	**289**	**72.3**
Slurry	**Low**	**384**	**263**	**F**	**268**	**F**	**219**	**H**	**250**	**65.1**
Slurry	High	724	332	B	371	A	294	E	332	45.9
Fert+slurry	High	546	371	A	374	A	321	C	355	65.0
Mean			269		267		226			

[a]Values not followed by same letter are significantly different at $P < 0.05$.
[b]NUE is nitrogen use efficiency, N uptake/N applied as a percent. Values greater than 100% can be due to atmospheric N deposition, biological N fixation and net soil N mineralization. Bolded rows received similar rates of total N.
Source: Adapted from: Bittman et al. (2023).

the 7 trial years, perhaps due to legacy N and the excellent hardiness of tall fescue.

There are a few other studies on the relative response among species to fertilizer and manure. In a New York State study, where semi-solid manure (14–19% DM) was applied by broadcasting, orchardgrass had lower yield but higher tissue N concentrations than tall fescue (Cherney et al. 2002). To optimize the efficiency of slurry N uptake by grass, it is important to appreciate specific crop nuances such as seasonal and total growth and preference or tolerance for N species, especially in high-input systems where over-application will not only reduce NUE but inevitably lead to environmental losses (see effects of long-term manure applications, below).

5 Assessing long-term manure and slurry application

Interpreting crop nutrient uptake and NUE of manure from short-term trials can be problematic. As a result of pre-trial soil disturbances, changes in the cropping system, trial establishment activities, etc., soil nutrient dynamics are perturbed and not in steady state. Hence, estimates of crop N uptake require adjustments for net soil N mineralization, which is often done by subtracting available soil N concentrations and crop N uptake on untreated control treatments (i.e. 'apparent N uptake'). This approach provides only an approximation of the available N in the absence of manure application, since it does not include the effects of N priming on mineralization in the treated soil. Also, crop N uptake includes both manure-derived readily available N (in the form of TAN) plus

N that is gradually released from the organic fraction in the manure (about 40-50% of the total slurry N; Table 1). Organic N mineralization is controlled by soil conditions, especially temperature and moisture, and other factors such as soil pH and aeration. Net mineralization of the manure organic fraction is also influenced strongly by the C:N ratio of the manure which, in turn, is affected by percent crude protein and fiber in the cattle diet and the amount and type of bedding used (Webb et al. 2013). In the cool soils of Denmark, the release of organic N from applied dairy slurry, in the 5 years after application, followed a decay curve of 17%, 17%, 8.5%, 4%, and 4% (Sørenson et al. 2017). In the milder climate of the Netherlands, these values were slightly higher in years 1 and 2 (21%) but similar thereafter. Over the 5-year study period, mineralization of organic N in cattle slurry was 51% compared to 71% for pig slurry.

Contrary to short-term agronomic trials, which are inevitably affected by rarely reported pre-trial conditions, long-term trials allow soil to reach a quasi steady state and more fully account for the mineralization of organic N derived from repeated applications of manure. Long-term trials also allow for monitoring of changes in SOC and in other soil minerals, like P and Cu. We evaluated the efficacy of N from dairy slurry relative to commercial fertilizer (and slurry + fertilizer) for tall fescue, between 1996 and 2002 at Agassiz, BC (Bittman et al. 2007). The slurry from commercial dairy farms was applied uniformly to replicated plots with low-emission trailing shoes at 19-cm spacing. Both slurry and fertilizer were applied in four equal doses each year: twice in cool weather (March and September) and twice in warm weather (June and August) to supply each of four grass harvests. The manure was applied at target (nominal) annual rates of 400 kg total-N ha^{-1} and 800 kg total-N ha^{-1} or 200 kg ha^{-1} and 400 kg ha^{-1} of TAN (actual rates in Table 5). The N fertilizer (NH_4-NO_3^- 'Fert' in Table 5) was broadcast at 200 kg N ha^{-1} and 400 kg N ha^{-1} with other

Table 5 Fate of N from dairy slurry and mineral fertilizer (Fert; NH_4-NO_3) applied to tall fescue 1996-2002 at Agassiz, BC

Treatment	Annual min/total N (nominal)	Apparent N recovery	Stored soil N	Unaccounted N
	Kg ha^{-1}	% of applied total-N		
Control	0	–	–	–
Fert-low	200/200	67.1	22.4	10.5
Fert-high	400/400	58.7	14.3	27.0
Slurry-low	200/400	42.0	32.0	26.0
Slurry-high	400/800	37.4	30.2	32.4
Alternate-high[a]	400/600	53.3	23.1	23.6

[a]Alternating applications of Fert-high and Slurry-high.
Note: Stored soil N was measured by soil sampling and nitrogen analysis.
Source: Adapted from: Bittman et al. (2007).

nutrients (and lime) supplemented according to soil test requirements. The experiment also included a treatment where applications of fertilizer at 100 kg N ha^{-1} were alternated with manure at 200 kg total-N ha^{-1} each application (annual total-N = 600 kg N ha^{-1}) and an unfertilized control. Over 7 years, about 60–67% of applied N was recovered from the Fert treatment, about 37–42% was recovered from slurry, and 53% was recovered from the alternate treatment with relatively lower N recoveries observed at the higher N application rates (Bittman et al. 2007) as reported by Webb et al. (2013).

These recovery rates were lower than expected based on concentrations of available TAN in the slurry; the fertilizer equivalency of the slurry at a comparable total N application (400 kg N ha^{-1} nominal) was 70% based on 42% recovery of slurry N vs 60% recovery of fertilizer N. Over 7 years, 14% of fertilizer N and 32% of slurry-applied total-N accumulated in the soil (Table 5). Nitrogen loss calculated from cumulative uptake plus cumulative soil storage vs N applied was lowest for the low fertilizer treatment but similar for the fertilizer and slurry treatments applied at the 400 kg N ha^{-1} nominal rate, due to lower uptake but greater soil storage for slurry (Table 5). It was deduced from this multi-year study that NH_4-NO_3 fertilizer, which emits <5% fertilizer N as NH_4 (Van der Weerden and Jarvis 1997), lost N mainly by leaching over the non-growing season (Kowalenko 1989) and that slurry N was lost by NH_3 emissions soon after application, especially in summer (as predicted by the ALFAM and ALFAM 2 models; Søgaard et al. 2002; Hafner et al. 2019) and by leaching and denitrification

Direct measurement of the suite of N losses over multiple years and variable weather conditions is difficult to implement because they are laborious and requires an assortment of specialized equipment. Also, losses can be missed when they occur near detection limits over long periods or as sudden bursts (e.g. following freeze-thaw events; Fig. 9), which are easily missed (Pelster et al. 2022; Hunt et al. 2019). In this way, N loss pathways inferred from long-term trials may be less precise but more robust than direct measurements.

Longer-term-treatment effects of this trial were re-examined after 15 years (Zhang et al. 2021). Mean annual yields for fertilizer- or slurry-treated grass ranged between 11.4 Mg dry matter ha^{-1} and 15.0 Mg dry matter ha^{-1}, while total N uptake over the entire trial duration ranged from 1530 kg N ha^{-1} (102 kg N ha^{-1} $year^{-1}$) for the control to 5565 kg N ha^{-1} (371 kg N ha^{-1} $year^{-1}$) for the alternating (Alt-high) treatment (Table 6). Overall, actual NUE (unadjusted for unfertilized plots) varied between 114.1% (Fert-low) and 48.6% (Slurry-high). At 400 kg total-N ha^{-1} (nominal) application rate, the NUE of the Slurry-low was similar to commercial fertilizer (Fert-high) at 77.6% and 74.5%, respectively. The high values for slurry can be attributed to the low emission slurry application and capture of mineralized legacy organic slurry N. High crop yields may also

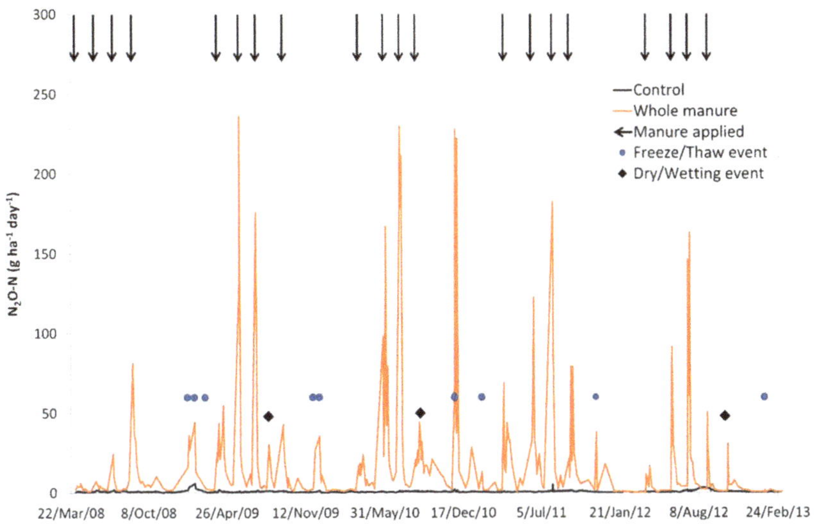

Figure 9 Daily nitrous oxide (N_2O-N) emissions measured between 10 April 2008 and 6 March 2013 (~59 months) on replicated perennial grass plots treated with dairy slurry (nominal 600 kg N ha^{-1} year^{-1}) in four equal doses. Day of manure application, freeze/thaw events and dry/wet cycles are depicted by arrows, circles, and diamonds, respectively. Source: Adapted from: Based on Hunt et al. (2019).

play a role. Notably, the Alt-high treatment had the highest yield and crop N recovery but a lower NUE (67.6%).

In contrast to the earlier study (Bittman et al. 2007), only 13–15% of the applied slurry N was stored in the soil compared with −1.9% and 8% of the low and high fertilizer N, respectively (Zhang et al. 2021; Table 6). The lower percentage of soil N storage after 15 years than after 7 years suggests that, over time, soil N pools move towards an equilibrium in which N mineralization rates approach N immobilization rates. The Alt-high treatment sequestered less N than the Slurry-low treatment (3.2% vs 13% of applied N, respectively) even though both treatments received the same amount of slurry N and produced similar amounts of biomass (14–15 Mg dry matter ha^{-1} year^{-1}). This unexpected result suggests that the added mineral N in the Alt-high treatment increased N mineralization rates (lower C:N ratio in soil), while possibly increasing the shoot to root ratio of the grass (i.e. lower inputs of root biomass N).

Nitrogen losses ranging from 36.2% to −12.2% (based on unadjusted NUE values) are probably underestimates because wet and dry atmospheric deposition of reduced and oxidized N species is not usually measured and may be as high as 30 kg N ha^{-1} year^{-1} where there is intensive livestock production (Hao et al. 2009; Table 6). The adjusted NUE, which subtracts the N uptake in unfertilized Control plots, corrects for N deposition but is likely an over-adjustment, due to higher biological N fixation in the Control plots than treated

Table 6 Responses to applications of mineral fertilizer (NH_4-NO_3) and dairy slurry on grass over a 15-year period: annual yield and N uptake by forage (total of four cuts per year); un-adjusted nitrogen use efficiency (NUE); adjusted NUE, total-N sequestered in soil over 15-year period (measured); changes in natural abundance δ15N in herbage. Un-adjusted NUE is defined as (total-N uptake by herbage)/(total-N applied); adjusted NUE is defined as ((total N uptake by treated herbage) – (total N uptake by control herbage))/(total-N applied)

	Applied Total-N kg ha^{-1}		DM yield t ha^{-1}	Uptake tot-Nkg ha^{-1}	NUE %Unadjust.	NUE %Adjust.	Sequest. N% of applied	lost N% of applied	Herbage δ15N (‰)	
Treatment	Annual	15 year	Annual	15 year	15 year	15 year	15 year	15 year	1996	2010
Control	0	0	5.89	1530	–	–	0	–	4.32	2.01
Fert-low	198	2970	11.36	3390	114.1	93.9	-1.9	-12.2	3.62	1.10
Fert-high	392	5880	11.11	4380	74.5	64.3	8.2	17.3	3.09	0.77
Slurry-low	379	5685	13.89	4410	77.6	67	13	19.4	5.99	5.20
Slurry-high	726	10890	14.34	5295	48.6	43.1	15.2	36.2	6.39	7.40
Alt[a]-high	549	8235	14.96	5565	67.6	60.3	3.2	29.2	4.72	4.27

[a]Alternating applications of Fert-high and Slurry-high.
Source: Adapted from: Zhang et al. (2021).

plots due to more volunteer white clover (10–15% ground cover in Controls compared to <5% in treated plots, unpublished data). Furthermore, micro-run-off may also transfer some N to low N plots and dry deposition of NH_3 may be greater on control plots due to steeper atmosphere–plant N gradients (Sutton et al. 1993). Thus, it is likely that the true NUE falls somewhere between the adjusted and unadjusted values but closer to the unadjusted values.

Determination of natural abundance $\delta^{15}N$ values of crop tissue can be used to make inferences about the source of N taken up by the plant. This is because fertilizer $\delta^{15}N$ value tends to be lower than that of manure, while urine $\delta^{15}N$ value tends to be lower than that of faeces. Even in the first year of the trial, differences in the sources of N inputs were reflected in the natural abundance $\delta^{15}N$ values of the harvested herbage and became more marked over time (Table 6; Zhang et al. 2021). For example, herbage $\delta^{15}N$ values in Fert-treated plots, especially at the high rate, declined towards fertilizer $\delta^{15}N$ value, particularly after ~10 years, showing that most of the N taken up by the crop in Fert plots was derived directly from fertilizer (not shown). By contrast, the herbage $\delta^{15}N$ values in the Slurry-low and Alt-high treatments remained largely unchanged through time, suggesting that the experimental field had received similar forms of N inputs prior to the trial (which was supported by the oral history for that field). The herbage $\delta^{15}N$ values in the Slurry-low and Alt-high plots were much lower than that of applied slurry (~11.9‰), reflecting preferential root uptake of urine N, which is more depleted in ^{15}N than faeces. Thus, the higher herbage $\delta^{15}N$ values in the Slurry-high treatments suggest that a greater proportion of N uptake came from faeces. Finally, the herbage $\delta^{15}N$ value in the control treatment (~2.01‰) probably reflects the uptake of N derived from biological N fixation and/or atmospheric deposition (Zhang et al. 2021).

In long-term dairy slurry trials conducted in northern Quebec, N accumulation in high clay soil was attributed to manure applied to forages grown in a rotation with cereal crops and soil storage of dairy slurry-derived N applied to arable crops was found to be limited (D'Amours et al. 2021). Canada-wide, low-soil C storage (5%) from dairy slurry compared to solid cattle manure, in mainly arable crops, likely reflects a low C:N ratio (Liang et al. 2021), which is consistent with higher C storage observed from dairy manure than from chemical fertilizer (Maillard et al. 2015, 2016).

Indeed, as mentioned above, the current interest in increasing SOC as a means of supporting climate change mitigation and improving soil health has renewed interest in the use of manure as a C-rich soil amendment. Numerous studies and meta-analyses have shown the potential for increased SOC with applications of solid manure (Gross and Glaser 2021; Li et al. 2021b), but research on the effects of manure slurry on SOC storage remains relatively scant (Maillard et al. 2014; Bolinder et al. 2020). Nevertheless, several Canadian studies have explored the longer-term impact of continuous slurry applications on SOC stocks and soil health. In the long-term trials described directly above, the SOC stock of the Slurry-high treatment (16.1 kg C m^{-2}) was significantly greater than the Control and Fert (13.2–13.7 kg C m^{-2}) treatments after 17 years, while the SOC stock of the Slurry-low treatment was intermediate (14.6 kg C m^{-2}; Maillard et al. 2015). This positive effect of high slurry application on SOC stock was attributed both to inputs of solid material in the manure (derived from undigested forage and wood chips used as bedding material) and to increased crop growth (which tended to be higher in the Slurry-high treatments than in the Slurry-low or Fert treatments) but note that the above-ground biomass was mostly harvested and removed. The increased SOC was concentrated in the top 20 cm of the soil; although there was limited mixing of the SOC into deeper soils, where it might be protected from decay, greater formation of C-rich organo-mineral complexes near the soil surface suggested that this newly formed SOC may be fairly stable. Similarly, a 21-year study comparing the effects of liquid dairy manure and fertilizer N applications on SOC stocks in combination with a cereal monoculture or cereal-perennial forage rotation found that applications of dairy manure increased SOC stocks in the cereal-perennial forage rotation treatment; again, treatment differences were concentrated in the top 20 cm of soil (Maillard et al. 2016). The authors proposed that SOC stocks were only observed in the forage rotation treatment because of reduced tillage frequency, and the addition of high-quality residue derived from red clover (*Trifolium pretense* L.). Together, these studies show that both direct inputs of manure-derived organic C and plant biomass growth following manure applications mediate the response of SOC stocks to applications of dairy manure. Further work is required to explore the 'stability' of dairy slurry-derived SOC.

Important questions also remain on how applications of dairy manure influence the composition and function of the soil biotic community. In a perennial grass sward dairy manure, whole or separated liquid fraction, applied at 400 kg total N ha^{-1} increased soil microbial biomass, microbial activity, and soil health enzyme indicators over control and mineral fertilizer, with no difference between the form of manure (Neufeld et al. 2017). However, the ratio of fungi:bacteria was significantly higher following applications of whole manure than after applications of the manure separated liquid fraction or commercial fertilizer, suggesting that organic C-rich manure can alter the structure of the soil microbial community with possible implications for soil C accumulation. In a separate trial, bacterial counts showed a rapid response to applications of whole dairy slurry compared to fertilizer, suggesting that soil bacteria may be capable of rapidly immobilizing manure N (Bittman et al. 2005). Microbivorous protozoa, bacterivorous nematodes, and fungivorous nematodes were also more abundant in soil treated with manure than in control or fertilized soil, indicating an enhanced microbial food web which may explain the relative lack of differences in microbial community size among treatments (Bittman et al. 2005; Forge et al. 2005b). Microbial processing of SOC contributes to the formation of C-rich organo-mineral complexes that are more resistant to decay (Lehmann and Kleber 2019), but the relationship between fungal (or bacterial) community composition and the formation of more stable forms of SOC remains an open question (Cotrufo et al. 2019). The response of the soil fungal community to applications of dairy manure may be mediated by available N and P (Ma et al. 2016), so applications of manure with different C:N and C:P ratios may elicit different pathways of SOC formation with varying resistance to mineralization.

Of note, the dairy slurry increased populations of the parasitic nematode, *Pratylenchus penetrans*, which contrast with some previous studies showing reductions but may be related to lower doses (Forge et al. 2005a). Slurry manure increased the population of a single species of carabid beetles (*Pterostichus melanarius*) which consumed more onion maggot (*Delia antiqua*), but decreased indicators of carabid diversity (Raworth et al. 2004).

6 Dairy manure and farm management

Over the long term, we estimate a NUE of 70% for slurry applied with a low-emission trailing shoe applicator on a high-producing grass in a seasonally moist climate. The NUE could be further improved, e.g. with shallow injection (Webb et al. 2010) or other methods that facilitate slurry-N infiltration and stabilize NH_3 (separation, rolling tines, acidification). Our estimate of NUE is higher than national efficiency values in the EU, which are mostly under 60% (reviewed by Webb et al. 2013). Across farms and crops, there will be a range of NUEs, depending on soil attributes, weather conditions, and the ability of farmers to manage their crop productivity (variety selection, pest control, irrigation) and soil productivity (compaction and conservation of soil C) and even to set manure/fertilizer rates (which requires some degree of guessing and intuition by farmers to anticipate weather and revenue).

There are soil testing and decision-support tools intended to help farmers improve their crop and nutrient management practices. A fairly recent innovation is real-time N modelling tools such as ADAPT-N® that can improve NUE (in maize) by guiding in-season adjustments of fertilizer rates and timing (van Es et al. 2023), although at present most tools have not been calibrated for N from manure. Advances in visual modelling platforms, like STELLA/ ITHINK©, are being developed to customize N models for a larger array of soils, crops, and nutrient inputs, including manure (e.g. Dion et al. 2020; Dorais et al. 2019; Hirsch 2007).

Because there are large differences in management skills and farm challenges among farmers, on-farm efficiencies across the sector can be gained by training, education, and mentorship programs. Living Labs have been implemented across Canada, some on dairies in BC and Quebec, to help farmers identify and adopt best management practices, such as in-crop testing, variable nutrient rates, cover cropping, efficient manure use, and nutrient budgeting. The objectives of these initiatives are to help mitigate climate effects and reduce other possible negative environmental impacts associated with milk production while maintaining farm production and profits (Longchamps et al. 2023).

To ensure efficient use of N inputs on a dairy farm, N losses from slurry must be staunched from barns and manure storages, and from grazing animals that have low NH_3 emission losses but, paradoxically, poor nutrient distribution which leads to inefficiencies (Webb et al. 2013). Increased adoption of measures, such as manure storage covers and manure acidification can reduce air pollution and preserve manure N (Sutton et al. 2022). However, these measures may not be sufficient for farms that have limited land to produce feed and accommodate manure nutrients. For all dairy farms, particularly those with a limited land base, it is necessary to address nutrient leakages throughout their operation while also balancing farm nutrients by matching inputs and outputs, as depicted in the 'leaky pipe' in Fig. 10. The leaks in this pipe, which apply to both N and P although only N is shown in Fig.10, occur throughout the farm: from wasted feed to inefficient handling and utilization of animal wastes and to crop harvesting and storage. Whole-farm models, such as the Integrated Farm System Model, can be used to simulate nutrient flows and test different management scenarios on dairy farms (e.g. changes in fertilizer or feed, or response to climate change; Rotz et al. 2021). These approaches must also account for spatial pollution swapping; e.g.,production of imported feeds also comes with an environmental cost that occurs elsewhere, infrequently represented in whole farm models, though it is likely to be included in life cycle analyses.

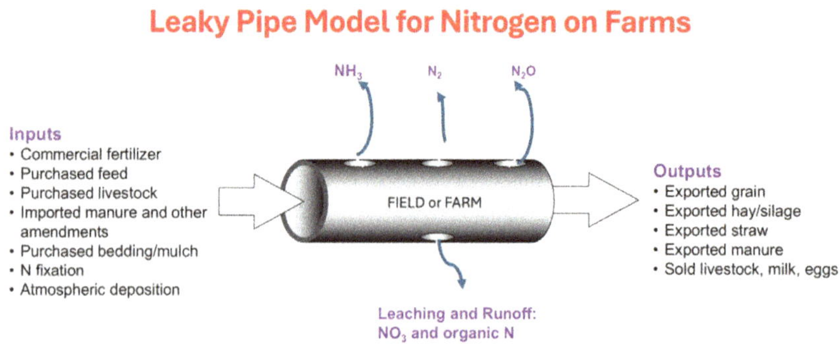

Figure 10 The leaky pipe model of N budgets for field or farm systems. Nitrogen surplus will result in leaks that can be shifted between pathways (e.g. NH_3 to NO_3) but not staunched unless surpluses are reduced.

The counterintuitive environmental benefits of growing maize on dairy farms was first reported on the DeMarke model dairy farm, where researchers demonstrated that replacing some grass with maize helped reduce farm N surplus and NO_3 leaching, and coincidentally, irrigation water consumption (Aarts et al. 2000). Reducing N surplus does not guarantee environmental protection since there can still be poor practices in sensitive areas, but improved environmental sustainability cannot be attained without constraining surplus nutrients at the field, farm, regional, and national levels. Nutrient surpluses on farms can be mitigated by reducing nutrient imports through increasing on-farm production, reducing livestock numbers, and/or adding land; increasing export of products, including wastes or feed, can also be used to balance whole farm-nutrient budgets. Commercial pilot dairy farms, perhaps the forerunners of the Living Lab concept, have been used to encourage participatory farm management as a means of reducing the environmental impacts of dairy production (Oenema et al. 2011). As well, 'dairy farmlet' trials have been employed in New Zealand to assess the effect of changing farm practices, e.g. grazing management, on GHG emissions (Van der Weerden et al. 2018).

A research framework was initiated at the Agassiz Research and Development Centre in 2017 to help integrate nutrient and crop management at the field and farm levels. This framework, called 'semi-virtual farmlets', is intended to provide a more complete and integrated assessment of advancing farm nutrient sustainability; the framework combines (i) field testing of best management practices for nutrient use and crop production; (ii) crop and soil sampling and analysis; and (iii) cow nutrition models (Li 2019; Li et al. 2021a). Slurry management measures for reducing nutrient losses are tested in the field while crop management measures are integrated into the design to help improve NUE and reduce losses. Manure management measures involving manure separation and precision application are combined with improved cropping measures, such as double cropping and relaxed harvesting to produce more crop with less inputs (Fig. 11). The crops are harvested and tested for key nutritional quality factors pertinent to dairy cattle. In this hybrid approach, the on-farm feed produced, rather than actually feeding to animals, are balanced to meet cattle requirements with minimum supplemental feed importation using the feeding model: Cornell Net Carbohydrate and Protein System feeding model (4 AMTS, Agricultural Modeling and Training Systems LLC, Groton, NY) (Fig. 12; Koenig et al. 2023). This ongoing 'dairy farmlet' trial has demonstrated the potential for reduced N losses and elimination of P fertilizer. Farm feed production was improved by optimizing maize and grass allocation, and improving the nutritional balance of feed, thereby reducing the need to purchase feed supplements, lowering N excretion rates, and freeing up land to grow additional needed crops that could further replace imports. With

Figure 11 Field portion of 'semi-virtual farmlet' trial showing Replicate 1 with corn and grass management plots. Farmlet 3 is identified. Photo by S. Bittman and D.E. Hunt, Agriculture and Agri-Food Canada.

this framework, the potential for pollution swapping has been identified, and measures have been adopted to mitigate these unintended consequences.

The progressive nutrient and crop measures integrated within these semi-virtual farmlet trials are intended to mitigate nutrient leakages and identify and address within the farming system any unintended negative consequences of adopting new practices. These efforts demonstrate that the sustainability of maize and grass production on dairy farms can be improved individually and in concert by sharing and optimizing resources, including land allocation for crops having complimentary and contrasting attributes in terms of plant physiology, nutrient requirements, seasonality, yield and feed quality, and different environmental risks.

7 Future trends in manure management

While there has been incremental progress on farm nutrient management, increasing societal pressures are driving more ambitious approaches to manure nutrient use. These pressures include concerns about animal production, including dairy, for its disproportionate environmental impact within the agri-food sector; the expectation that agriculture will contribute to climate mitigation; and an emerging understanding that nutrient accumulation across landscapes, especially in urban and peri-urban areas, must be rationalized (Bittman and Hunt 2024). Food production activities near cities help preserve greenspace and farmland; these farms support local food supply and safeguard food security,

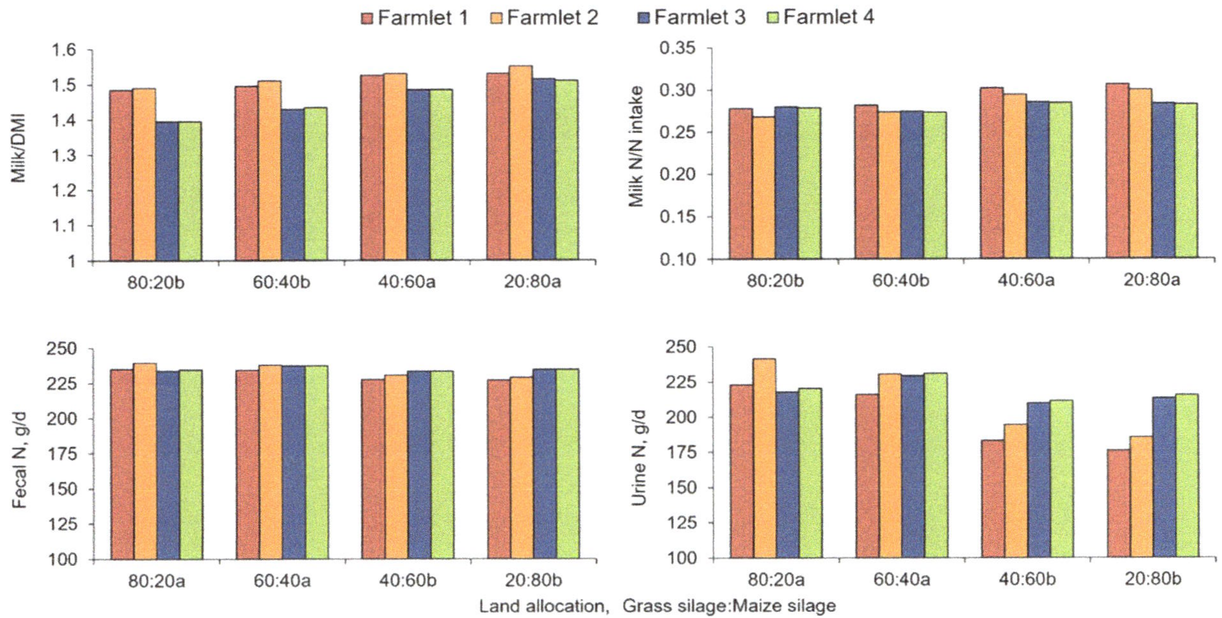

Figure 12 Effect of soil and crop measures and ratio of land allocation (grass silage:maize silage) on some production indicators determined in a 'semi-virtual' dairy system study, The study combines field experiments (Fig. 11) and simulated animal production using the Cornell Net Carbohydrate and Protein System model in an integrated assessment of farm management practices. Farmlet 1 reflects local farming practices, while Farmlets 2-4 employ improved soil/nutrient measures, improved nutrient and crop practices, and advanced measures, respectively. The indicators shown are milk per dry matter intake, milk N per N intake (= milk NUE), and excretion of faecal and urine N. The study shows that milk production and N use efficiency can be significantly improved while reducing urine N. Crop ratios (across farmlets) not followed by the same letter are significantly different at $P < 0.05$. Source: Adapted from: Koenig et al. (2023).

but it cannot be overlooked that these regions are becoming termini for farm-food nutrient streams and no strategy has been devised in Canada to return nutrients to remote feed sources. This question has been articulated in the USA as Manuresheds, an attempt to visualize the flow of feed, food, and waste nutrients across land, landscapes, and geographies (Kleinman et al. 2022). In peri-urban areas and regions with large animal production, soils are likely to be replete with N, and even more so with P and other non-volatile nutrients. At best, this nutrient accumulation reflects a wasted non-renewable resource but, at worst, it is a legacy problem for future farmers and citizens. At present, processing manure to produce marketable products is expensive – in part because it requires removal of large amounts of water from manure (e.g. via filtration, centrifugation, reverse osmosis) or manure nutrients from water (e.g. NH_3 stripping).

Historically, manure has been used across the world to produce energy, mostly via direct combustion of cattle dung. Increased commitments to climate mitigation are now driving industrial-scale biogas production, especially from dairy slurries with economic supports for energy bio-CH_4 production and environmental CH_4 mitigation. More recently, tropospheric CH_4 release has been blamed for ozone pollution. In some countries and jurisdictions, biogas is limited by supplies of feedstock; to fill these gaps, crops such as silage maize may be grown, though this practice may have unintended consequences for the environment and for food security. Co-digestion of dairy slurry with maize silage may improve biogas yield, but the use of organic wastes must be carefully controlled so that digestate residues can be safely returned to the food production stream. Companies may be reticent to use unregulated waste materials, such as household wastes or biosolids, as feedstock or may relegate digestate to non-agricultural uses, such as forestry or production of dedicated biomass crops, such as switchgrass (*Panicum virgatum*) or *Miscanthus*. Reusing manure for non-agricultural uses may be justified at the larger scale, but it does not respect the circularity of nutrients, i.e. farm-to-fork-to-farm. Energy production will be in direct conflict with food production if agricultural land is used to produce energy crops rather than feed crops. Feeding crop wastes to biodigesters rather than to cattle and other livestock comes with similar philosophical problems. Furthermore, family-run dairy farms may be too small to run biodigesters effectively and economically (VanderZaag et al. 2023); they must be near other farms to capture sufficient feedstock, although artificial intelligence may help to operate small digestors, much as robotics and GPS have helped improve the efficiency of other family farm operations.

8 Conclusion

Using manure on crop land is a long-established practice and is of great value when there is limited access to mineral N and P fertilizers. It remains important across all farms to use manure to reduce mineral fertilizer inputs and improve soils; direct application of manure to soils remains the simplest and best way of using this troublesome resource. Manure is bulky and heavy, irregular and inconsistent, with an inconvenient nutrient composition that does not match crop needs. Manure nutrients are also prone to losses as gases, solutes, and suspensions, and manure is a source of pathogens. Solving the manure conundrum so that it contributes value to farms, rather than merely posing a disposal issue is an incremental process with synergies, trade-offs, paradoxes and, often, costs. Integrated approaches for using manure are needed to fully exploit this resource, which is needed to help Canadian dairy farms reach their goal of 'net zero C'.

Dairy manure can be used to replace mineral fertilizers and thereby diminish C footprints. Manure C, properly harnessed, sequesters C in soils and, since much of the SOC stored in soils following manure application is derived from primary production (i.e. crop photosynthesis), this additional stored C may cancel some of the unavoidable C inputs on dairy farms. There are now new and better ways to use the C in dairy manure to produce cattle bedding, mulches, soil amendments, and even as substrates for energy production, especially biomethane, while reusing nutrients.

Feed self-sufficiency and land conservation are important, and inter-related, aspirations because they lead to better cropping and feeding practices which, in turn, improve the production efficiency of dairy crops, animals, and farms. With the staggering complexity and demands of modern dairy farming, it is difficult for farmers to unravel the manure conundrum. This will require a community effort, where resources, knowledge, and expertise can be shared and advanced. A model for such an effort might be the 'Collective systems with centralized feed and manure management systems' in Japan, as described by Toda et al. (2020) where several farms share a centralized feed and manure management system. In these shared systems, manure handling and crop production is managed by specialized companies, enabling the dairy farmers to focus on animal management, feeding, and milking.

Worldwide, farmers are working to solve the manure conundrum on dairy farms in the context of their diverse production systems (Ghourley and Misselbrook; book chapter in review).

9 Acknowledgments

This chapter was supported by a contribution from the Dairy Research Cluster 4 partners (Dairy Farmers of Canada and Agriculture and Agri-Food Canada) under the Sustainable Canadian Agricultural Partnership AgriScience Program. Funding was also provided by Agriculture and Agri-Food Canada's Agricultural Climate Solutions' BC Living Lab (Project J-002979).

10 References

Aarts, H.F.M., et al., 2000. Nitrogen (N) management in the 'De Marke' dairy farming system Nutr. *Cycling Agroecosyst*, 56, pp.231–240.

Ackerman, J., et al., 2018. Sustainable re-use of dairy cow manure as bedding and compost: nutrients and self-heating potential. *Canadian Biosystems Engineering*, 60, pp.6.1–6.7.

Asman, W.A., et al., 1998. Ammonia: emission, atmospheric transport and deposition. *The New Phytologist*, 139(1), pp.27–48.

Bach, A., et al., 2020. Symposium review: decomposing efficiency of milk production and maximizing profit. *Journal of Dairy Science*, 103, pp.5709–5725.

Barthelmie, R.J. and Pryor, S.C., 1998. Implications of ammonia emissions for fine aerosol formation and visibility impairment: a case study from the Lower Fraser Valley, British Columbia. *Atmospheric Environment*, 32, pp.345–352.

Bhandral, R., et al., 2009. Enhancing soil infiltration reduces gaseous emissions and improves N uptake from applied dairy slurry. *Journal of Environmental Quality*, 38(4), pp. 1372–1382.

Bittman, S. and Hunt, D., 2024. Multilevel integration of agricultural systems: the case of a modern peri-urban region. In R.Y. Yada, et al. (eds) *Future Food Systems* (pp. 31–42). Academic Press, London.

Bittman, S., et al., 2005a. Responses of the bacterial and fungal biomass in a grassland soil to multi-year applications of dairy manure slurry and fertilizer. *Soil Biology and Biochemistry*, 37(4), pp.613–623.

Bittman, S., et al., 2005b. Surface-banding liquid manure over aeration slots: a new low-disturbance method for reducing ammonia emissions and improving yield of perennial grasses. *Agronomy Journal*, 97(5), pp.1304–1313.

Bittman, S., et al., 2006. A. Land application of solid and liquid fractions from sedimented dairy slurry. In DIAS Report 122, Proceedings of the 12th Ramiran Inter-national Conference, Arhus, Denmark, 11–13 September 2006; Peterson, S.O., Ed.; Ministry of Food, Agriculture and Fisheries, Danish Institute of Agricultural Sciences: Tjele, Denmark, pp. 16–159.

Bittman, S., et al., 1999. Surface-banded and broadcast dairy manure effects on tall fescue yield and nitrogen uptake. *Agronomy Journal*, 91(5), pp.826–833.

Bittman, S., et al., 2007. Agronomic effects of multi-year surface-banding of dairy slurry on grass. *Bioresource Technology*, 98(17), pp.3249–3258.

Bittman, S., et al., 2011. Removing solids improves response of grass to surface-banded dairy manure slurry: a multiyear study. *Journal of Environmental Quality*, 40(2), pp.393–401.

Bittman, S., et al., 2012. Precision placement of separated dairy sludge improves early phosphorus nutrition and growth in corn (Zea mays L.). *Journal of Environmental Quality*, 41(2), pp.582–591.

Bittman, S., et al., 2014a. Nitrogen recovery from long term applied dairy slurry by three forage grasses: a long-term trial. In -C.S.C. Marques dos Santos Cordovil (ed) *The Nitrogen Challenge: Building a Blueprint for Nitrogen Use Efficiency and Food Security*, pp. 259–260. Proceedings of the 18th Nitrogen Workshop. Lisbon, Portugal, 30 June–3 July. Instituto Superior Agronomia Press, Lisbon, Portugal.

Bittman, S., et al., 2014b. In E. Taylor and A. McMillan (eds) *Effects of Agriculture on Air Quality in Canada. Air Quality Management: Canadian Perspectives on a Global Issue*. Springer, London, pp. 237–259.

Bittman, S., et al. 2017. Potential for mitigating atmospheric ammonia in Canada, 2017. *Soil Use and Management*, 33(2), pp. 263–275.

Bittman, S., et al., 2023. Inferences on soil nutrients and carbon following application of dairy manure from long term grass experiments in a temperate climate xxv international Grassland congress proceedings. 14–19 May 2023, Covington, Kentucky, pp.1469–1472. https://doi.org/10.52202/071171-035680.

Bolinder, M.A., et al., 2020. The effect of crop residues, cover crops, manures and nitrogen fertilization on soil organic carbon changes in agroecosystems: a synthesis of reviews. *Mitigation and Adaptation Strategies for Global Change*, 25, pp.929–952.

Cade-Menun, B.J., et al., 2015. Stratification of phosphorus forms from long-term conservation tillage and poultry litter application. *Soil Science Society of America Journal*, 79, pp. 504–516.

Chen, G., et al., 2017. Changes in runoff chemistry and soil fertility after multiple years of cattle winter bale feeding on annual cropland on the Canadian Prairies. *Agriculture, Ecosystems and Environment*, 240, pp.1–13.

Cherney, D.J., et al., 2002. Orchardgrass and tall fescue utilization of nitrogen from dairy manure and commercial fertilizer. *Agronomy Journal*, 94(3), pp.405–412.

Clark, S.C., et al., 2021. Persistent nitrate in alpine waters with changing atmospheric deposition and warming trends. *Environmental Science & Technology*, 55(21), pp.14946–14956.

Cotrufo, M.F., et al., 2019. Soil carbon storage informed by particulate and mineral-associated organic matter. *Nature Geoscience*, 12, pp.989–994.

Cox, S.E., et al., 2018. Concentrations of nutrients at the water table beneath forage fields receiving seasonal applications of manure, Whatcom County, Washington, autumn 2011–spring 2015: U.S. Geological Survey Scientific Investigations Report 2018-5124, 41 p. https://doi.org/10.3133/sir20185124.

D'Amours, E., et al., 2021. Combining perennial grass–legume forages and liquid dairy manure contributes to nitrogen accumulation in a clayey soil. *Canadian Journal of Soil Science*, 101(3), pp.378–388.

Dell, C.J., et al., 2012. Low-disturbance manure incorporation effects on ammonia and nitrate loss. *Journal of Environmental Quality*, 41(3), pp. 928–937.

Dion, P.P., et al., 2020. NLOS-OG: a management tool for nitrogen fertilization in organic greenhouses. *Acta Hortic*, 1296, pp.1093–1098. https://doi.org/10.17660/ActaHortic.2020.1296.138.

Dorais, M., et al., 2019. NLOS-OG: a management tool for nitrogen fertilization in organic greenhouses. In *International Symposium on Advanced Technologies and Management for Innovative Greenhouses: GreenSys2019 1296* (pp. 1093–1098).

Duplessis, M., et al., 2021. Toward precision feeding regarding minerals: what is the current practice in commercial dairy herds in Québec, Canada? *Animals*, 11, 1320.

Embertson, N. 2016. Protecting Puget Sound watersheds from agricultural pollution using a progressive manure Application Risk Management (ARM) system. https://www.eopugetsound.org/articles/protecting-puget-sound-watersheds-agricultural-pollution-using-progressive-manure.

Federolf, C.P., et al., 2015. Optimizing nitrogen and phosphor use efficiencies from liquid manure by slurry injection to reduce environmental pollution. *Procedia Environmental Sciences*, 29, pp.227–228.

Feng, X., et al., 2022. Evaluation of near-infrared reflectance and transflectance sensing system for predicting manure nutrients. *Remote Sensing*, 14(4), p.963.

Forge, T.A., et al., 2005a. Impacts of sustained use of dairy manure slurry and fertilizers on populations of Pratylenchus penetrans under tall fescue. *Journal of Nematology*, 37(2), p.207.

Forge, T.A., et al., 2005b. Responses of grassland soil nematodes and protozoa to multi-year and single-year applications of dairy manure slurry and fertilizer. *Soil Biology and Biochemistry*, 37(10), pp.1751–1762.

Franzluebbers, A., et al., 2021. Integrated crop-livestock systems: lessons from New York, British Columbia, and the south-eastern United States. *Frontiers of Agricultural Science and Engineering*, 8(1), pp.81–96.

Gross, A. and Glaser, B., 2021. Meta-analysis on how manure application changes soil organic carbon storage. *Scientific Reports*, 11(1), p.5516.

Hafner, S.D., et al., 2019. A flexible semi-empirical model for estimating ammonia volatilization from field-applied slurry. *Atmospheric Environment*, 199, pp.474–484.

Hansen, J.C., et al., 2004. Phosphorus speciation in manure-amended alkaline soils. *Journal of Environmental Quality*, 33(4), pp. 1521–1527.

Hao, X., et al., 2009. Spatial pattern of ammonia sorption by soil and vegetation downwind of a beef feedlot. *Agriculture, Ecosystems & Environment*, 132(1–2), pp.39–47.

Harrigan, T.M., 2010. Liquid manure hauling capacity of custom applicators using tank spreader systems. *Applied Engineering in Agriculture*, 26(5), pp.729–741.

He, Z. and Waldrip, H.M. 2015. Composition of whole and water-extractable organic matter of cattle manure affected by management practices. *Labile Organic Matter – Chemical Compositions, Functions, and Significance in Soil and the Environment*, 62, pp. 41–60.

He, Z., et al., 2015. Potential traceable markers of organic matter in organic and conventional dairy manure using ultraviolet-visible and solid-state 13C nuclear magnetic resonance spectroscopy. *Organic Agriculture*, 5, pp.113–122.

He, Z., et al., 2009. Characterization of plant-derived water extractable organic matter by multiple spectroscopic techniques. *Biology and Fertility of Soils*, 45, pp.609–616.

Hirsch, H.R., 2007. Applicability of the NLOS model for predictions of soil water movement and nitrogen transport in an agricultural soil, Agassiz, BC. WWU Graduate School Collection. 441.https://cedar.wwu.edu/wwuet/441.

Hunt, D.E. and Bittman, S., 2021. Precision injection of dairy sludge on crop yield and N and P uptake in juvenile and mature no-till silage corn. *Agronomy*, 11(2), p.370.

Hunt, D.E., et al., 2016. Improving nitrogen and phosphorus response of corn (Zea mays L.) to dairy slurry by precision injection: benefits and risks. In Proceedings of the 2016 International Nitrogen Initiative Conference. Solutions to improve nitrogen use efficiency for the world, Melbourne, Australia. www.ini2016.com.

Hunt, D.E., et al., 2019. Year-round N2O emissions from long-term applications of whole and separated liquid dairy slurry on a perennial grass sward and strategies for mitigation. *Frontiers in Sustainable Food Systems*, 3, p.86.

Johnson, K.N., et al., 2011. Effect of dairy manure slurry application in a no-till system on phosphorus runoff. *Nutrient Cycling in Agroecosystems*, 90, pp.201–212.

Kelln, B., et al., 2012. Effects of beef cow winter feeding, pen manure and compost on soil nitrogen and phosphorus amounts and distribution, soil density and crop biomass. *Nutrient Cycling in Agroecosystems*, 92(2), pp.183–194.

Kleinman, P.J., et al., 2022. Envisioning the manureshed: toward comprehensive integration of modern crop and animal production. *Journal of Environmental Quality*, 51(4), pp.481–493.

Koenig, K.M., et al., 2023. Effects of sustainable agronomic intensification in a forage production system of perennial grass and silage corn on nutritive value and predicted milk production of dairy cattle. *Journal of Dairy Science*, 106(1), pp.274–293.

Kowalenko, C.G., 1989. The fate of applied nitrogen in a Fraser Valley soil using 15N in field microplots. *Canadian Journal of Soil Science*, 69(4), pp.825–833.

Kowalenko, C.G., et al., 1989. An evaluation of the T-sum method for efficient timing of spring nitrogen applications on forage production in south coastal British Columbia. *Canadian Journal of Plant Science*, 69(4), pp.1179–1192.

Kulesza, S.B., et al., 2016. Injecting poultry litter into orchardgrass hay. *Communications in Soil Science and Plant Analysis*, 47(11), pp.1389–1397.

Kumaragamage, D. and Akinremi, O.O., 2018. Manure phosphorus: mobility in soils and management strategies to minimize losses. *Current Pollution Reports*, 4, pp.162–174.

Lalor, S.T.J., et al., 2011. Nitrogen fertilizer replacement value of cattle slurry in grassland as affected by method and timing of application. *Journal of Environmental Quality*, 40(2), pp.362–373.

Landry, H., et al., 2011. Development and evaluation of subsurface application technology for solid organic fertilizers. *Applied Engineering in Agriculture*, 27(4), pp.533–549.

Lapen, D.R., et al., 2008. Effect of liquid municipal biosolid application method on tile and ground water quality. *Journal of Environmental Quality*, 37(3), pp.925–936.

Lau, A., et al., 2003. Odor measurements for manure spreading using a subsurface deposition applicator. *Journal of Environmental Science and Health, Part B*, 38(2), pp.233–240.

Lehmann, J. and Kleber, M., 2019. The contentious nature of soil organic matter. *Nature*, 528, pp.60–68.

Li, Y., 2019. *Forage Crop Nitrogen Recovery and Nitrogen Field-Losses Determined on Semi-Virtual Dairy Farms Under Integrated Nutrient and Crop Management Scenarios* (Dissertation). University of British Columbia.

Li, C., et al., 2021a. Integrated farm management systems to improve nutrient management using semi-virtual Farmlets: agronomic responses. *Environmental Research Communications*, 3(7), p.075009.

Li, B., et al., 2021b. Responses of soil organic carbon stock to animal manure application: a new global synthesis integrating the impacts of agricultural managements and environmental conditions. *Global Change Biology*, 27(20), pp.5356–5367.

Liang, B.C., et al., 1996. Characterization of water extracts of two manures and their adsorption on soils. *Soil Science Society of America Journal*, 60, pp.1758–1763.

Liang, C., et al., 2021. Manure-induced carbon retention measured from long-term field studies in Canada. *Agriculture, Ecosystems & Environment*, 321, p.107619.

Longchamps, L., et al., 2023. Farmer-led on-farm experimentation enhanced with digital agronomy. In A. Rejeb, et al. (eds) *Precision Agriculture'23* (pp. 137–143). Wageningen Wageningen Academic.

Ma, X., et al., 2016. Shifts in microbial biomass and community composition in subtropical paddy soils under a gradient of manure amendments. *Biology and Fertility of Soils*, 52, pp.755–787.

Maillard, É. and Angers, D.A. 2014. Animal manure application and soil organic carbon stocks: a meta-analysis. *Global Change Biology*, 20, pp.666–679.

Maillard, É., et al., 2015. Carbon accumulates in organo-mineral complexes after long-term liquid dairy manure application. *Agriculture, Ecosystems & Environment*, 202, pp.108–119.

Maillard, É., et al., 2016. Greater accumulation of soil organic carbon after liquid dairy manure application under cereal-forage rotation than cereal monoculture. *Agriculture, Ecosystems & Environment*, 233, pp.171–178.
McDowell, R.W., et al., 2008. A comparison of phosphorus speciation and potential bioavailability in feed and feces of different dairy herds using ^{31}P nuclear magnetic resonance spectroscopy. *Journal of Environmental Quality*, 37, pp.741–752.
Mederle, M. and Bernhardt, H., 2017. Analysis of influencing factors and decision criteria on Infield-Logistics of different farm types in Germany. *Agricultural Engineering International: CIGR Journal*, 19(2), pp.139–148.
Misselbrook, T.H., et al., 2022. Field application of organic and inorganic fertilizers. In M.A. Sutton (ed) *Nitrogen Opportunities for Agriculture, Food and Environment* (pp. 89–114). UK Centre for Ecology and Hydrology.
Møller, H.B., et al., 2000. Solid–liquid separation of livestock slurry: efficiency and cost. *Bioresource Technology*, 74(3), pp.223–229.
Möller, H.B., et al., 2002. Separation efficiency and particle size distribution in relation to manure type and storage conditions. *Bioresource Technology*, 85, pp.189–196.
Neufeld, K.R., et al., 2017. Long-term alternative dairy manure management approaches enhance microbial biomass and activity in perennial forage grass. *Biology and Fertility of Soils*, 53, pp.613–626.
Ni, J.Q. and Lim, T.T., 2022. Manure Characteristics, Testing, and Sampling. ABE-166-W. Purdue University Extension. https://extension.purdue.edu/extmedia/ABE/ABE-166-W.pdf.
Noack, S.R., et al. 2012. Crop residue phosphorus: speciation and potential bioavailability. *Plant and Soil*, 359, pp.375–385.
Oenema, J., et al., 2011 Participatory farm management adaptations to reduce environmental impact on commercial pilot dairy farms in the Netherlands NJAS-Wageningen *Journal of Life Sciences*, 58, pp.39–48.
Ouatahar, L., et al., 2021. Modelling the effect of feeding management on greenhouse gas and nitrogen emissions in cattle farming systems. *Science of the Total Environment*, 776, p.145932.
Pedersen, I.F., et al., 2020. Row-injected cattle slurry can replace mineral P starter fertiliser and reduce P surpluses without compromising final yields of silage maize. *European Journal of Agronomy*, 116, p.126057.
Pedersen, J., et al., 2021. Effect of reduced exposed surface area and enhanced infiltration on ammonia emission from untreated and separated cattle slurry. *Biosystems Engineering*, 211, pp.141–151.
Pelster, D.E., et al., 2022. Ratio of non-growing season to growing season N2O emissions in Canadian croplands: an update to national inventory methodology. *Canadian Journal of Soil Science*, 103(2), pp.344–352.
Piepel, M.F. and Olfs, H.W., 2023. Development of a physicochemical test kit for on-farm measurement of nutrients in liquid organic manures. *Agriculture*, 13(2), p.477.
Pogue, S.J., et al. 2025. Impact of climate change on the nitrogen budget of a dairy farm in the Fraser Valley, British Columbia, Canada. Agricultural Systems, 227, pp. 104333.
Raworth, D.A., et al., 2004. Effects of dairy slurry application on carabid beetles in tall fescue, British Columbia, Canada. *Agriculture, Ecosystems & Environment*, 103(3), pp.527–534.

Published by Burleigh Dodds Science Publishing Limited, 2026.

Reeves III, J.B., 2007. The present status of "quick tests" for on-farm analysis with emphasis on manures and soil: what is available and what is lacking?. *Livestock Science*, 112(3), pp.224–231.

Rotz, A., et al., 2021. Environmental assessment of United States dairy farms. *Journal of Cleaner Production*, 315, p.128153.

Sampedro, J., et al., 2023. Marginal damage of methane emissions: Ozone impacts on agriculture. *Environmental and Resource Economics*, 84(4), pp.1095–1126.

Schröder, J.J., et al., 2015. Maize yields benefit from injected manure positioned in bands. *European Journal of Agronomy*, 64, pp.29–36.

Sheppard, S., et al., 2011a. Ecoregion and farm size differences in dairy feed and manure nitrogen management: a survey. *Canadian Journal of Animal Science*, 91(3), pp.459–473.

Sheppard, S., et al., 2011b. Modelling monthly NH3 emissions from dairy in 12 Ecoregions of Canada. *Canadian Journal of Animal Science*, 91(4), pp.649–661.

Sherman, J., et al., 2021. Impacts of low disturbance liquid dairy manure incorporation on alfalfa yield and fluxes of ammonia, nitrous oxide, and methane. *Agriculture*, 11(8), p.750.

Singh, A.K., et al., 2020. A review: effect of bedding material on production, reproduction and health and behavior of dairy animals. *International Journal of Livestock Research*, 10(7), pp.11–20.

Sommer, S.G., et al., 1991. Effects of temperature, wind speed and air humidity on ammonia volatilization from surface applied cattle slurry. *The Journal of Agricultural Science*, 117(1), pp.91–100.

Sommer, S.G., et al., 2003. Processes controlling ammonia emission from livestock slurry in the field. *European Journal of Agronomy*, 19(4), pp.465–486.

Søgaard, H.T., et al., 2002. Ammonia volatilization from field-applied animal slurry–the ALFAM model. *Atmospheric Environment*, 36(20), pp.3309–3319.

Sørensen, P., et al., 2017. Empirical model for mineralisation of manure nitrogen in soil. *Soil Research*, 55(6), pp.500–505.

Sutton, M.A., et al., 1993. The exchange of atmospheric ammonia with vegetated surfaces. I: unfertilized vegetation. *Quarterly Journal of the Royal Meteorological Society*, 119(513), pp.1023–1045.

Sutton, M.A., et al., 2022. Nitrogen opportunities for agriculture, food & environment. UNECE guidance document on integrated sustainable nitrogen management. https://unece.org/environment-policy/publications/guidance-document-integrated-sustainable-nitrogen-management.

Toda, M., et al., 2020. Nitrogen balance and use efficiency on dairy farms in Japan: a comparison among farms at different scales. *Environmental Research Communications*, 2, 125001. https://doi.org/10.1088/2515-7620/abce4a.

Toor, G.S., et al., 2005. Establishing a link between phosphorus forms in dairy diets, feces and manure. *Journal of Environmental Quality*, 34, pp. 1380–1391.

Van der Weerden, T.J. and Jarvis, S.C., 1997. Ammonia emission factors for N fertilizers applied to two contrasting grassland soils. *Environmental Pollution*, 95(2), pp.205–211.

Van der Weerden, T., et al., 2018. The effects of system changes in grazed dairy farmlet trials on greenhouse gas emissions. *Animals*, 8(12), p.234.

VanderZaag, A.C., et al., 2013. Towards an inventory of methane emissions from manure management that is responsive to changes on Canadian farms. *Environmental Research Letters*, 8(3), p.035008.

VanderZaag, A., et al., 2023. Farm size distribution in the Canadian livestock sector. *Canadian Journal of Animal Science*, 103(2), pp.218–222.

van Es, H., et al., 2023. Adapt-N®(Yara International). In D. Cammarano, et al. (ed) *Precision Agriculture: Modelling* (pp. 201–208). Cham: Springer International Publishing.

Van Kessel, J.S. and Reeves III, J.B., 2000. On-farm quick tests for estimating nitrogen in dairy manure. *Journal of Dairy Science*, 83(8), pp.1837–1844.

van Vliet, L.J.P., et al., 2006. Aerating grassland before manure application reduces runoff nutrient loads in a high rainfall environment. *Journal of Environmental Quality*, 35(3), pp. 903–911.

Webb, J., et al., 2010. The impacts of manure application methods on emissions of ammonia, nitrous oxide and on crop response–a review. *Agriculture, Ecosystems & Environment*, 137(1–2), pp. 39–46.

Webb, J., et al., 2013. An assessment of the variation of manure nitrogen efficiency throughout Europe and an appraisal of means to increase manure-N efficiency. *Advances in Agronomy*, 119, pp.371–442.

Wu, Z., et al., 2000. Milk production, reproductive performance and fecal excretion of phosphorus by dairy cows fed three amounts of phosphorus. *Journal of Dairy Science*, 83, pp.1028–1041.

Wyer, K.E., et al., 2022. Ammonia emissions from agriculture and their contribution to fine particulate matter: a review of implications for human health. *Journal of Environmental Management*, 323, p.116285.

Zhang, H., et al., 2018. Availability of phosphorus after long-term whole and separated slurry application to Perennial Grass prior to Corn Silage. *Journal of Environmental Quality*, 47(4), pp.893–901.

Zhang, H., et al., 2019. Animal-based organic amendments and their potential for excessive nitrogen leaching and phosphorus loading. *Agronomy Journal*, 111(5), pp.2207–2217.

Zhang, H., et al., 2021. Nitrogen dynamics after low-emission applications of dairy slurry or fertilizer on perennial grass: a long term field study employing natural abundance of δ15N. *Plant and Soil*, 465(1), pp.415–430.

Chapter 12

Methane utilisation on dairy farms

Sigrid Kusch-Brandt, University of Southampton, UK and University of Applied Sciences Ulm, Germany

1 Introduction

Livestock breeding has a significant impact on the environment, including its contributions to climate change (Xu et al. 2021). Greenhouse gas (GHG) emissions in the form of methane (CH_4), nitrous oxide (N_2O), and ammonia (NH_3) are among the unavoidable outputs of livestock production systems (Cheng et al. 2022; Tubiello et al. 2021), and CH_4 as a GHG is of particular concern in this context (Prathap et al. 2021). Different livestock breeding systems have very different levels of GHG emissions, with those from dairy systems being particularly high. Average methane emission rates per animal unit (AU) and day (d) were reported to be 281–323 g CH_4/(d*AU) for dairy cows, 161–194 g for beef heifers, 56–118 g for beef steers, 16–24 g for pigs, and 12–13 g for poultry layer houses (Liu and Powers 2014; Liu S. et al. 2021). Even higher variations and emission levels were reported for dairy cows by Broucek (2014), who found 151–497 g CH_4/(d*AU). Overall, livestock breeding systems are responsible for about 14.5% of the global anthropogenic GHG emissions,

http://dx.doi.org/10.19103/AS.2025.0157.24

and beef and dairy cattle production together account for more than 60% of this contribution (Gerber et al. 2013; Menghistu et al. 2021). Consequently, there is an urgent need to reduce the net GHG emissions of livestock-related businesses, including dairy farms.

On dairy farms, GHG emissions in the form of methane from the livestock have two main sources, namely enteric fermentation and manure/slurry handling. Methanogenic microorganisms play an important role in the enteric fermentation in the rumen of ruminants, and this process naturally generates methane as a by-product of the digestion of an animal's feed. This methane is belched or exhaled by the animal. In addition, the handling of manure and slurry contributes to the production of CH_4 and its release into the atmosphere. Under anaerobic conditions, i.e. in the absence of free oxygen, microbial consortia decompose organic material in the stored animal excreta/manure/slurry into CH_4 and other components. How much CH_4 is produced from manure/slurry and how much of it is emitted to the atmosphere as GHG depends on many factors, including the composition of the manure/slurry, the collection and handling of the manure/slurry, the storage conditions and the techniques used to apply manure/slurry to agricultural land. Enteric fermentation and anaerobic decomposition of manure/slurry during storage are both anaerobic fermentation processes that convert organic matter into CH_4 and other components in the absence of free oxygen. To lower direct GHG emissions from dairy farms, there are several approaches. They focus on reducing CH_4 production and emission from enteric fermentation and from manure/slurry handling, and research in this area is ongoing. These approaches are discussed in other chapters of this book. In those contexts, methane is addressed as an undesired gaseous emission, and any reduction in its release to the atmosphere contributes to improving the environmental performance of dairy farms by lowering the site's GHG emissions. While such efforts reduce unwanted GHG emissions, they are not sufficient to achieve net-zero GHG emissions from dairy farms.

Anaerobic digestion (AD) is a biotechnological solution to break down organic (biogenic) matter into an energy-rich gas (referred to as 'biogas') and a stabilised semi-liquid or liquid residue (referred to as 'digestate'). The obtained biogas can either be used directly or upgraded to almost pure methane, which is referred to as 'biomethane'. Anaerobic digestion for biogas/biomethane production is based on a series of biochemical processes implemented in engineered systems and in an oxygen-free ('anaerobic') environment. The biochemical processes are carried out in several steps by microbial consortia that contain different groups of specialised microorganisms. The relevant biochemical processes also occur naturally in anaerobic environments, e.g. as enteric fermentation in the rumen of ruminants and as unwanted degradation of stored livestock manure/slurry. The naturally occurring processes are deliberately intensified and accelerated in engineered AD systems to obtain the methane-rich (and therefore energy-rich) biogas. Although the basic biochemical processes are the same as those that occur naturally, the term 'anaerobic digestion' usually refers to the engineered anaerobic process.

Deployment of AD through engineered systems to produce, capture and utilise methane-rich biogas from organic residues adds to dairy farms a renewable energy production scheme that can achieve negative net GHG emissions. This represents an important pathway towards achieving net-zero GHG emissions for the dairy sector, and it also offers the opportunity for an additional source of income for dairy farms.

This chapter provides an overview of how AD for biogas and biomethane production and the utilisation of the methane can be implemented on dairy farms. Sections 2–4 describe the production and utilisation of the methane-rich gas from dairy substrates. How this contributes to achieving net-zero GHG emissions on dairy farms is explained in Section 5. While the AD processes themselves and the gas valorisation routes are rather well documented in the available literature, there is a relative lack of literature on how an AD scheme can be successfully integrated into the overall management of a dairy farm, and therefore a separate part of this chapter (Section 6) is devoted to this aspect. Finally, Sections 8 and 9 provide conclusions, outline future trends and suggest additional sources of information.

2 Anaerobic digestion as a solution to produce renewable methane on dairy farms

2.1 The biogas process

Biogas production through AD is an outcome of a series of interlinked biochemical processes, namely hydrolysis, acidogenesis, acetogenesis and methanogenesis (Anukam et al. 2019; Náthia-Neves et al. 2018; Wellinger et al. 2013). The key processes are carried out synergistically by microbial communities in a wet environment (usually with a minimum moisture content of around 30%) and in the absence of free oxygen (Atelge et al. 2020; Meegoda et al. 2018; Náthia-Neves et al. 2018).

Depending on the feedstock used and the process conditions applied, the biogas composition can show some variations, but it typically contains 50–70% CH_4 per unit volume of gas (as dry gas), 30–45% carbon dioxide (CO_2), and small amounts of other constituents such as nitrogen, sulphur compounds (e.g. hydrogen sulphide (H_2S)), ammonia, halogenated compounds, and other volatile organic compounds (Calbry-Muzyka et al. 2022). As energy-rich component, CH_4 is the target biogas component, but the occurrence of relevant quantities of CO_2 and gas impurities is unavoidable in the context of the metabolic processes of the microorganisms involved. AD converts only a part of the feedstock's carbon into methane, while some is released as gaseous CO_2, and a further part remains in the digestate, i.e. the liquid or semi-liquid residue of the digestion process. The AD digestate itself is a very valuable organic fertiliser and soil improver (Czekała et al. 2022), and therefore, an additional benefit of AD systems.

Different types of biomasses can be converted into biogas and digestate through AD, such as agricultural residues, food waste, organic waste from industry, separately collected biowaste from households and sewage sludge (Kougias and Angelidaki 2018; WBA 2019). The use of energy crops, i.e. crops grown specifically to serve for energy generation rather than as food or feed, is also possible and exists in practice, but their cultivation remains critically debated due to sustainability implications of water demand, fertiliser application and land use change (Duarah et al. 2022). Organic substrates with high water content and low lignification levels are particularly well suited for AD treatment. Carbohydrates, proteins, and fats serve as valuable biomass constituents for biogas production (Ekstrand et al. 2022), while lignin cannot be microbiologically converted under anaerobic conditions. The lignification of biomass also complicates microbial access to the degradable biomass components. The AD particularly benefits from feedstocks such as livestock slurry, food waste or grease, i.e. materials that can be pumped and stirred and/or are easily hydrolysable, whereas structure-rich solid substrates are more suitable for composting and would usually require extensive pre-treatment for any integration into AD (Wellinger et al. 2013).

It is essential to ensure favourable process conditions for the AD microbial consortium. This includes adequate availability of macro- and micro-nutrients to avoid process inhibition (Song et al. 2020). Key macro-nutrients are carbon (C), nitrogen (N), phosphorus (P), and sulphur (S), and a ratio of C:N:P:S around 450:15:5:3 is considered adequate for sufficient nutrient supply (Kaiser et al. 2008). A target C:N ratio between 20:1 and 30:1 is assumed compatible with the aim of ensuring stability of the AD process (Kaiser et al. 2008). If the C:N ratio exceeds the favourable range, nitrogen deficiency will cause malfunctioning of the microbial metabolism, while a C:N ratio below this range risks process inhibition by ammonia, in particular below a minimum C:N ratio of 10:1 (FNR 2016). If the sulphur concentration is high, the occurrence of elevated H_2S concentrations in the digester quickly causes inhibition (Vu et al. 2022). Deficiency in micro-nutrients can be a complex challenge if the feedstock is unfavourably composed. Important micro-nutrients are cobalt (Co), copper (Cu), manganese (Mn), molybdenum (Mo), nickel (Ni), selenium (Se), tungsten (W), and tin (Zn), but others can also affect the stability and efficiency of the process (Bardi et al. 2023; Rasapoor et al. 2020; Song et al. 2020). Where necessary, co-digestion of different substrates and supplementation of trace elements are effective and well-established measures to mitigate nutrient imbalances (Afroze et al. 2023; Bardi et al. 2023; Karki et al. 2021).

Anaerobic digestion feedstocks should be free of any contamination that could cause process inhibition or toxification. For example, industrial residues may contain intolerable levels of heavy metals (Huang et al. 2022). Livestock residues may contain unsuitable detergents from cleaning activities or residues from animal medications. Even low concentrations of antibiotics in animal manure/slurry cause critical inhibitions of the AD process (Steinmetz and Gressler 2019) and must therefore be avoided.

During AD, a favourable pH level must be maintained. While hydrolysis benefits from acidic conditions, acetogens and methanogens prefer a rather neutral pH value between 6.5 and 7.5, but not higher than 8.0 (FNR 2016). The pH of the feedstock has an influence on the pH in the digester, but it is mainly due to intermediate metabolic products that the pH in the digester changes during the process. Long-chain fatty acids and volatile fatty acids (VFAs) released during hydrolysis, acidogenesis, and acetogenesis can cause a critical drop in pH (acidification of the reactor content) (Song et al. 2020). Conversely, excess ammonia production during acidogenesis risks a too high pH increase. Ammonia itself inhibits methanogens at elevated pH because in such a state a relevant share of the ammonia is present as inhibitory free ammonia (NH_3) (Jiang et al. 2019).

2.2 Anaerobic digestion plant types

A variety of different AD plant types are state of the art and can be found in full-scale applications. Table 1 provides an overview of the main AD plant types that are commonly used to digest agricultural feedstock or biogenic wastes. The expression 'digester' refers to the AD reactor employed, i.e. the air-tight vessel or other equipment in which the AD process takes place; as such, 'digester' and 'reactor' are synonymous.

2.3 Anaerobic digestion plant types on dairy farms

On dairy farms, most AD plants represent continuous, wet digestion systems, based on the CSTR type (Table 1), and are either one- or two-stage solutions. Figure 1 shows a typical solution for a small to medium dairy farm (e.g. a site with 200 or fewer dairy cows). Two-stage systems are more likely to be found on large dairy farms, or at sites where significant quantities of energy-rich co-substrates (e.g. food waste, energy crops) are received. At dairy farms with solid manure, dry digestion, for example in batch digesters, has also been implemented (Kusch 2007). Mesophilic processing is more widely applied on farms for manure and slurry. Where thermophilic AD is implemented, this is usually at sites with

Table 1 Common types of AD plants (source: author's own elaboration, based on Kusch (2007), Kougias and Angelidaki (2018), Kranert et al. (2012), Nsair et al. (2020), Suryawanshi et al. (2010))

Digestion temperature	In terms of AD temperature, *mesophilic digestion*, conducted around 30–38°C, is most frequently applied. In countries with colder seasons, the AD temperature is controlled by technical solutions (heating). In some countries (tropical countries), AD plants are operated in the mesophilic range without temperature control, at ambient temperatures (ca. 20–45°C). Compared to thermophilic processes, mesophilic AD benefits from the greater number of microorganisms present in this temperature range, which makes the process more tolerant to changes and shocks. AD schemes that employ *thermophilic digestion*, with a temperature maintained at around 48–57°C, are also common. Compared to mesophilic AD, the higher temperature provides better sanitation and faster reaction rates, resulting in shorter required retention times and smaller reactors. However, the thermophilic process is less resilient against shocks or disturbances, and it requires more energy to heat the reactor. AD plants operating at *psychrophilic temperatures* (<20°C) are limited to low-tech applications. At such low temperatures, the degradation of organic material is slow and long residence times are required to achieve degradation of the feedstock.
Operation mode: continuous batch, fed-batch	In a *continuous* (or more accurately semi-continuous) system, new substrate is added as input to the reactor on a regular basis. This results in continuous biogas production along with continuous (or more accurately semi-continuous) removal of a corresponding quantity of digestate. Large scale plants will usually adopt such a mode of operation. Changes in the feedstock composition can negatively affect the continuity of biogas production and should therefore be avoided or carefully monitored and controlled. In *batch systems*, distinct batches of feedstock are treated one after the other. The entire batch is loaded into the reactor at once, digested over a pre-defined period of time, and then unloaded. The biogas production varies over the treatment period of a batch, so several reactors (at least three) need to be operated off-set for commercial operation. In *fed-batch mode,* feedstock is gradually added to the digester until the space is used up, and then the reactor is emptied, and a new cycle begins. Like in the batch mode, biogas production changes over time.
Total solids (TS) content	So-called *wet digestion plants* typically operate at TS < 12%. They are most frequently applied in agriculture, including at dairy farms. When treating solid materials (feedstock with high TS), the water content of the reactor content must be adjusted by adding liquid material (e.g. water) or by recirculating liquid digester effluent. For the digestion of solid feedstocks, so-called *dry digestion plants* are typically operated at TS > 20%. Although referred to as 'dry', some moisture needs to be present for microbial activity to occur. The water content within the digester is usually not set to a specific level but is a result of the digesting substrates. Dry digestion plants exist in agriculture, including for the treatment of solid dairy manure or horse manure, but they are more common for the processing of separately collected biowaste from households.

	Although the TS content is used as a criterion to distinguish between wet and dry digestion, there exists no strict threshold and some variations can be found in the literature. Occasionally, a third type is introduced, namely *semi-dry digestion*, which refers to AD processes operated between 12% and 20% TS.
Transport and homogenisation of reactor content	Most AD plants employ the *continuous stirred tank reactor* (CSTR) concept. The digester content is continuously (or in many cases semi-continuously) stirred by appropriate technical equipment, which homogenises the material. This is directly applicable to feedstocks that can be pumped and stirred (i.e. with elevated water content), whereas solid feedstocks require the addition of liquid. The stirring means that not all matter has the same residence time in the reactor (different particles have different retention times). An average retention time is given, but some of the feedstock will leave the reactor after a very short digestion with incomplete degradation, while some will remain in the digester longer than the calculated average. For solid or semi-solid feedstocks, *plug flow digesters* can be found in practice. Their inlet and outlet are at opposite ends of the longitudinal installation, and feeding is typically done semi-continuously with a thick substrate (TS content ca. 15%). Material advances through the digester whenever new substrate is added. Compared to CSTR solutions, there is less variation in the particle retention time, and a minimum retention time is more reliably achieved. However, this solution is not applicable for wet digestion.
One- or two-stage (multi-stage) systems	By far the most AD plants are *one-stage systems*, consisting of one single reactor for the digestion process. The reactor is usually followed by a storage tank. In *two-stage systems* (or in rare cases multi-stage systems with more than two reactors), the AD process conditions can be optimised for the different groups of microorganisms involved in the AD process chain. The first stage usually focuses on achieving a rapid liquefaction of the feedstock, while the second stage converts soluble matter into biogas. Compared to single-stage systems, this sequence of separate processes is faster and more stable, but the investment and maintenance costs are significantly higher.

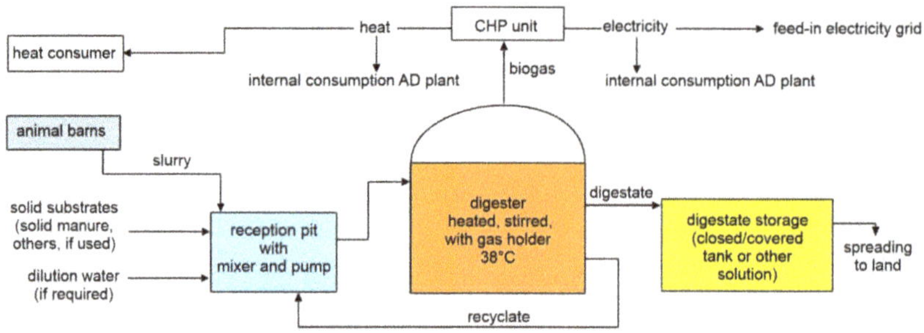

Figure 1 Schematic representation of a typical anaerobic digestion (AD) plant to produce methane-rich biogas at a small to medium-sized dairy farm: continuous stirred tank reactor (CSTR) concept, one-stage, mesophilic, wet digestion plant, equipped with a combined heat and power (CHP) unit (source: author's own elaboration).

co-digestion (further feedstocks are digested together with the manure/slurry), in particular sites with substrates that require hygienisation, such as food waste.

The AD plants associated with dairy farms are usually operated on the individual farm, i.e. as a decentralised solution – either digesting only the farm's residues, or underpinning biogas generation by adding energy-rich co-substrates. However, there also exist solutions where the feedstock is collected from several farms and treated together in a centralised, large-scale AD plant, for example in Denmark (Booker Nielsen 2022; Raven and Gregersen 2007). The solutions implemented in practice to a large degree reflect the framework set by policy and legislation. These determine which AD solution can be economically viable and obtain the required permits. For example, if agricultural residues destined for external AD are classified as waste, these materials will fall under waste management regulations and transport to other farms will require appropriate provisions. Shifts in energy and environmental policies can make previously attractive AD schemes unviable, which has been observed for the centralised plants in Denmark (Raven and Gregersen 2007). As a result, solutions that are common in some countries may not exist in others, and the common plant types may also change within countries and regions over time.

3 Specificities of dairy manure and slurry as feedstock for methane production

Dairy manure and slurry are common feedstocks for AD. Their characteristics depend on several factors, such as the digestibility, protein and fibre content of the feed, animal age, barn management practices and choice of bedding material (Atandi and Rahman 2012; Zhu et al. 2021). This effects biogas and methane yields in AD. Table 2 shows average values and compares them with other common AD feedstocks.

Due to the rather low energy density of dairy manure and slurry, the potential methane yield is rather limited. The relatively high moisture content (low TS content) of animal slurry further contributes to a low energy density of this feedstock, and therefore to low methane production potential per digester volume. In addition, the digestion of dairy manure/slurry is a relatively slow process. Without pre-treatment, only a fraction of the organic matter contained in dairy manure and slurry is converted to biogas because a significant fraction of the organic matter (about 40 to 60%) is lignocellulosic material (Khan and Ahring 2021; Li et al. 2021), which is difficult for the AD microbial community to degrade (Section 2.1). As a result, there is a significant discrepancy between the theoretical methane yield based on the feedstock constituents and the actual methane yield achieved in practice (Atandi and Rahman 2012). Pre-treatment can increase the biogas yield to some extent, but the additional costs and efforts in most cases hinder a viable implementation at plants that mainly process feedstocks with a rather low energy density (Section 6). These limitations may reduce the attractiveness of AD schemes to be operated on the basis of dairy manure and slurry alone (Atandi and Rahman 2012).

However, dairy AD feedstock also has advantages. The AD of dairy manure and slurry is a steady and stable process – one that is resilient against shocks or disturbances, with a very low risk of process instability or failure. Dairy manure/slurry brings a particularly high buffering capacity to the AD process (Wellinger et al. 2013). The buffering effect unfolds mainly due to the high alkalinity of

Table 2 Physiochemical characteristics and biogas yield of selected biogenic materials, including dairy manure and slurry (TS: total solids; VS: volatile solids; FM: fresh material) (Source: author's compilation based on data reported in FNR (2016))

Substrate	TS (%)	VS (% of TS)	Biogas yield (Nm^3/t FM)	CH_4 yield (Nm^3/t FM)	CH_4 yield (Nm^3/t VS)
Cattle manure	20–25	68–76	60–120	33–36	130–330
Cattle slurry	6–11	75–82	20–30	11–19	110–275
Pig slurry	4–7	75–86	20–35	12–21	18–360
Poultry manure			130–270	70–140	200–360
Maize silage	28–35	85–98	170–230	89–120	234–364
Grass silage	25–50	70–95	170–200	93–109	300–338
Molasses	80–90	85–90	290–340	210–247	261–355
Brewer spent grains	20–25	70–80	105–130	62–112	295–443

Note: The biogas and methane yields are based on validated measurements campaigns and indicate ultimate yields of standardised laboratory testing. They are reported in 'norm cubic metres' (Nm^3), which refers to a condition of 0°C temperature and 1013 mbar pressure.

the feedstock, which lowers the VFA/alkalinity ratio throughout the digestion (Srisowmeya et al. 2021), and therefore the risk of VFA accumulation. This makes dairy manure/slurry an ideal candidate for co-digestion with energy-rich, rapidly hydrolysable biomasses such as food waste. Such co-digestion has several advantages for the substrates. The buffer capacity reduces the risk of process inhibition when digesting the rapidly hydrolysable biomass, e.g. food waste (Abbas et al. 2023). From the dairy farm perspective, the main objective of such co-digestion solutions is to increase biogas production per unit volume of AD reactor. Co-digestion also ensures a more balanced nutrient composition in the digester.

In terms of nutrients, dairy manure and slurry generally have a favourable composition and provide the required macro- and micro-nutrients for the AD process (FNR 2016; Wellinger et al. 2013). However, the carbon to nitrogen (C:N) ratio of dairy manure and slurry is below the optimal range (Section 2.1) and can often be lower than the minimum recommended range for AD of 10:1 (FNR 2016). Atandi and Rahman (2012) report an average C:N ratio of 9:1 for dairy manure. This excess nitrogen risks process inhibition by ammonia. Co-digestion with carbon-rich substrates such as food waste or energy crops is a widely used and effective strategy to mitigate this risk (Karki et al. 2021; Ma et al. 2020). Deficiency of micro-nutrients should not be a common challenge, but Wang et al. (2021) reported an increased methane yield from cattle manure when a mixture of iron (Fe), Co and Ni was added. This suggests that micro-nutrients could also be monitored.

Foam formation can occur in the digester. Proper mixing, adjustment of the organic loading, and biological defoaming methods are effective measures to combat this (Atandi and Rahman 2012; Yang et al. 2021).

4 Utilisation of the methane-rich gas

Methane, as the combustible and energy-rich component of biogas, can be utilised in different ways, such as electricity generation, heat supply, chilling/cold supply, cooking fuel, or vehicle fuel (Köttner et al. 2008; Scarlat et al. 2018). Biogas can be used directly for such purposes, or it can be upgraded to biomethane to increase the methane content of the gas. Biogas upgrading removes the non-combustible CO_2 and other impurities to achieve a CH_4 content of typically 95–99% (Golmakani et al. 2022), but at least 90% (Sahota et al. 2018) per unit volume of gas. Removing the non-combustible components significantly increases the calorific value of the gas. The calorific value of a gaseous fuel is commonly expressed as its lower heating value (LHV), also referred to as 'net calorific value', which is the amount of heat released by complete combustion minus the heat of water vapourisation in the combustion product (the LHV differs from the higher heating value in that it includes the latent heat of water). Biomethane typically has a LHV that is comparable to natural gas, i.e. around 36 MJ/m^3 (megajoules per cubic metre), whereas the LHV of biogas is between 16 and 28 MJ/m^3, depending on the gas composition (IEA 2020). Both biogas and biomethane can be used as versatile energy carriers for different applications, but only biomethane can directly replace natural gas in all applications (e.g. as a fuel in natural gas vehicles).

On agricultural sites, it is most common to valorise biogas into electricity and heat through a combined heat and power (CHP) unit. Depending on the legal and economic framework, the electricity produced is either used on site or it is supplied to the electricity grid. For example, some countries have guaranteed feed-in tariffs for electricity from renewable sources, which makes it more favourable to supply the electricity to the grid rather than use it on site. Where delivery to the grid is an economically preferential option, the technical grid requirements need to be respected. Connection is usually to the low or medium voltage distribution grid. The heat is usually used on site to heat the biogas digester and to meet the heat needs of the farm and residential areas, but it is also common to supply surplus heat to nearby users to generate an additional income. Surplus heat can also be used to provide chilling/cooling services, for example to nearby hospitals, the residential buildings or livestock stables (Agonafer et al. 2022; Köttner et al. 2008).

One cubic metre (1 m^3) of biogas with a CH_4 content of 55% has the energy content of 0.55 litres of heating oil. The use of biogas in a gas boiler solely for heat production is only an option for very small AD plants, where very low quantities of biogas are available. This is sometimes the case in Europe and North America, but it is more common in regions where micro-scale AD plants are more frequent (e.g. Asia). For typical AD plants in Europe and North America, it is not economically viable to invest only in heat production infrastructure because even for very small AD plants the energy content of the biogas is significantly higher than what could be exploited as heat for the whole farm, and a possible economic benefit that can be leveraged through heat is generally significantly lower than the potential economic benefit of other energy forms (electricity, vehicle fuel). The use of biogas for household applications such as cooking is mainly applied in economically developing countries at very small and technically simple AD plants.

Upgrading biogas to biomethane is a solution that is applied more and more in practice, including on agricultural sites. This solution is generally more suitable for larger AD plants, as the economic viability of small plants is more difficult to achieve due to the significant investment required to establish the facilities for cleaning and upgrading the biogas to biomethane. A variety of biogas upgrading technologies are now state of the art (Ahmed et al. 2021). The most common are water scrubbing and membrane separation, which together provide around 60% of the global biomethane today (IEA 2020), while chemical scrubbing, pressure swing adsorption and organic-physical scrubbing are also operated in practice (Khan et al. 2021).

Under intense research are CO_2 biomethanation (also referred to as 'CO_2 biomethanisation') solutions that convert the CO_2 from biogas into additional CH_4, thereby increasing the overall methane yield (Chatzis et al. 2024; Zhang et al. 2023). This introduces a valorisation scheme for the non-combustible CO_2 and increases the amount of renewable energy produced in the form of CH_4. CO_2 is an unavoidable output of the biogas process from biogas (Section 2.1), but as an incombustible gas it is usually left unvalorised and emitted to the atmosphere – both where biogas is used directly and where biogas is upgraded to biomethane. In CO_2 biomethanisation, CO_2 is transformed into CH_4 by microbiological activity under the addition of exogenous hydrogen (H_2). This can be done either in-situ within the existing AD reactor or ex-situ in an additional reactor. In ex-situ applications, CO_2 is first separated from the biogas stream and then added together with H_2 to an additional reactor for microbial methanisation (Mahla et al. 2022; Thapa et al. 2023). A distinct advantage of in-situ arrangements is that they do not require an additional bioreactor; however, compared to the more robust ex-situ solution, they are much more susceptible to process disturbance and much more difficult to control. In-situ CO_2 biomethanisation is not yet at commercial scale, while the first full-scale ex-situ CO_2 biomethanisation plant was commissioned in 2023 in Denmark and is operational (Xirostylidou et al. 2024).

There exist also research efforts to convert biogas to hydrogen (H_2) as an alternative (and carbon-free) energy carrier (Kumar et al. 2022). The target energy carrier then is no longer CH_4 but H_2. This is a solution that assumes hydrogen will be the main energy carrier in the future energy system. Several technologies can be applied to convert biogas into H2, including steam reforming, dry reforming and plasma-catalysis (Boscherini et al. 2023).

5 How methane utilisation obtained through anaerobic digestion on dairy farms contributes to net-zero

'Net-zero' refers to a setting in which carbon neutrality is achieved. In such a setting, the 'net greenhouse gas emission' to the environment is zero, because an equivalent amount of GHG is removed for any quantity released. The 'net greenhouse gas emission' is the calculated balance between the GHG released in a specific setting and the GHG captured or removed (or avoided) in relation to that setting (e.g. a dairy farm). If more GHG is released than removed, the net GHG emission is positive, which accelerates climate change. Conversely, if more GHG is removed than is released, the calculated net GHG emission is negative, which makes a positive contribution to climate change mitigation and is referred to as 'climate positive'. As described in Section 1, livestock breeding operations at dairy farms emit significant amounts of GHGs, which cannot be reduced to zero. To achieve 'net-zero', operations with negative net GHG emissions need to be added, i.e. climate positive activities that remove or avoid more GHG than they emit themselves.

There are differences between AD sites and schemes in terms of their net GHG emissions (Bachmaier et al. 2013; Liebetrau et al. 2013). While AD schemes that focus on energy crops may struggle to achieve carbon neutrality (Bachmaier et al. 2013), biogas production based on organic agricultural residues such as animal manure can reliably be expected to have negative net GHG emissions if properly operated (Lijó et al. 2017), thus contributing towards reaching 'net-zero' for the agricultural business concerned, e.g. a dairy farm.

5.1 Methane production and utilisation on dairy farms contributes to net-zero through different mechanisms

The net GHG emissions of an AD installation are influenced by several factors. Among the most important ones are the following (Bachmaier et al. 2013; Kodba et al. 2023; Meyer-Aurich et al. 2012):

- Types of feedstocks used for AD (e.g. organic residues or purposely grown energy crops).
- Procedures applied to source and handle the substrates to be digested (e.g. transport distances, transport means and logistics, storage of feedstocks).
- The technical AD equipment implemented at a site and its operation (e.g. AD plant type, reactor size, digestion temperature, loading of material into the reactor, and unloading).
- Valorisation of the generated biogas (e.g. upgrading of biogas to biomethane, utilisation in CHP unit).
- Handling of the digestate (e.g. covered or uncovered storage, storage duration, techniques for applying digestate to agricultural land).

Some of the related activities cause GHG emissions themselves, for example transport of feedstock with diesel-fuelled vehicles, and it depends on the overall balance whether the net GHG emission is positive or negative (Kodba et al. 2023). For energy crops in particular, it is not necessarily the case that AD schemes will actually achieve negative net GHG emissions, and consequently the individual case would need to be carefully analysed under this lens. In contrast, the use of residues and wastes, including from livestock breeding, can reliably be expected to achieve negative net GHG emissions if digestion is carried out in state-of-the-art AD plants and if good agricultural practices are adopted to manage the materials both before and after digestion (Lijó et al. 2017; Pucker et al. 2013), including the appropriate choice of equipment to apply digestate to agricultural land (Korba et al. 2024). Digestate storage in uncovered facilities, i.e. without gas collection, has been identified as one of the most significant CH_4 emission sources at AD plants in practice (Lehtoranta et al. 2024), and consequently uncovered digestate storage is not appropriate and should no longer be used.

Schemes that produce methane on dairy farms through the digestion of organic residues and the use of this methane as a renewable energy source contribute to 'net-zero' through several different and important mechanisms. For dairy farms, the following key mechanisms can be highlighted:

- Substitution of fossil fuels with renewable methane.
- Establishment of a sound manure and slurry management.
- Reduced need for synthetic fertilisers and improved carbon sequestration in soil through AD digestate.

Biogas/biomethane produced by AD substitutes fossil fuels such as natural gas, oil, and coal as an energy carrier. Biogas and biomethane are carbon-based energy carriers as well and, just like fossil fuels (e.g. natural gas), they release CO_2 when used to generate electricity and heat, or as a transport fuel. However, the implications of this CO_2 in the context of climate change and its impact on the natural carbon cycle are very different. While fossil fuels release additional CO_2 into the atmosphere when burned, the use of biogas/biomethane does not. The CO_2 released when biogas/biomethane is used energetically is biobased – it comes from biomass. During its growth phase, this biomass has incorporated the respective amount of carbon by removing CO_2 from the atmosphere (Golmakani et al. 2022; Rafiee et al. 2021). Therefore, substitution of fossil-based energy by biomass-derived energy displaces CO_2 emissions.

Compared to the use of solid biomass energy carriers, such as wood-based materials, the use of biogas or biomethane has additional advantages in terms of climate change mitigation potential. The production of biogas and biomethane is not linked to deforestation. Deforestation is a key driver of climate change (Kunte and Bhat 2024). When biogas or biomethane replaces solid, wood-based biomass fuels, the risk of deforestation is reduced.

Diverting farm residues, including animal manure and slurry, for treatment in a controlled and enclosed environment avoids uncontrolled and fugitive methane emissions associated with the storage and handling of the materials. Consequently, AD of manure and slurry from dairy farms avoids or reduces GHG emissions from the livestock systems by implementing AD-related practices that aim to store and handle the organic materials in a way that avoids any degradation of the material prior to the actual AD in the biogas reactor.

In the digestate, the remaining organic material is present in a stabilised form. Compared to manure and slurry without AD, the digested material has altered biochemical properties. This may mean fewer emissions of fugitive methane or other GHGs (e.g. NH_3, N_2O) during the further handling, especially when the organic material is applied to agricultural land as fertiliser and soil conditioner (Kovačić et al. 2022), although contradictory results have been reported in the literature and the knowledge base remains somewhat fragmented and unclear, especially with regard to gaseous ammonia emissions from digested versus undigested animal slurry (Pedersen and Hafner 2023). The digestate application techniques and the equipment used also have an influence on GHG emissions (Korba et al. 2024).

The fertiliser value of AD digestate is higher compared to undigested manure and slurry, and the risk of environmental pollution is lower. An important component of the environmental concerns associated with the spreading of undigested manure/slurry is nitrate. Undigested animal manure/slurry contains nitrogen mainly in organic form (30–50% of the nitrogen) and as ammonia. Some plants can take up ammonia directly, others after it has been converted to nitrate in the soil, while organic nitrogen must first be mineralised to be available for plant growth. Depending on the time of application and the plant growth stage, there is a risk that significant amounts of the nutrients may not be taken up by plants. In AD, most of the organic nitrogen has already been converted and around 60–80% of the nitrogen is present as ammonia (Banks et al. 2007), which minimises nitrogen losses and makes the fertiliser effect more predictable.

The use of AD digestate as a fertiliser on agricultural land reduces or eliminates the need for synthetic fertilisers. Substitution of synthetic fertilisers means less need to extract primary resources through mining or other activities, and it avoids the GHG emissions associated with the production, market distribution and application of such fertilisers (Walling and Vaneeckhaute 2020). Digestate from AD also generally has soil-improving properties when used in agriculture, which increases agricultural productivity and improves the soil's ability to store and sequester carbon (Czekała et al. 2022). However, knowledge gaps remain regarding the effects of long-term and repeated applications of digestate on soil biota and consequently on soil health (van Midden et al. 2023).

5.2 The untapped methane potential on dairy farms remains huge

Although AD with biogas production can be considered state of the art (Kougias and Angelidaki 2018), with many thousands of AD facilities in operation worldwide (Kabeyi and Olanrewaju 2022; WBA 2019), the currently untapped biogas/biomethane potential is huge. Only about 6% of the biogas potential of wastes and residues was exploited in 2018 (IEA 2020), and the untapped biogas potential would be even higher if some energy crops were also considered. Estimates by the World Biogas Association (WBA 2019) suggest that AD of realistically collectable livestock manure/slurry alone could make available renewable energy of 2600–3800 TWh (terawatt hours), based on a biomethane potential of 250–370 bcm CH_4 (billion cubic metres methane), which corresponds to a GHG reduction potential of 930–1260 Mt CO_2-eq (mega tonnes CO_2 equivalents). These data take into account that not all occurring livestock manure/slurry can realistically be collected; for example, if cows are kept free range, those manure quantities cannot be considered available for AD. Global GHG emissions in 2023 amounted to 53.0 Gt CO_2-eq (giga tonnes CO_2 equivalents) (Crippa et al. 2024). Thus, biogas/biomethane production through AD of the currently available and realistically collectable livestock manure/slurry has the potential to reduce global GHG emissions by 1.8% to 2.4%. When adding co-substrates to the digestion of manure and slurry, the GHG reduction potential is boosted.

The existing literature is not sufficiently detailed to clarify exactly how much of the mentioned GHG mitigation potential is associated with dairy farms and how much is related to other livestock systems (e.g. pigs, poultry). More detailed assessments would be required to quantify this precisely, but the available data base (WBA 2019) suggests that dairy units contribute at least half of this potential. Overall, it can be concluded for dairy farms worldwide that AD of their manures and slurries, with biogas/biomethane production along with the energetic use of the generated methane, holds the potential to reduce global GHG emissions by at least 1%. In addition, AD plants operated at dairy farms could use other biomasses from the farm or take in additional substrates for processing, which significantly increases their possible contribution to GHG reduction.

6 Integration of anaerobic digestion into the dairy farm management

The production of methane by AD and its utilisation on the dairy farm or by external schemes provides an additional revenue stream for the farm, but the economic viability needs to be carefully assessed before decisions are made. Investment in an AD plant is a long-term commitment and some changes in farm management may also be required. If the AD unit is appropriately integrated into the overall farm operation, the interactions between AD, livestock breeding and other farm enterprises are usually positive. Some of the benefits that AD brings to dairy farms are the production of electricity and heat, the supply of a digestate with high fertiliser and soil improvement value, a reduction in odour nuisance and a reduced need for storage capacity for untreated manure/slurry (Köttner et al. 2008). Other potential benefits include collaborations with nearby farms, local communities and other stakeholders, for example to co-develop and cooperatively set up local energy solutions (electricity, heat), which strengthens the wider societal role and agency of dairy farmers (Gadirli et al. 2024; Niang et al. 2022).

A number of factors related to livestock management need to be considered when planning and implementing an AD plant, and this may require changes to the established operational practices on the farm. As a general issue, animal manure and slurry should be used as fresh as possible to avoid loss of organic matter and thus methane production potential (Wellinger et al. 2013). It is advantageous to have the livestock breeding units and the biogas plant in close spatial proximity. Animal slurry should be loaded into the AD reactor as soon as possible, avoiding any significant intermediate storage. In some regions, it is still a common practice to store slurry under the barns, which is not suitable in a farm concept that includes AD. Any available storage capacity under the barns also cannot be used to store the effluent from the AD plant, i.e. the digestate, as there is a risk that some degassing of the material will continue during digestate storage (Awiszus et al. 2018), leading to unfavourable or even dangerous air conditions within the building.

Facilities for appropriate storage of the digestate must be provided, for example covered tanks (Section 5.1), and sufficient storage capacity must be available, as digestate cannot normally be spread on agricultural land throughout the whole year due to legal restrictions. Transport of liquid digestate quickly becomes uneconomical over longer distances and is therefore usually only an option for shorter distances. If digestate cannot be used on the farm, installing solid-liquid separation equipment can be a viable solution, as it provides a solid fraction with higher concentrations of valuable digestate components (Kovačić et al. 2022), making transport over longer distances more profitable. Drying of digestate (Czekała et al. 2022) or upgrading of digestate to a refined fertiliser (Chozhavendhan et al. 2023) can also be considered, but all such solutions come with additional costs.

Barn cleaning may require adjustments to existing procedures. If equipment is newly installed, some factors should already be considered. Dilution of manure and slurry during cleaning should be avoided. This happens, for example, at sites where flushing systems are used to remove the manure from the barns. Flush systems ensure very effective cleaning and are less labour intensive than other solutions, but they are not an optimal choice for dairy farms with AD systems (Köttner et al. 2008). Any dilution of the manure/slurry increases the volume flow of the feedstock and would therefore either require larger AD facilities or it will reduce the residence time of the material in a given AD reactor. Dilution also increases the risk of stratification within the digester

(Burke 2001), with fibrous components, e.g. straw, floating on top of the digester content and forming a thick mat, which makes this biomass less well available for degradation and can also cause technical problems. Therefore, from an AD perspective, water-free systems, such as scraper systems, which collect the dairy manure by scraping it into a sump without changing its consistency, are more favourable than flushing systems for cleaning the barns (Köttner et al. 2008). The use of detergents and disinfectants in the livestock breeding units must take into account possible negative effects on the microbial communities in the AD system to avoid inhibition of the biogas production.

Any medication of the animals in the livestock breeding units must factor in the potential impact on the microbial communities in the AD plant. Antibiotics or other medications can cause severe inhibition of the biogas process (Koniuszewska et al. 2021; Ma et al. 2021). Contaminated manure and slurry may need to be handled separately and with great care, and it may need to be excluded from the AD unit.

The litter (bedding) material used should not only enhance the welfare of the animals in the stable but must also be compatible with the requirements of the AD plant. After its use in the stable, the bedding material is fed to the AD plant together with the animal excreta as a mixture. It is technically hardly feasible to separate litter material from animal excreta in an economically viable way. In addition, appropriately selected bedding material serves as an additional AD feedstock and increases the methane yield of the biogas plant. Sand is problematic in AD and should be excluded. Sand can clog pipes (loading of reactor, transport of material) and damage equipment, including mixing equipment, through abrasion. In addition, sand retained in the manure/slurry is released from the organic material during its digestion and its precipitation gradually reduces the effectively available digester volume. The installation of dedicated sand removal equipment can improve the situation to some extent in cases where sand cannot be avoided, but the AD feedstock obtained from such schemes is suboptimal in terms of quality and quantity. Not all the sand can be removed, while some of the organic material is removed together with the sand and therefore lost for AD.

Woody bedding materials, such as wood shavings or sawdust, are generally less technically demanding in the AD unit than sand, but such lignified materials are poorly degradable in a biogas plant and block reactor space without contributing significantly to the biogas production. Therefore, woody litter materials are not an optimal choice when operating an AD plant on the dairy farm. Pre-treatment solutions are potentially applicable to facilitate the degradation of woody materials in AD (Müller et al. 2022), but they require additional investment and labour, making them a possible option only for large-scale AD plants.

Straw is generally a good choice, but prior to its usage as bedding material, the straw size should be reduced through chopping to facilitate microbial degradation during AD and thus increase the biogas yield of the plant (Kusch 2007). Other materials compatible with AD are less commonly used as animal bedding and include paper-based materials, recycled manures, hemp, and other field biomasses (Manni et al. 2023).

Where animals are kept outdoors (free range) during some or all of the seasons, this needs to be taken into account. Animal excreta deposited on pasture by cows on free range is mostly degraded aerobically, which reduces the unwanted methane production linked to storage of manure/slurry from housed animals. However, excreta deposited on pasture cannot contribute to a digester. Therefore, the total amount of manure and slurry available to serve as AD feedstock is adversely affected where animals are kept free range, as is the continuity of feedstock availability (Nleya et al. 2023). A greatly reduced amount of manure or slurry can only be collected during such periods, and in some cases no collection at all is possible. Dairy cows will be in for milking for a certain period of the day, and it is plausible to assume that this will enable the collection of 20% of the manure/slurry (Köttner et al. 2008). This may make continuous feeding of the AD plant impossible. Supplementation with other substrates is one solution to facilitate continuous feeding of the AD system on such farms.

Supplementing methane production from dairy manure and slurry with energy-rich co-substrates, such as energy crops or food waste, is an effective strategy at many sites to achieve or improve the economic viability of the AD installation (Bywater and Kusch-Brandt 2022). However, site-specific analysis is required, and it is also possible that the intake of additional substrates may reduce AD profitability. For example, if digestion of energy crops is planned, this will require additional investment in feedstock storage, which may undermine overall profitability. It is also necessary to understand and consider the technical and regulatory implications of each substrate. For example, if accepting food waste from nearby canteens, schools or industrial sites brings the AD site under animal by-products regulations, pasteurisation will need to be implemented (Zhang et al. 2020).

7 Case studies

There is no shortage of case studies in the existing literature on the production and use of biogas and biomethane on farms. Hundreds of case studies have been documented and are available. For example, Task 37: Energy from Biogas has compiled a collection of case studies on its website (https://task37.ieabioenergy.com/case-success-stories/).

It is important to note that successful projects are much more likely to be documented for a wider audience than failed projects. Stakeholders involved in unsuccessful initiatives are rarely interested in publishing these experiences themselves, and they may not give their consent for others to do so. As a result, there is some bias in the literature in terms of published case studies. The learning from case studies will therefore be somewhat limited and may be unbalanced.

Biogas and biomethane solutions in the agricultural sector are site-specific solutions. They take into account and reflect the individual characteristics of each farm. What has worked on one site may not be compatible with the needs of another. The AD can be a source of additional income for a dairy farm, but it can also put a farm into debt. The production and use of biogas and biomethane should not be seen as a panacea for a financially struggling farm. A carefully prepared, site-specific assessment and detailed pre-planning are required for each dairy farm.

Smaller dairy farms will generally find it more difficult to achieve economic viability of an AD installation, but there is no threshold for farm size as many factors influence the cost-benefit outcome. Economic viability on a global scale is not documented in the literature (Lisowyj and Wright 2020). For the US dairy sector, manure/slurry from at least 3,000 cows has been recommended to ensure economically viable operation of an AD system (Lauer et al. 2018), while 500 cows may be considered a minimum (Yiridoe et al. 2009). For US farms of 250 dairy cows or less, economic modelling showed a negative cash flow for all digesters considered (Klavon et al. 2013). Costs and revenue streams associated with AD vary widely between countries and regions. Some countries provide greater tariff support for smaller AD systems, such as Germany and Italy, making AD more attractive for smaller dairy farms (Lisowyj and Wright 2020). For Ireland, small-scale AD systems were found to be profitable for dairy herds of at least 100 cows (O'Connor et al. 2020). In Germany, profitable small-scale AD systems exist on farms of around 70 cows, often with some addition of on-farm co-substrates such as grass (Eggemann et al. 2023). Many case studies, e.g. for the UK (Bywater and Kusch-Brandt 2022; Köttner et al. 2008), document the positive effect on economic viability of adding some co-substrates to dairy manure/slurry.

8 Conclusion and future trends

Anaerobic digestion with subsequent utilisation of methane as a renewable energy carrier is a key pathway for establishing environmentally sound manure management on dairy farms and moving towards 'net-zero' GHG emissions from the dairy sector. The AD is currently the most promising way to drastically reduce gaseous emissions from agriculture, in particular GHG emissions (CH_4,

N_2O) from manure management. Most importantly, the methane-rich energy carrier produced can displace fossil energy sources and thus cut away the associated GHG emissions. Biogas and biomethane from dairy manure and slurry alone can reduce global GHG emissions by at least 1% and likely more. The total GHG emission reduction potential of AD plants on dairy farms is significantly higher than this, as co-substrates (e.g. food waste) can be added to the manure and slurry.

Anaerobic digestion on dairy farms has additional benefits. Compared to undigested manure/slurry, AD produces a digestate with improved fertiliser value and soil amelioration quality, which enhances plant nutrient uptake, reduces leaching losses, and improves soil carbon storage capacity. Furthermore, AD minimises the odour nuisance associated with animal manure and slurry.

Biogas and biomethane production on dairy farms can be considered state of the art, with thousands of plants in operation worldwide. However, the untapped potential for AD remains huge, with more than 90% of the potential estimated to be unutilised today (Section 5.2). Generating the confidence of practitioners to adopt AD in practice is a key factor in promoting more widespread adoption of this technology (Kusch-Brandt et al. 2023); this refers to the availability of appropriate information, the reliability of technologies, and the credibility and continuity of policies. Appropriate and reliable policies are essential to encourage investment in AD and to enable profitable installations, including at farm scale (Bywater and Kusch-Brandt 2022).

Much of the current research around biogas and biomethane focuses on how to create more value from the processed materials and the infrastructure used (Kusch-Brandt et al. 2023). This includes, but goes well beyond, further research into the performance of substrates under different AD regimes. There is scope to identify and develop more effective utilisation schemes for biogas and biomethane. For example, the regional clustering of several AD plants with collection of their gas through a dedicated gas grid allows for economies of scale and novel types of gas upgrading processes (Stockl and Lemmer 2023). A particularly vibrant area of research is the biomethanisation of CO_2 to methane (Thapa et al. 2023; Tommasi et al. 2024). More broadly, the valorisation of CO_2 from biomethane production as a resource, for example to supply carbon-based materials to the chemical industry, is attracting increasing attention and interest (Akash et al. 2023; Leiman and Seelmann 2024). Biomethanisation and most other CO_2 valorisation strategies are linked to the usage of hydrogen (Tommasi et al. 2024). Therefore, exploring synergies between biomethane production and a future hydrogen economy is an emerging and promising line of research.

Agricultural activities, including biogas and biomethane production, are increasingly seen as key elements in building a bioeconomy, particularly a circular bioeconomy. For example, the development of a regional bioeconomy concept for the Alb-Donau-Kreis district in southern Germany has given a priority role to agriculture and on-farm AD throughout the district (Kusch-Brandt et al. 2024). The implementation of biorefinery solutions at agricultural AD facilities to obtain a range of products from manure or other biomasses, rather than just methane and digestate, is a further evolving concept (Bolzonella et al. 2023). More generally, the role of methane utilisation and CO_2 valorisation in establishing a robust circular economy is underexplored.

9 Where to look for further information

There are several handbooks on biogas and biomethane available. Among others, they provide comprehensive information on the biological processes that take place during AD, which is not covered in detail in this chapter. A particularly authoritative and frequently referenced handbook is that compiled by Wellinger et al. (2013):

- Wellinger, A., et al. (2013), *The Biogas Handbook: Science, Production and Applications,* Woodhead Publishing, Cambridge, UK.

The research landscape in the field of biogas and biomethane production and utilisation, including institutions involved, is vast and spans the globe. Excellence in research is demonstrated by many of the institutions and research groups. It is inevitable that there will be omissions of other excellent researchers and institutions in the highlighting of individual institutions. Nevertheless, some key research institutions and initiatives can be listed as examples:

- The State Institute of Agricultural Engineering and Bioenergy (https://la-bioenergie.uni-hohenheim.de/en/labioenergy), associated with the University of Hohenheim, has a long tradition of supporting dairy farms to improve their practices, including through better management of manure, slurry, and other residues. Anaerobic digestion of agricultural residues and energy crops, biomethane production, and biomethane utilisation are some of the key research areas today. At one of the university's research farms, the institute has set up a full-scale AD plant running on cow manure and co-substrates, to conduct research at the practice scale.
- The American Biogas Council has set up the Biogas Research Directory (https://americanbiogascouncil.org/resources/biogas-research-directory/), which structures information related to more than 1200 completed biogas-related research projects.
- The Environmental Biotechnology Network (EBNet) (https://ebnet.ac.uk/) has continued the work of the former AD Network and has established

several AD-related working groups. It integrates the expertise and research capabilities from several academic institutions, including those established at the University of Southampton over several decades in this field (www.southampton.ac.uk/research/groups/water-environmental-engineering-group).

Task 37: Energy from Biogas (https://task37.ieabioenergy.com/) is an international group of experts that operates under the Technology Cooperation Programme of the International Energy Agency to explore and address the challenges related to the economic and environmental sustainability of biogas and biomethane. Task members come from both academia and the private sector. Their work covers the entire AD value chain, from feedstock collection to biogas and biomethane utilisation. Feedstocks addressed include agricultural residues (e.g. manure and crop residues), energy crops, biowaste from households, industrial organic wastes, and others. A wealth of authoritative technical reports and case studies are publicly available on the website.

There are many associations that not only actively promote the production and utilisation of biogas and biomethane, but also provide publications and data on the sector. Some have a national or regional focus, while others are transnational or not specific to any place. Two transnational associations that regularly review the situation and trends in the biogas and biomethane sector are the following:

- The European Biogas Association (EBA) (https://www.europeanbiogas.eu/) addresses the sustainable deployment of biogas and biomethane mainly in the context of the European continent. More than 300 national associations and other organisations contribute to its work. Policy assessments across Europe and technology analysis are among its regular activities. Recent publications are freely available through the Info Hub on the organisation's website. The website also contains a section on R&D projects, reflecting some of the key lines of research and development in the field of biogas and biomethane production and utilisation.
- The World Biogas Association (WBA) (https://www.worldbiogasassociation.org/) works with various stakeholders in the biogas and biomethane sector worldwide, including national associations, AD plant operators, biogas and biomethane industry players, academic and research institutions.

10 References

Abbas, Y., et al. (2023), 'Co-digestion of cow manure and food waste for biogas enhancement and nutrients revival in bio-circular economy', *Chemosphere*, 311 (Part 2), 137018. https://doi.org/10.1016/j.chemosphere.2022.137018.

Afroze, N., et al. (2023), 'Effects of trace elements on digester performance and microbial community response in anaerobic digestion systems', *Environmental Technology*, 44(27), 4157–4172.

Agonafer, T. D., et al. (2022), 'Biogas-based trigeneration system: a review', *Results in Engineering*, 15, 100509.

Ahmed, S. F., et al. (2021), 'Biogas upgrading, economy and utilization: a review', *Environmental Chemistry Letters*, 19, 4137–4164. https://doi.org/10.1007/s10311-021-01292-x.

Akash, S., et al. (2023), 'Biotechnology to convert carbon dioxide into biogas, bioethanol, bioplastic and succinic acid using algae, bacteria and yeast: a review', *Environmental Chemistry Letters*, 21, 1477–1497. https://doi.org/10.1007/s10311-023-01569-3.

Anukam, A., et al. (2019), 'A review of the chemistry of anaerobic digestion: methods of accelerating and optimizing process efficiency', *Processes*, 7, 504. https://doi.org/10.3390/pr7080504.

Atandi, E. and Rahman, S. (2012), 'Prospect of anaerobic co-digestion of dairy manure: a review', *Environmental Technology Reviews*, 1(1), 127–135. https://doi.org/10.1080/09593330.2012.698654.

Atelge, M. R., et al. (2020), 'Biogas production from organic waste: recent progress and perspectives', *Waste and Biomass Valorization*, 11, 1019–1040. https://doi.org/10.1007/s12649-018-00546-0.

Awiszus, S., et al. (2018), 'Gas-releasing potential of biogas digestate', *Biofuels*, 12(6), 689–695. https://doi.org/10.1080/17597269.2018.1519764.

Bachmaier, H., et al. (2013), 'Changes in greenhouse gas balance and resource demand of biogas plants in southern Germany after a period of three years', *Waste Management & Research*, 31(4), 368–375.

Banks, C. J., et al. (2007), 'Potential of anaerobic digestion for mitigation of greenhouse gas emissions and production of renewable energy from agriculture: barriers and incentives to widespread adoption in Europe', *Water Science & Technology*, 55(10), 165–173.

Bardi, M. J., et al. (2023), 'Opportunities and challenges of micronutrients supplementation and its bioavailability in anaerobic digestion: a critical review', *Renewable and Sustainable Energy Reviews*, 186, 113689. https://doi.org/10.1016/j.rser.2023.113689.

Bolzonella, D., et al. (2023), 'Toward the transition of agricultural anaerobic digesters into multiproduct biorefineries', *Processes*, 11(2), 415. https://doi.org/10.3390/pr11020415.

Booker Nielsen, M. (2022), 'Identifying challenges and drivers for deployment of centralized biogas plants in Denmark', *Sustainability*, 14(13), 8021. https://doi.org/10.3390/su14138021.

Boscherini, M., et al. (2023), 'New perspectives on catalytic hydrogen production by the reforming, partial oxidation and decomposition of methane and biogas', *Energies*, 16(17), 6375. https://doi.org/10.3390/en16176375.

Broucek, J. (2014), 'Production of methane emissions from ruminant husbandry: a review', *Journal of Environmental Protection*, 5, 1482–1493. https://doi.org/10.4236/jep.2014.515141.

Burke, D. A. (2001), *Dairy waste anaerobic digestion handbook*, Environmental Energy Company, Olympia, USA.

Bywater, A. and Kusch-Brandt, S. (2022), 'Exploring farm anaerobic digester economic viability in a time of policy change in the UK', *Processes*, 10(2), 212. https://doi.org/10.3390/pr10020212.

Calbry-Muzyka, A., et al. (2022), 'Biogas composition from agricultural sources and organic fraction of municipal solid waste', *Renewable Energy*, 181, 1000–1007. https://doi.org/10.1016/j.renene.2021.09.100.

Chatzis, A., et al. (2024), 'Biological methanation (BM): a state-of-the-art review on recent research advancements and practical implementation in full-scale BM units', *Energy Conversion and Management*, 314, 118733. https://doi.org/10.1016/j.enconman.2024.118733.

Cheng, M., et al. (2022), 'Climate change and livestock production: a literature review', *Atmosphere*, 13(1), 140. https://doi.org/10.3390/atmos13010140.

Chozhavendhan, S., et al. (2023), 'Current and prognostic overview on the strategic exploitation of anaerobic digestion and digestate: a review', *Environmental Research*, 216(Part 2), 114526. https://doi.org/10.1016/j.envres.2022.114526.

Crippa, M., et al. (2024), *GHG emissions of all world countries*, Publications Office of the European Union (JRC138862), Luxembourg, Luxembourg. https://doi.org/10.2760/4002897.

Czekała, W., et al. (2022), 'Biogas plant operation: Digestate as the valuable product', *Energies*, 15, 8275. https://doi.org/10.3390/en15218275.

Duarah, P., et al. (2022), 'A review on global perspectives of sustainable development in bioenergy generation', *Bioresource Technology*, 348, 126791. https://doi.org/10.1016/j.biortech.2022.126791.

Eggemann, L., et al. (2023), 'The ecological potential of manure utilisation in small-scale biogas plants', *Applied Energy*, 331, 120445. https://doi.org/10.1016/j.apenergy.2022.120445.

Ekstrand, E. M., et al. (2022), 'Identifying targets for increased biogas production through chemical and organic matter characterization of digestate from full-scale biogas plants: what remains and why?', *Biotechnology for Biofuels and Bioproducts*, 15(1), 16. https://doi.org/10.1186/s13068-022-02103-3.

FNR (2016), *Leitfaden Biogas - Von der Gewinnung zur Nutzung*, 7th edition, Fachagentur Nachwachsende Rohstoffe, Gülzow, Germany.

Gadirli, G., et al. (2024), 'Fundamentals, operation and global prospects for the development of biogas plants–A review', *Energies*, 17(3), 568. https://doi.org/10.3390/en17030568.

Gerber, P. J., et al. (2013), *Tackling climate change through livestock – A global assessment of emissions and mitigation opportunities*, Food and Agriculture Organization of the United Nations (FAO), Rome, Italy.

Golmakani, A., et al. (2022), 'Advances, challenges, and perspectives of biogas cleaning, upgrading, and utilisation', *Fuel*, 317, 123085. https://doi.org/10.1016/j.fuel.2021.123085.

Huang, Z., et al. (2022), 'Effects of heavy metals and antibiotics on performances and mechanisms of anaerobic digestion', *Bioresource Technology*, 361, 127683. https://doi.org/10.1016/j.biortech.2022.127683.

IEA (2020), *Outlook for biogas and biomethane: prospects for organic growth*, International Energy Agency, Paris, France. Available online: https://www.iea.org/reports/outlook-for-biogas-and-biomethane-prospects-for-organic-growth (accessed 20 February 2024).

Jiang, Y., et al. (2019), 'Ammonia inhibition and toxicity in anaerobic digestion: a critical review', *Journal of Water Process Engineering*, 32, 100899. https://doi.org/10.1016/j.jwpe.2019.100899.

Kabeyi, M. J. B. and Olanrewaju, O. A. (2022), 'Biogas production and applications in the sustainable energy transition', *Journal of Energy*, 2022(1), 8750221.

Kaiser, F., et al. (2008), *Sicherung der Prozessstabilität in landwirtschaftlichen Biogasanlagen*, 2nd edition, Bayerische Landesanstalt für Landwirtschaft (LfL), Freising, Germany.

Karki, R., et al. (2021), 'Anaerobic co-digestion: current status and perspectives', *Bioresource Technology*, 330, 125001.

Khan, M. U. and Ahring, B. K. (2021), 'Improving the biogas yield of manure: effect of pretreatment on anaerobic digestion of the recalcitrant fraction of manure', *Bioresource Technology*, 321, 124427. https://doi.org/10.1016/j.biortech.2020.124427.

Khan, M. U., et al. (2021), 'Current status of biogas upgrading for direct biomethane use: a review', *Renewable and Sustainable Energy Reviews*, 149, 111343. https://doi.org/10.1016/j.rser.2021.111343.

Klavon, K. H., et al. (2013), 'Economic analysis of small-scale agricultural digesters in the United States', *Biomass and Bioenergy*, 54, 36–45. https://doi.org/10.1016/j.biombioe.2013.03.009.

Kodba, A., et al. (2023), 'Analysis of specific greenhouse gas emissions savings from biogas production based on agricultural residues and industrial by-products', *Energies*, 16(9), 3721. https://doi.org/10.3390/en16093721.

Köttner, M., et al. (2008), *Economic modelling of anaerobic digestion/biogas installations in a range of rural scenarios in Cornwall*, Cornwall Agri-Food Council, Truro, UK.

Koniuszewska, I., et al. (2021), 'The effect of antibiotics on mesophilic anaerobic digestion process of cattle manure', *Energies*, 14, 1125. https://doi.org/10.3390/en14041125.

Korba, J., et al. (2024), 'Digestate application methods and rates with regard to greenhouse gas emissions and crop conditions', *Agronomy*, 14(2), 336. https://doi.org/10.3390/agronomy14020336.

Kougias, P. G. and Angelidaki, I. (2018), 'Biogas and its opportunities–a review', *Front. Environ. Sci. Eng.*, 12, 14. https://doi.org/10.1007/s11783-018-1037-8.

Kovačić, D., et al. (2022), 'Digestate management and processing practices: a review', *Applied Sciences*, 12(18), 9216. https://doi.org/10.3390/app12189216.

Kranert, M., et al. (2012), 'Anaerobic digestion of waste'. In Karagiannidis, A. (Ed), *Waste to Energy - Green Energy and Technology*, Springer, London, UK, pp. 107–135. https://doi.org/10.1007/978-1-4471-2306-4_5.

Kumar, R., et al. (2022), 'Overview of hydrogen production from biogas reforming: technological advancement', *International Journal of Hydrogen Energy*, 47(82), 34831–34855. https://doi.org/10.1016/j.ijhydene.2022.08.059.

Kunte, G. and Bhat, V. (2024), 'Deforestation, climate change and the sustainability of agriculture: a review', *Journal of Resources and Ecology*, 15(1), 140–150. https://doi.org/10.5814/j.issn.1674-764x.2024.01.012.

Kusch, S. (2007), *Methanisierung stapelbarer Biomassen in diskontinuierlich betriebenen Feststofffermentationsanlagen*, Herbert Utz Verlag, Munich, Germany.

Kusch-Brandt, S., et al. (2023), 'Unlocking the full potential: New frontiers in anaerobic digestion (AD) processes', *Processes*, 11(6), 1669. https://doi.org/10.3390/pr11061669.

Kusch-Brandt, S., et al. (2024), 'The circular bioeconomy as a regional task', *Waste Management & Research*, 42(11), 949–952. https://doi.org/10.1177/0734242X241277462.

Lauer, M., et al. (2018), 'Making money from waste: the economic viability of producing biogas and biomethane in the Idaho dairy industry', *Applied Energy*, 222, 621–636. https://doi.org/10.1016/j.apenergy.2018.04.026.

Lehtoranta, S., et al. (2024), 'The implications of management practices on life cycle greenhouse gas emissions in biogas production', *Journal of Environmental Management*, 366, 121884. https://doi.org/10.1016/j.jenvman.2024.121884.

Leiman, P. and Seelmann, C. (2024), *Leitfaden biotechnologisches CO2-Recycling*, Umwelttechnik BW, Stuttgart, Germany.

Li, L., et al. (2021), 'Strategies to boost anaerobic digestion performance of cow manure: laboratory achievements and their full-scale application potential', *Science of The Total Environment*, 755(Part 1), 142940. https://doi.org/10.1016/j.scitotenv.2020.142940.

Liebetrau, J., et al. (2013), 'Analysis of greenhouse gas emissions from 10 biogas plants within the agricultural sector', *Water Science and Technology*, 67(6), 1370–1379.

Lijó, L., et al. (2017), 'Eco-efficiency assessment of farm-scaled biogas plants', *Bioresource Technology*, 237, 146–155. https://doi.org/10.1016/j.biortech.2017.01.055.

Lisowyj, M. and Wright, M. (2020), 'A review of biogas and an assessment of its economic impact and future role as a renewable energy source', *Reviews in Chemical Engineering*, 36(3), 401–421. https://doi.org/10.1515/revce-2017-0103.

Liu, S., et al. (2021), 'Rethinking methane from animal agriculture', *CABI Agriculture and Bioscience*, 2(1), 1–13.

Liu, Z. and Powers, W. (2014), 'Greenhouse gases emissions from multi-species animal operations and potential diet effects', *Transactions of the ASABE*, 57(1), 219–227. https://doi.org/10.13031/trans.57.10246.

Ma, G., et al. (2020), 'Methane yields during anaerobic co-digestion of animal manure with other feedstocks: a meta-analysis', *Science of The Total Environment*, 728, 138224. https://doi.org/10.1016/j.scitotenv.2020.138224.

Ma, J., et al. (2021), 'Effects of different antibiotic operation modes on anaerobic digestion of dairy manure: Focus on microbial population dynamics', *Journal of Environmental Chemical Engineering*, 9(4), 105521. https://doi.org/10.1016/j.jece.2021.105521.

Mahla, P. K., et al. (2022), 'Biomethanation: advancements for upgrading biomethane using biogas technologies. In Verma, P. (Ed), *Industrial Microbiology and Biotechnology*, Springer, Singapore, Singapore, pp. 487–504. https://doi.org/10.1007/978-981-16-5214-1_17.

Manni, K., et al. (2023), *Report on bedding materials: analysis of the current bedding material situation and assessment of the near-future development outlook in Finland*, Natural Resources and Bioeconomy Studies 89/2023, Natural Resources Institute Finland, Helsinki, Finland. Available online: https://jukuri.luke.fi/bitstream/handle/10024/553840/Luke-luobio_89_2023.pdf?sequence=1&isAllowed=y (accessed 31 October 2024).

Meegoda, J. N., et al. (2018), 'A review of the processes, parameters, and optimization of anaerobic digestion', *Int. J. Environ. Res. Public Health*, 15, 2224. https://doi.org/10.3390/ijerph15102224.

Menghistu, H. T., et al. (2021), 'Greenhouse gas emission and mitigation potential from livestock production in the drylands of Northern Ethiopia', *Carbon Management*, 12(3), 289–306. https://doi.org/10.1080/17583004.2021.1921620.

Meyer-Aurich, A., et al. (2012), 'Impact of uncertainties on greenhouse gas mitigation potential of biogas production from agricultural resources', *Renewable Energy*, 37(1), 277–284. https://doi.org/10.1016/j.renene.2011.06.030.

Müller, R., et al. (2022), 'Biogas production from dairy cattle residues: Definition of the pretreatment approach through a bibliometric analysis of publications and patents'. In Sinharoy, A. and Lens, P. N. L. (Eds), *Renewable Energy Technologies for Energy Efficient Sustainable Development. Applied Environmental Science and Engineering for a Sustainable Future*, Springer, Cham, Switzerland, pp. 121–137. https://doi.org/10.1007/978-3-030-87633-3_5.

Náthia-Neves, G., et al. (2018), 'Anaerobic digestion process: technological aspects and recent developments', *International Journal of Environmental Science and Technology*, 15, 2033–2046. https://doi.org/10.1007/s13762-018-1682-2.

Niang, A., et al. (2022), 'How do local actors coordinate to implement a successful biogas project?', *Environmental Science & Policy*, 136, 337–347. https://doi.org/10.1016/j.envsci.2022.06.019.

Nleya, Y., et al. (2023), 'Opportunities and challenges for anaerobic digestion of farm dairy effluent', *ChemBioEng Reviews*, 10(6), 924–940. https://doi.org/10.1002/cben.202300021.

Nsair, A., et al. (2020), 'Operational parameters of biogas plants: a review and evaluation study', *Energies*, 13(15), 3761. https://doi.org/10.3390/en13153761.

O'Connor, S., et al. (2020), 'Economic and environmental analysis of small-scale anaerobic digestion plants on Irish dairy farms', *Energies*, 13(3), 637. https://doi.org/10.3390/en13030637.

Pedersen, J. and Hafner, S. D. (2023), 'Ammonia emissions after field application of anaerobically digested animal slurry: Literature review and perspectives', *Agriculture, Ecosystems & Environment*, 357, 108697. https://doi.org/10.1016/j.agee.2023.108697.

Prathap, P., et al. (2021), 'Towards sustainable livestock production: Estimation of methane emissions and dietary interventions for mitigation', *Sustainability*, 13(11), 6081. https://doi.org/10.3390/su13116081.

Pucker, J., et al. (2013), 'Anaerobic digestion of agricultural and other substrates – implications for greenhouse gas emissions', *Animal*, 7(s2), 283–291. https://doi.org/10.1017/S1751731113000840.

Rafiee, A., et al. (2021), 'Biogas as an energy vector', *Biomass and Bioenergy*, 144, 105935. https://doi.org/10.1016/j.biombioe.2020.105935.

Rasapoor, M., et al. (2020), 'Recognizing the challenges of anaerobic digestion: critical steps toward improving biogas generation', *Fuel*, 261, 116497.

Raven, R. P. J. M. and Gregersen, K. H. (2007), 'Biogas plants in Denmark: successes and setbacks', *Renewable and Sustainable Energy Reviews*, 11(1), 116–132. https://doi.org/10.1016/j.rser.2004.12.002.

Sahota, S., et al. (2018), 'Review of trends in biogas upgradation technologies and future perspectives', *Bioresource Technology Reports*, 1, 79–88. https://doi.org/10.1016/j.biteb.2018.01.002.

Scarlat, N. et al. (2018), 'Biogas: Developments and perspectives in Europe', *Renewable Energy*, 129(Part A), 457–472. https://doi.org/10.1016/j.renene.2018.03.006.

Song, H., et al. (2020), 'Comparison of variable and constant loading for mesophilic food waste digestion in a long-term experiment', *Energies*, 13, 1279. https://doi.org/10.3390/en13051279.

Srisowmeya, G., et al. (2021), 'Improving process stability, biogas production and energy recovery using two-stage mesophilic anaerobic codigestion of rice wastewater with cow dung slurry', *Biomass and Bioenergy*, 152, 106184. https://doi.org/10.1016/j.biombioe.2021.106184.

Steinmetz, R. L. R. and Gressler, V. (2019), 'Impact of antibiotics on biogas production'. In Treichel, H. and Fongaro, G. (Eds), *Improving Biogas Production*, Biofuel and Biorefinery Technologies book series, vol. 9, Springer, Cham, Switzerland, pp. 181–198. https://doi.org/10.1007/978-3-030-10516-7_8.

Stockl, A. and Lemmer, A. (2023), 'Entwicklung von innovativen Konzepten zur Clusterung von Bestandsbiogasanlagen für die Bereitstellung von Biomethan'. In Kuratorium für Technik und Bauwesen in der Landwirtschaft (KTBL) (Ed), *Biogas in der Landwirtschaft - Stand und Perspektiven*, KTBL, Darmstadt, Germany, pp. 352–354.

Suryawanshi, P. C., et al. (2010), 'Mesophilic anaerobic digestion: first option for waste treatment in tropical regions', *Critical Reviews in Biotechnology*, 30(4), 259–282.

Thapa, A., et al. (2023), 'Ex-situ biomethanation for CO2 valorization: state of the art, recent advances, challenges, and future prospective', *Biotechnology Advances*, 108218.

Tommasi, M., et al. (2024), 'Advancements in CO2 methanation: a comprehensive review of catalysis, reactor design and process optimization', *Chemical Engineering Research and Design*, 201, 457–482. https://doi.org/10.1016/j.cherd.2023.11.060.

Tubiello, F. N., et al. (2021), 'Greenhouse gas emissions from food systems: building the evidence base', *Environmental Research Letters*, 16(6), 065007. https://doi.org/1088/1748-9326/ac018e.

van Midden, C., et al. (2023), 'The impact of anaerobic digestate on soil life: a review', *Applied Soil Ecology*, 191, 105066. https://doi.org/10.1016/j.apsoil.2023.105066.

Vu, H. P., et al. (2022), 'Hydrogen sulphide management in anaerobic digestion: a critical review on input control, process regulation, and post-treatment', *Bioresource Technology*, 346, 126634. https://doi.org/10.1016/j.biortech.2021.126634.

Walling, E. and Vaneeckhaute, C. (2020), 'Greenhouse gas emissions from inorganic and organic fertilizer production and use: A review of emission factors and their variability', *Journal of Environmental Management*, 276, 111211. https://doi.org/10.1016/j.jenvman.2020.111211.

Wang, K., et al. (2021), 'Binary and ternary trace elements to enhance anaerobic digestion of cattle manure: Focusing on kinetic models for biogas production and digestate utilization', *Bioresource Technology*, 323, 124571.

WBA (2019), *Global Potential of Biogas*, World Biogas Association, London, UK.

Wellinger, A., et al. (2013), *The Biogas Handbook: Science, Production and Applications*, Woodhead Publishing, Cambridge, UK.

Xirostylidou, A., et al. (2024), 'Biomethanation on demand: continuous and intermittent hydrogen supply on biological CO2 methanation', *Chemical Engineering Journal*, 495, 153677. https://doi.org/10.1016/j.cej.2024.153677.

Xu, X., et al. (2021), 'Global greenhouse gas emissions from animal-based foods are twice those of plant-based foods', *Nature Food*, 2, 724–732. https://doi.org/10.1038/s43016-021-00358-x.

Yang, P., et al. (2021), 'Foaming mechanisms and control strategies during the anaerobic digestion of organic waste: a critical review', *Science of The Total Environment*, 779, 146531. https://doi.org/10.1016/j.scitotenv.2021.146531.

Yiridoe, E. K., et al. (2009), 'Nonmarket cobenefits and economic feasibility of on-farm biogas energy production', *Energy Policy*, 37(3), 1170–1179. https://doi.org/10.1016/j.enpol.2008.11.018.

Zhang, Y., et al. (2020), 'Effect of pasteurisation on methane yield from food waste and other substrates in anaerobic digestion', *Processes*, 8(11), 1351. https://doi.org/10.3390/pr8111351.

Zhang, Y., et al. (2023), 'Validation of two theoretically derived equations for predicting pH in CO2 biomethanisation', *Processes*, 11(1), 113. https://doi.org/10.3390/pr11010113.

Zhu, Q.-L., et al. (2021), 'Bioenergy from dairy manure: technologies, challenges and opportunities', *Science of The Total Environment*, 790, 148199. https://doi.org/10.1016/j.scitotenv.2021.148199.

Chapter 13

Integrating dairy farming and crop production

Susanne Wiesner, Technical University of Denmark, Denmark; Shabda Gajbhiye, University of Wisconsin, USA; and Laxmi Prasad, North Dakota State University, USA

1 Introduction

We have tried to persuade ourselves that there is such a thing as an effect that is not also, in turn, a cause, by essentially ignoring effects we do not wish to see.

Duncan, Colin A. M., *The Centrality of Agriculture: Between Humankind and the Rest of Nature*

The year is 2026, just 4 years shy of the Paris Agreement's set goal to reduce our greenhouse gas emissions (GHG) by nearly 50% from 2010 levels to keep global warming below 1.5°C with a global net zero CO_2 emissions goal by 2050 (Soergel et al., 2021). Crop-livestock agricultural systems like dairy farms and their parent producer corporations have similarly pledged to meet carbon neutrality (i.e. US Dairy Net Zero Initiative, Arla's Climate Ambition),

http://dx.doi.org/10.19103/AS.2025.0157.30

which involves emission reductions via improved nutrient management or feed additives, offsets via increased carbon storage on-farm using, for example, cover crops, or by outsourcing emissions via large-scale offset campaigns, typically coordinated by the distributors and parent companies (Hand et al., 2024). In the European Union (EU), member nations are seeking an increase in agricultural sustainability as part of a 'Common Agricultural Policy' (CAP) and protect water resources as part of the Nitrates Directive (Kathage et al., 2022), seeking climate resilience and energy efficiency of agricultural production systems, to ensure food security (Nowak & Różańska-Boczula, 2024). For example, several EU member states require farmers to use winter cover crops as part of EU CAP's 'Ecological Focus Area' greening obligations (Kathage et al., 2022). Some countries have implemented a carbon tax on livestock emissions to curb the negative pressures on our climate (Blandford, 2024). For example, Denmark has announced plans to implement a carbon tax on livestock emissions starting in 2030, representing a pioneering step toward curbing the negative pressures on our climate (Blandford, 2024).

Livestock agriculture, including dairy, is often cited as a major contributor to global CO_2 and methane (CH_4) emissions, largely due to emissions from enteric fermentation, nutrient management, and soil disturbance. Enteric fermentation involves the digestion of fibers high in cellulose and other carbohydrates, which are broken apart to gain energy in the form of volatile fatty acids, nutrients, and organic acids in symbiosis with microorganisms in the ruminants' gut system (Thacharodi et al., 2024). In the process, greenhouse gases such as CO_2 and CH_4 are produced as byproducts, and water is lost through evaporative demands from the skin and respiratory system of dairy cows (Zhou et al., 2023). However, while livestock emissions are significant, in the following we argue that removing livestock from agricultural production might have more negative effects, which could reduce the overall resilience of global agricultural systems to climate change. To adapt our agricultural systems to changes in climate, it is crucial to understand their overall energy efficiency by incorporating all inputs, internal processes, and outputs into comprehensive lifecycle assessments. This chapter specifically focuses on integrated crop dairy livestock systems (ICLS), which combine crop production with dairy herd feed and nutrient management. Integrated crop dairy livestock systems offer the advantage of recycling nutrients on-farm, thereby reducing the need for imported nutrients and promoting circularity. By optimizing nutrient requirements, these systems can potentially lower energy demands. We will compare ICLS to agricultural production systems that do not integrate dairy crop production to provide a comprehensive analysis of which systems are best suited to achieve global sustainability goals.

2 A brief history of dairy livestock agriculture

Ruminant animals evolved in the late Eocene approximately 37.8–33.9 million years ago and make up six families today, which include modern dairy cattle species *Bos taurus* (Sánchez et al., 2015). Animal husbandry and humans have been an integral part of ancient and modern human societies for at least the past 15 000 years, with the domestication of dogs (Newsome, 2018) and goats and sheep around 8000 years ago (Serpell, 2021). Dairying stretches back at least 6000 years, shown by evidence found in the skeletal remains of our ancestors from Kenya and Sudan (Bleasdale et al., 2021). In Western Europe, the first people to drink milk were early farmers and pastoralists. Ancient artwork in Egypt, dating back to more than 2000 BC, depicts cattle and references to milk consumption, highlighting its significance in Egyptian culture (Wilkin et al., 2020). Dairy production systems were essential for the success of human civilizations, providing food during the dormant season, as well as non-edible products that were indispensable for early human survival and adaptation, such as casein glue used in woodworking and bookbinding in ancient Egypt (Wilkin et al., 2020). Integrated crop dairy livestock systems were the norm in human agricultural history up to the Middle Ages, but policy and economic shifts have led to more specialized and intensive agricultural practices by essentially separating livestock from its feed source in the name of efficiency, which now is often associated with negative externalities, such as air and water pollution (Garrett et al., 2017).

3 The importance of dairy production in human diets

Recent outcries about the contribution of GHG emissions of livestock agriculture to global climate change, from both media outlets and scientific journals, are increasingly challenging the environmental sustainability of cohabitation of humans and ruminant livestock (Crippa et al., 2021). Many outlets are calling for a shift from livestock agriculture toward a primarily vegetarian diet ('plant-based') to reduce our GHG burden (Röös et al., 2017). While plant-based diets are often cited as substantially reducing GHG emissions compared to livestock agriculture, they require the input of external nutrients to grow crops. Nutrient management is needed to replace soil nutrient deficiencies after harvest, producing GHG emissions off- and on-site from synthetic fertilizer production, transport, and field emissions (Detzel et al., 2022; Zanten et al., 2023). The need to drastically reduce and/or remove atmospheric GHG emissions has resulted in proposals such as the abandonment of livestock agriculture. However, the concept to phase out meat and dairy agriculture to reduce our GHG burdens should be heavily scrutinized with respect to other large contributors, such as fossil fuel or built environment industries. While ruminant livestock agriculture has a significant GHG footprint, it also provides essential benefits such as energy and protein-rich nutrition through milk and meat, food preservation, and nutrient recycling via manure and urea (Scholtz et al., 2020). Instead of eliminating livestock agriculture, research shows that a balanced crop and livestock agriculture, particularly in the form of ICLS, with respect to nutrient management, can significantly improve environmental sustainability (Zanten et al., 2023). Several studies have found that while animal agriculture has a large environmental footprint including lower efficiency when compared to a vegan diet, the supplement of dairy products in the human diet has great potential to meet nutrient needs of a growing global population (Peters et al., 2016). While dairy production contributes to GHG emissions, it provides an efficient source of essential nutrients that are difficult to replace through plant-based alternatives without potentially compromising nutrient availability for the population. The US dairy industry contributes about 1.58% of total US GHG emissions while providing significant proportions of essential nutrients like calcium, essential fatty acids, Vitamin B12, and Vitamin D (Liebe et al., 2020; Månsson, 2008) which are crucial for human health, particularly during child development for bone strength, and to build immune function (Garcia et al., 2019; Gaucheron, 2011). Beyond nutrition, dairy farming sustains millions of people, particularly in rural areas, by generating employment and income for smallholder farmers (Brown et al., 2022). Additionally, dairy products are available in various forms, from fresh milk to powdered milk and cheese, enhancing food accessibility and in some cases shelf life, especially in regions with limited food preservation infrastructure. However, balancing affordability for consumers while ensuring fair earnings for farmers remains a challenge. While innovations in dairy production, such as plant-based alternatives and advancements in breeding and feed efficiency, can contribute to a more sustainable and resilient food system, responsibly managed dairy production remains a cornerstone of global food security. It provides essential nutrition, economic stability, and supports agricultural sustainability. Additionally, adopting a more targeted and balanced diet, coupled with efforts to reduce and reuse food waste, can effectively mitigate the negative impacts of agriculture on our climate, water resources, and overall health (Zanten et al., 2023).

4 Drivers of changes in crop-livestock agriculture

While likely well-intentioned, the regulation of food safety and quality standards to prevent harmful ingredients, such as pathogens, often related to the chemical and nutritional content of food, has led to the creation of more uniform food systems requiring standardized growth and production criteria. Interestingly, consumer preferences rarely made it into the debate (Spiekermann, 2011). Dairy products, such as milk, butter, and cheese were first standardized in Germany in 1897, which had significant effects on local producers and their traditional manufacturing traditions and the diversity of products (Spiekermann, 2011). The introduction of pasteurization was primarily driven by an increase in urbanization, which required longer transport routes for the raw product, resulting in a regulation by the US FDA (Food and Drug Administration) prohibiting the sale of raw milk in the USA and a nationwide ban on cross state raw milk sales (Weisbecker, 2007). While evidence show that regulations significantly reduced the occurrence of foodborne illnesses (notably from *Campylobacter* spp., *E. coli*, and *Salmonella* outbreaks), some research has also found health benefits such as protection against asthma, atopy, and respiratory illnesses, associated with the consumption of raw milk (Klerk & Robinson, 2022). Food standardization has also resulted in the specialization of dairy agriculture, including the separation of cropland and livestock, which often does not leave room for adaptive practices to changes in environmental or market conditions. Increasing populations, agricultural yields, and the introduction of food science have formed the basis for industrially processed food, which also led to the availability of more diverse food options from around the globe, as well as a significant decrease in prices, at least in the developed world. Furthermore, food standardization has led to a separation of production, processing, markets, and consumption, connected by a variety of long transport routes, which can lower their adaptive capacity (Bazargani & Deemyad, 2024; Petersen-Rockney et al., 2021). Additionally, quality standards have created barriers for developing countries to access export value chains, particularly due to high compliance costs. As a result, many smallholder farms have been excluded from these markets. Only a few developing economies have successfully adapted by creating their own standards, such as Chile GAP (Chile Good Agricultural Practices) and Kenya GAP (Kenya Good Agricultural Practices), which have allowed them to maintain or increase their role in international trade, despite the challenges posed by these private standards (Henson & Humphrey, 2010). The standardization of food markets, particularly franchising in the USA, further caused a decline in the local adaptive capacity, as any change during the production or processing would affect the 'similarity of the end product' (Cox & Mason, 2007). In India, the traditional practice of local fresh milk supply is gradually being replaced by processed milk. This shift is driven by the increasing demand for standardized and safe milk products, which has led to the growth of modern milk supply chains. While this transition offers benefits such as improved food safety and extended shelf life, it also poses challenges for smallholder dairy farmers who may struggle to meet the stringent quality standards required for processed milk (Lyngkhoi et al., 2022). Interestingly, research found that global food supplies have become similar in composition particularly due to a greater reliance on a few globally dominant crops such as wheat, rice, maize, and oil crops. The growing interdependence between countries could become detrimental when considering climate change adaptation, which will be dependent on local conditions (Khoury et al., 2014).

5 Types of dairy crop-livestock systems

Dairy crop-livestock systems are integrated farming systems in which crops and livestock are managed together to enhance productivity, sustainability, and economic viability (Moraine et al., 2017; Ryschawy et al., 2012). These systems leverage the interdependence between crops and livestock to optimize resource use, improve soil fertility, and increase farm resilience. Cropping systems commonly associated with dairy farming, particularly in the USA, include corn (*Zea mays* L.), soybeans (*Glycine max* L.), and alfalfa (*Medicago sativa* L.), providing feed for dairy cattle while integrating with manure management strategies (Martin et al., 2017). Corn silage is a high-energy feed that complements dairy diets, while alfalfa crops offer high-protein forage essential for milk production. Soybeans contribute as both a feed ingredient and a rotational cash crop that in some cases can enhance soil fertility (Crusciol et al., 2014; Yang et al., 2024), but in some cases have been shown to increase GHG emissions and reduce soil C accumulation (Wiesner et al., 2022). In regions like North America, Australia, and Europe, wheat (*Triticum* spp.), barley (*Hordeum vulgare*), and sorghum (*Sorghum bicolor* ssp. L.) are often associated with dairy systems, particularly when the price of corn is too high (Mundia et al., 2019). Generally speaking, but depending on local agricultural practices and crop availability, dairy ICLS often include grasses, legumes, cereals, and roughages to feed the herds (Martin et al., 2017). In addition, pasture grasses are also integral to dairy farming, providing essential forage and contributing to sustainable agricultural practices (Bell & Moore, 2012). Common pasture grasses include perennial ryegrass (*Lolium perenne* L.), known for its high digestibility and suitability for grazing; orchard grass (*Dactylis glomerata*), valued for its rapid growth and resilience; and Kentucky bluegrass (*Poa pratensis*), appreciated for its palatability and adaptability to various soil types (Hearn et al., 2021; Küsters et al., 2021). These grasses supply necessary nutrients to dairy cattle and support soil health through their root systems, which help prevent erosion and improve soil structure (Teague & Kreuter, 2020). Integrating diverse forage crops and pasture grasses into dairy farming systems enhances feed availability, promotes animal health, and contributes to the overall sustainability of agricultural practices.

Comparing the nutrient management needs and intensity of these cropping systems versus pasture-based grazing systems reveals significant differences. Row crops like corn typically require precise fertilization schedules, often supplemented with manure to meet their high nitrogen and phosphorus demands (Khalafi et al., 2021; Rayne & Aula, 2020). In contrast, pasture-grazing systems rely more on nutrient cycling through animal manure deposition, reducing the need for external inputs while promoting soil health through continuous vegetation cover (Teague & Kreuter, 2020). The management intensity also varies. For example, row cropping systems require frequent soil testing, fertilization, tillage, and harvesting operations, whereas managed grazing systems emphasize rotational grazing plans, pasture rest periods, and controlled stocking rates (Hyland et al., 2018). Understanding these differences helps dairy farmers optimize nutrient use efficiency while balancing productivity and sustainability in their operations.

Typical dairy production systems encompass tie-stall, free-stall, which can include different types of bedded pack (i.e. sand, water mats, etc.), drylot, and pasture-grazing operations. Dairy ICLS can incorporate these types individually or in combination. In integrated confinement systems, feed is cultivated on the farm, harvested after the growing season, and converted into silage to be fed to the herd in the following year, ensuring continuous milk productivity (Campbell-Arvai, 2009). Other barn confinement systems rely on purchased feed for milk production (Brito et al., 2022). Tie-stall barns aim to enhance the efficiency of monitoring cow health, particularly through feed intake, reducing labor intensity. Furthermore, this type of herd management can enhance milk production efficiency in comparison to cows housed in free-stall barns (Simensen et al., 2010), possibly as they expend less metabolic energy on movement (Oliveira et al., 2024). However, tie-stall barns are frequently linked to a higher incidence of animal health issues (Beaver et al., 2021). In the USA, a large proportion (~70%) of tie-stall cows are allowed pasture access at least for part of the day (Beaver et al., 2021). Barriers for farmers to transition from tie-stall to free-stall farm is usually cost and labor intensity from the transition to manage the herd as a group, instead of on an individual basis. In the USA, approximately 39% of farmers still use tie-stalls, while the number rises to 74% in Canada (Beaver et al., 2021). Drylot facilities can help keep production costs low (7% of total costs) but are best maintained in areas with annual rainfall below 500 mm. However, these facilities offer fewer benefits to the herd, including limited climate control and increased health problems such as bovine mastitis, compared to pasture-based systems, which provide better animal welfare and lower overall costs (Hawkins et al., 2020). Considerations of bedding are important for animal health, as a dairy cow spends most of the day lying down and resting, while only ~5 hours are spent on eating, hence providing ample comfort should improve health, as well as milk productivity.

All confinement systems share the requirement of multiple daily feed and fodder deliveries to the stalls. This, coupled with energy-intensive field management activities such as planting, chemical and nutrient applications, and harvesting, contributes to the overall carbon footprint. Additionally, the transport of manure, while having a smaller carbon footprint (around 1–3% of total manure management emissions), still adds to the overall emissions. In contrast, emissions from rumination represent approximately 38–54% of the annual greenhouse gas budget from dairy systems (Basarab et al., 2012; Hospers et al., 2022; Wang et al., 2024b). As a result, pastoral systems often carry lower carbon footprints and costs compared to confinement systems, due to the lower energy needs from feed production, machinery, and electricity use, as well as reduced diesel consumption from farm management (Hawkins et al., 2020; Ilyas et al., 2019). Pasture-based grazing has become an attractive management for improved sustainability in ICLS agriculture among researchers (Herron et al., 2022), however only 4–20%, depending on the nation of national dairy operations can be considered grazing, and often only for part of the year (FAO, IDF, IFCN 2014), with declining trends in places like Europe (Dasselaar et al., 2020). Further, grazing is more common on small farms (<100 cows) than large ones due to cost and labor constraints. National socio-economic factors like policy, land area, tradition, and pricing models (quota vs. market) also influence dairy farm size and type (Wróbel et al., 2023).

A review of dairy farming literature shows that government policies, economies of scale, and sustainability challenges significantly impact the industry. For example, policies promoting environmental sustainability and animal welfare can influence farm sizes and production methods (Bravo-Ureta et al., 2021). In Organization for Economic Co-operation and Development countries, dairy farm sizes are influenced by government policies and subsidies, as well as crop insurance policies (Requena-i-Mora & Barbeta-Viñas, 2024; Sumner, 2014). Furthermore, intensive livestock operations often operate as 'franchises,' in connection with larger processing and marketing operations (i.e. agribusiness), hampering the ability of smallholder farms to compete (Bonanno et al., 2018). Typically, larger farms receive more subsidies and insurance payouts compared to smaller family farms, and they also exhibit higher economic productivity due to economies of scale and better resource management (Sumner, 2014). Nevertheless, smallholder family farms make up the bulk of global food production, contributing 50–70% of the supply, but they often struggle with economic viability and food insecurity, particularly in lower-income countries. Despite their potential for more sustainable farming practices, these farms face challenges due to limited resources and global market dynamics (Giller et al., 2021). In Brazil, dairy farming is influenced by socio-economic factors such as land availability and pricing models (Tonet et al., 2023). While in New Zealand, tariffs on fertilizer and livestock feed imports incentivize the adoption of mixed crop-livestock systems (Garrett et al., 2017). Around the globe, small-scale family farms are common, and government programs aimed at supporting these farmers impact the structure and productivity of the dairy sector (Novo et al., 2013). In countries like India and Kenya, dairy farming is influenced by traditional practices and socio-economic conditions. Smallholder dairy farms are prevalent, and cooperative milk collections play a crucial role in their support. Nevertheless, challenges with market access, including the distance to milk processors, perception of risk, and payment systems affect commercial farm preservation (Wangu et al., 2021). For instance, low milk prices and resource fluctuations hinder commercial dairy production of smallholder dairy farms in Ethiopia (Duguma, 2022). All these examples highlight that financial, political, and socioeconomic drivers are important considerations when transitioning dairy production systems to greater sustainability and net zero emissions.

6 Benefits of integrated crop-livestock systems

Net zero dairy production must address the sustainable management of agricultural systems in a holistic way, which includes healthy soils, reducing erosion and land use conversion (Bilotto et al., 2024), as well as biodiversity, including plant, animal, and genetic, but also market diversity, to increase the resilience to climate and economic shocks. An integration of livestock on crop farms has the benefit of allowing for improved nutrient cycling, as manure and urea droppings add nutrients back into the system (Garrett et al., 2017). This can reduce the reliance on energy-intensive synthetic fertilizer, while also saving emissions from reduction in production and transport. Crop agriculture often relies on imported synthetic fertilizers, as harvests lead to soil nutrient deficiencies that need to be replaced (Ren et al., 2023).

In contrast, intensified dairy agriculture has led to an overproduction of manure, which often exceeds the land base available to farmers to effectively manage and recycle the accumulated nutrient slurries (Kleinman et al., 2022). In dairy agroecosystems, livestock manure management (collection, processing, storage, transport, and land application) is a longstanding major challenge (Sharara et al., 2018). Advances in engineering, research, and capital investment on manure handling equipment provide solutions to some of these challenges (Aguirre-Villegas et al., 2014; Holly et al., 2017; Sharara et al., 2017; Sharara et al., 2018). However, land applications are associated with a risk of losing manure nutrients to the environment through runoff, leaching, and emissions (Lewis & Makarewicz, 2009; Shibin Liu et al., 2020; Srinivasan et al., 2006). Nutrient losses lead to fish kills, eutrophication, degraded freshwater quality, and many other environmental issues (Kakade et al., 2021). Farmers are often forced to overapply manure to fields to reduce the burden on their storage facilities. Various manure management strategies are used, including collection methods (gutters, scrapers), storage systems (liquid, solid), and land application methods (liquid manure spreaders, composting). Research shows that larger farms are more likely to use liquid or slurry manure systems

(Aguirre-Villegas & Larson, 2017; Niles et al., 2019). Liquid manure storage facilities can increase GHG emissions due to the anaerobic conditions that promote methane production. However, these facilities also offer the advantage of allowing for more controlled and timely nutrient applications, ensuring that crops receive nutrients when they are most needed. In the USA, manure handling can vary significantly from farm to farm, with fewer producers using advanced techniques such as manure injections or digesters to recapture energy (Niles et al., 2022). Many research studies point out that the separation between crops and animal production of intensive livestock systems has led to an imbalance in soil nutrients, concentrated where animals are present and lacking where crops are grown (Flynn et al., 2023; Kleinman et al., 2022). As a result, ideas surrounding a 'manureshed' have been developed, which aim to balance nutrient surpluses and deficiencies by assessing nutrient needs per land area, as a function of animal head counts and manure production (Flynn et al., 2023; Spiegal et al., 2020). The development of a 'manureshed' seeks to overcome this imbalance by recycling nutrients among farm operations to balance production, thereby mimicking traditional crop-livestock agricultural systems, but at a larger scale and in co-operation with other farmers (Flynn et al., 2023). However, major hurdles for its widespread implementation are coordination of producers, special analysts, and industries, as target areas are subject to variability over time based on nutrient needs, crops grown, as well as environmental conditions (Flynn et al., 2023). Furthermore, nutrient needs vary depending on the nutrient species, and effective nutrient management strategies are required to balance these needs and avoid overapplication of any one nutrient, which may require more adaptive approaches to model nutrient flows across systems (Dell et al., 2022). Using efficient collection and redistribution systems could significantly reduce nutrient loss (Aarons et al., 2023). While grazed pastures often have more patchy nutrient distribution, the nutrient extraction has been found to be comparable to fertilizer management (Aarons et al., 2023), thus reducing the need for fertilizer production and transport. Patchy nutrient distribution may also be improved using better grazing management, such as rotational grazing.

Another major source of concern is the amount of water livestock agriculture uses. Agricultural irrigation is the largest human water consumer, accounting for about 70% of global human freshwater withdrawal and about 90% of global human freshwater consumption. It is estimated that growing crops for livestock consumes approximately 41% of the total agricultural water use (Heinke et al., 2020). Additional water is also used during further processing of animal products, such as cheese production. In terms of water withdrawals, livestock production accounts for 5% globally, with greater extraction in India, China, and the Eastern USA, due to the high density of animals (Huang et al., 2018). Milk product consumption alone contributes approximately 7% to the global water footprint (Hoekstra & Mekonnen, 2012). Research has shown that the reintegration of cropping systems with livestock has the potential to reduce water and air pollution and optimize nutrient cycling. Sustainable dairy agriculture should consider water and nutrient footprints, in addition to carbon footprints to better understand sustainable practices. Mixed crop-livestock systems could provide a more efficient and conservative way to meet future nutritional needs while providing biodiversity and more resilient food ecosystems to weather extremes (Kronberg et al., 2021), while also providing risk management during variable weather and economic market situations.

7 Greenhouse gas emissions from crop-livestock systems

Animal agriculture is a larger contributor to agricultural GHG emissions compared to production systems aimed at 'plant-based diets,' largely due to emissions associated with manure storage and application and enteric fermentation. Methane emissions from enteric fermentation differ significantly between nations, with the USA and Canada leading (117 and 128 kg/head/year), followed by EU nations with 109 kg/head/year (Lauerwald et al., 2024), and lowest emissions from India (56 kg/head/year) accounting for 50–67% of agricultural emissions (Mech et al., 2023), while enteric CH_4 emissions from smallholder dairy farms in Ethiopia are estimated to be significantly less (34 kg/head/year) and even lower compared to enteric CH_4 emissions from wild bison (Feyissa et al., 2023). These examples highlight the large variability in agricultural GHG emissions based on management. As one data mining study points out, the best predictor for enteric CH_4 emissions is dry matter intake (DMI); hence, emissions are a function of productivity rather than breed (Evangelista et al., 2024). Manure is the second largest contributor to on-farm GHG emissions of dairy agricultural systems, which are primarily composed of N_2O and CH_4 emissions during manure storage, as well as from field emissions after application. Manure is responsible for approximately 7% of agricultural CH_4 and N_2O emissions but varies based on manure storage and field application (Owen & Silver, 2017). To understand the impact agriculture has on global climate change, it is imperative to distinguish between anthropogenic and biogenic sources of GHG emissions.

Biogenic sources include ruminant grazing animals that emit CH_4 and CO_2 as part of their natural digestion and metabolism. For example, just a few decades ago, an estimated 30 million bison (*Bison bison* L.) were roaming and grazing in North America which also emitted GHGs to the atmosphere, while offering a valuable food and spiritual source for Native American tribes (Kelliher & Clark, 2010). Recent research has shown that ruminant emissions from bison amount to roughly ~81 g ± 28 g of CH_4 per bison per day (Stoy et al., 2020) or roughly 72 kg of CH_4 per head per year (Kelliher & Clark, 2010), which amounts to more than double the emissions from Ethiopian dairy cows, and ranging from ~50 to 100% of a modern-day dairy cow in Europe and North America (Thacharodi et al., 2024). Additionally, CH_4 has a relatively low lifetime in the atmosphere (~12 years) compared to CO_2 (Yenice, 2024). These results suggest that GHG emissions could be considered as part of the natural biogenic carbon cycle, and that focus should be put on improving management, including feed rations and soil conservation, to reduce GHG emissions in the short term, rather than eliminating livestock from the human diet (Shule Liu et al., 2021). Considering anthropogenic CH_4 sources from the burning of fossil fuels and emissions from landfills, livestock agricultural accounts for approximately 16% of global CH_4 emissions, of which dairy cattle account for approximately 30% (Tseten et al., 2022; Yenice, 2024). Yenice (2024) points out that a typical dairy cow emits approximately 1/3 of CO_2 equivalent (CO_2-eq) compared to an average passenger car per year, showing that while we need to reduce agricultural emissions to meet Paris climate targets, other anthropogenic sources are likely more important targets.

A study by Hristov (2012) showed that present-day animal husbandry may exceed historic wild ruminant GHG emissions by only 14%. Nevertheless, modern milk productivity in the USA has increased by roughly 59% compared to 1944, while the number of US milking cows decreased by about 64% from 1944 to 2007 and 16% from 1992 to 2012 (Keyserlingk et al., 2013). The global dairy cow population is estimated to be ~1.5 billion, with India and Brazil leading the chart with roughly 190 million in India (Mathkari, 2025), 214 million head in Brazil, while herd sizes in the USA and Europe are only 9.3 million and 19.7 million, indicating increasing trends in the Global South (Kozicka et al., 2023). Nevertheless, most agricultural research surrounding livestock agriculture focuses on eliminating or reducing ruminant GHG emissions, while land management often plays a secondary role as a strategy to offset emissions. In contrast to ICLS, where nutrients are recycled on-farm, industrial fertilizer production is a linear system that requires the constant supply of fossil fuels, which is predicted to become scarce in the coming decades (Kronberg et al., 2021). Synthetic fertilizers contribute approximately 2–5% to the global GHG budget (~11% of all agricultural emissions), with the largest share of emissions coming from field applications (~59%) and fertilizer production (~39%) and the rest being transport (Gao & Serrenho, 2023; Menegat et al., 2022). In contrast ruminant CH_4 emissions account for approximately 5% of global annual GHG emissions (Thacharodi et al., 2024).

Abandoning livestock agriculture, as it is suggested by a subset of the scientific literature, may reduce GHG emissions, but it would also make us heavily dependent on artificial ways of distributing nutrients and managing vegetation. For instance, while grazing and mowing prairie or pasture ecosystems can be comparable, natural grazing has been shown to offer the greatest benefits for soil carbon storage and biodiversity in semi-natural grasslands. Grazing promotes more heterogeneous ecosystems and helps maintain a higher species richness through mechanisms like seedling emergence and micro-site availability, which are less pronounced in mowing systems (Tälle et al., 2016). To put this further into perspective, maintaining vegetation types in artificial ways (i.e. mowing, planting, harvesting, etc.) will require a sustainable energy source. To date, these practices are maintained by fossil fuel, directly or indirectly (Paris et al., 2022), which require continued investment in infrastructure (i.e. mining, transport, and the production of machinery), which is associated with even larger CO_2 footprints amounting to 40% of the global CO_2 emissions (Wang et al., 2024a).

In contrast, research has shown that under the right environmental conditions and ample moisture, grazing can serve as a carbon sink compared to confinement systems in which forage is imported to the barns, but especially in contrast to mowing management of grasslands which can be a net carbon source due to the large reduction in leaf area and when hay is exported from the premises (Koncz et al., 2017; Senapati et al., 2014; Soussana et al., 2010). Correct grazing management can provide significant benefits to soil conditions, as well as entail a more conservative biomass removal compared to mowing, which leads to enhanced carbon sequestration and improved soil health (Koncz et al., 2017). Grass areas that get repeatedly mowed (for forage or to control biomass in urban areas) tend to have high N_2O emissions due to the application of fertilizer and pesticides and compaction of the soil (Bellarby et al., 2013; Foley et al., 2011). In contrast to mowing treatments, grazed pastures can offset soil GHG and enteric fermentation emissions via perennial plant CO_2 uptake and improved carbon sequestration due to the recycling of carbon from ingested biomass (Koncz et al., 2017; Senapati et al., 2014). Additionally, grazing diets for dairy livestock have been shown to reduce GHG emissions, with methane production being 17% lower for cattle fed grass through a cut-and-carry system and 39% lower for those allowed to graze, compared to cattle fed a grain-based total mixed ration (TMR) (Cameron et al., 2018).

A plethora of recent lifecycle analyses have focused on mitigating or offsetting on-farm emissions on an annual basis. However, biogenic processes occur on several different timescales, from hours to weeks, when considering plant physiological processes or the turnover rates of soil microbial organisms. For example, under nutrient-poor conditions, soil microbes can have generation times ranging from 120–180 days, whereas in environments with more readily available nutrients, turnover can occur within a matter of days (Hobbie & Hobbie, 2013). Other biogenic processes occur over several months (i.e. crop growth) to decades or even centuries when considering trees that may be present on a farm setting (Nair et al., 2010) or when considering national economies

(Ajani et al., 2013). When aiming for net zero, agricultural systems should be evaluated over multiple timescales, as different processes, such as nutrient cycling and decomposition, do not follow the same seasonal patterns (Bortolli et al., 2024). For example, nutrient availability has been shown to vary based on nutrient species, which can affect residue or litter decomposition and with it the subsequent crop, and differences in soil organic carbon accumulation, which occurs on timescales much larger than a year (Dean et al., 2012). Further, the global warming potential (GWP) varies significantly by gas species. For example, while CO_2 has a near-linear relationship with global warming over centuries, gases like CH_4 have a higher short-term GWP but dissipate more quickly (Rogelj et al., 2016). Moreover, modelling research indicates that while corn-based diets can reduce enteric methane emissions by 8%, a holistic assessment shows that the GHG emissions from converting pastures to grain systems may take 44 years to offset the non-recurrent emissions (Middelaar et al., 2013). Furthermore, while seasonal cycles, including growing seasons, occur on a roughly semiyearly scale at latitudes above 30 degrees north and south, as a function of temperature, tropical agricultural systems are more continuous and primarily affected by precipitation seasonality (Minoli et al., 2022), highlighting the need for localized adaptation strategies. To overcome growing season limitations in cooler climate, global dairy operations have adopted different strategies to maintain milk production throughout the year, which include silage production, or the import of feed (Lee et al., 2013). It should be noted that silage production and storage are also associated with GHG emissions, with anaerobic fermentation during silage production releasing up to 10% of total CH_4 emissions from agricultural systems and contributing to emissions ranging from 0.15–0.45 tons of CO_2 per ton of silage produced (Deeken et al., 2024).

8 Impacts of climate change on integrated dairy crop-livestock systems

The impact of climate change on integrated dairy crop-livestock systems will depend on its trajectory, climate region, extreme weather, and intensity, as well as socio-economic and national and international policy drivers. Climate change will affect agricultural systems in multiple ways, by altering rain patterns and intensity, soil fertility and crop yields, temperature, changes in soil microbiomes, as well as heat stress in animals. Likewise, pests and pathogens will likely increase in occurrence due to warmer temperatures, reductions in winter frosts, and the expansion of agricultural production into newly suitable regions (i.e. climate-driven agricultural frontiers). The increased spread of agricultural products globally and the shift toward more intensified agricultural systems (such as monocropping) may further exacerbate vulnerability to pests and diseases

(Hannah et al., 2020). The trend of removing trees from agricultural fields and paddocks to enhance management and crop productivity has resulted in a lack of shaded areas for grazing livestock (England et al., 2020; Schütz et al., 2011). In livestock agriculture, heat stress can lead to a reduction in feed intake by up to 20%, increasing metabolic expenses by 20%, and consequently lowering milk productivity, with studies showing up to a 35% decline in milk yield during periods of high temperature and humidity (Summer et al., 2018; Wankar et al., 2021). Additionally, rising global temperatures have increased energy costs, as barns require improved ventilation systems to maintain optimal animal body temperatures. Climate change indirectly affects livestock systems by altering crop productivity, which can be stressed by changes in water availability or heatwaves, leading to reduced productivity and forage quality. Further, typical monocropping species, such as corn, soybean, and wheat, are more susceptible to herbivory and pathogen outbreaks, creating significant economic losses and food insecurities. For example, soybean rust and wheat blotch have caused yield losses of more than 50% in severe outbreaks (Singh et al., 2023).

Integrated crop dairy livestock systems, particularly systems that incorporate grazing, have the potential to withstand these pressures, as diversity can foster healthier soil microbiomes that can withstand pathogen pressures (Singh et al., 2023), as well as improve the resilience of plants to changes in short-term water availability (Franzluebbers & Hendrickson, 2024). Research has shown that crop-livestock systems are more resilient to climate change compared to more specialized monocropping systems. In an analysis of soybean and beef integrated systems in Brazil, gross margins were found to be 23% higher on average compared to confinement systems, even under future climate scenarios. The integrated system showed greater resilience, with higher productivity and profitability, despite soybean yield reductions due to climate stress (Peterson et al., 2020). Other strategies include improving cow genetics via selective breeding, as well as optimizing the timing of feed intake to minimize negative impacts on milk productivity and heat stress (Cheng et al., 2022; Toledo et al., 2022). Heat stress in cows can further impact immune functions fighting pathogens and parasites, which can increase diseases such as mastitis, with negative economic implications for farmers. This is projected to lead to a reduction in milk yield by up to 2.8% by 2050 in some regions and an increase in health-related issues, with significant economic implications for farmers, including costs related to veterinary care (Guzmán-Luna et al., 2022).

It is important to note that climate change will likely impact producers' risk management strategies, which have traditionally focused on sustaining yields, particularly of monocropping systems. For example, several nations have public or private agricultural insurance programs that provide coverage for yield losses from, for example, natural disasters. However, such insurances have been shown to hamper with climate change adaptation as their availability reduces the demand for agricultural innovations by almost a quarter. Further, insurance policies might become unaffordable if climate change increases the frequency of weather extremes (Miao, 2020). Climate change has also been shown to significantly affect the price volatility of commodities, especially for energy and fertilizers, which in turn drive food price inflation, which can have negative implications for the sustainability of conventional farming systems, as well as global food security (Iliyasu & Sanusi, 2024). Many farmers already use futures and options contracts to hedge their risk against price fluctuations. However, larger farms generally have better means to employ multiple risk management strategies compared to smallholder farmers (Birthal et al., 2021). Nevertheless, crop and confinement crop-livestock systems are subject to much greater volatility compared to grazing systems, due to their dependence on fertilizer, chemical herbicides and fungicides, and feed imports (Khatri et al., 2024).

9 Strategies to achieve net zero and climate resilience

Several strategies to offset, capture or reduce GHG emissions from livestock agricultural systems have been proposed (Caradus et al., 2024). Historically, research has focused on improving dairy breeds for greater milk production and feed conversion efficiency, which was believed to reduce the GHG footprint per unit of milk produced (Oltenacu & Broom, 2010). However, such studies often overlook the negative impacts of high-protein and nitrogen-rich diets on the ecosystem, particularly in intensive systems, leading to higher nitrogen excretion and subsequently higher ammonia emissions (NH_3), which contribute to both GHG emissions and environmental degradation (Otálora et al., 2024). Other studies suggest that feed quality, particularly the digestibility of grazed pastures, can affect CH_4 emissions on a dry matter intake basis. For example, higher quality forages with lower fiber content, such as hay or silage, improve DMI efficiency and reduce the proportion of energy lost as CH_4. In contrast, poor-quality feed, with higher fiber levels, increases CH_4 production and reduces DMI (Min et al., 2022; Ormston et al., 2025). While genetic selection can lead to reductions in enteric emissions by up to 24% by 2050 (de Haas et al., 2021), improving feed efficiency might be more effective. Improving feed efficiency could lead to reductions in CH_4 emissions by 10–20% in the short term, depending on the selected traits (Manzanilla-Pech et al., 2022). Other studies suggest that cows with low CH_4 emissions can be selected based on their DNA, as the rumen microbiota, particularly methanogenic archaea, is influenced by the host animal's genetic profile (Roehe et al., 2016). However, this idea has been disputed by other research studies, which found that while genetic selection for low emitting cows is possible, the host genetics and the rumen microbiome contribute independently to methane emissions, meaning the rumen microbiome is not directly modulated by the cow's genetic makeup (Difford et al., 2018).

In recent years, much research has focused on feed or dietary additives (making up <1% of dry matter intake) such as the red seaweed Asparagopsis (Vijn et al., 2020), and microbial treatments to reduce CH_4 emissions (Hodge et al., 2024), both of which alter the microbial populations in the rumen. However, several metanalyses also found large variability in the results of CH_4 reductions. These studies point out that more *in vivo* experiments are required to understand the true effects of seaweed on enteric fermentation, particularly considering consistent sources of seaweed production that could be scaled (Lean et al., 2021; Vijn et al., 2020), Further, large quantities of seaweed, as in half of the current global seaweed production, would be necessary to meet reduction targets for US cattle herds alone (Vijn et al., 2020). Worse yet, smallholder farms are likely unable to participate due to the high cost of this feed additive. Additionally, the active ingredient in seaweed to inhibit methanogenesis, bromoforms, has been identified as a probable carcinogen to humans (Caradus et al., 2024; Patra et al., 2017).

While enteric fermentation is difficult to control, larger success has been achieved via more efficient manure collection, storage, recycling, and application operations, including manure treatments such as acidification which can reduce NH_3 and CH_4 emissions by up to 93% and 87%, respectively (Yan et al., 2024). Further, biofiltration technologies that focus on capturing GHG emissions from manure storage facilities exist with a removal efficiency of up to 85% (Melse & Werf, 2005; Pratt et al., 2012). However, their implementation is largely dependent on legislation and funding. Several European countries have implemented legislation to control and manage manure storage facilities, including manure storage covers to reduce ammonia emissions and nitrate leaching as well as volumes by restricting water inputs from rainfall (Köninger et al., 2021; Varma et al., 2021). These regulations also require waterproof facilities to prevent nitrate leaching into soil and groundwater, which can have significant positive effects on overall nutrient recovery and the environment (Agga et al., 2022; Sobhi et al., 2021). Additionally, the EU has introduced the European Climate Law in 2021 which sets legally binding EU-wide targets for GHG emissions reduction, including a 55% net reduction by 2030 and climate neutrality by 2050, while providing flexibility for member states to determine their own national measures to contribute to these collective goals (Tvarnø, 2022). For example, in Denmark, several regulations are in place that have reduced nitrogen leaching by 45% by restricting manure application to periods with crop demand for nutrients and mandating catch crops such as grass and winter cereals (Sommer & Knudsen, 2021). Just recently, biofilters, which have been successfully used to collect and reduce GHG emissions from landfills, are now tested to capture GHG emissions from manure storage, with efficiencies reaching nearly 100% for CH_4 and 96–100% for trace gases, making them a promising technology for reducing GHG emissions from agricultural waste (Duan et al., 2021). While anaerobic digesters offer a way to reduce GHG emissions and potential for reuse of energy, their adoption has been limited (with less than 2% dairy farms in California utilizing this technology), due to the significant installation cost and required changes in manure collection conditions (Khoshnevisan et al., 2021; Mashad et al., 2023).

Other strategies to reduce and offset emissions include the coexistence of dairy cattle with perennial vegetation, including pastures, shrublands, hedgerows, or trees, which could also include continuous green cover in the form of cover crops. However, results from cover crop experiments are varied and often show limited effects on carbon sequestration, nutrient retention, and nutrient runoff (Bolinder et al., 2020; Hu et al., 2023; Jian et al., 2020; Muhammad et al., 2019; Nouri et al., 2022). More promising approaches include silvopasture and agroforestry systems, where animals are integrated with trees and grasslands on the same land, which can reduce farming inputs (Brick et al., 2022). These systems are often part of tropical agricultural systems, as well as arid and semiarid climate zones, providing both environmental and economic benefits to farmers, including nutritional security (Raj et al., 2022). Integrated agroforestry livestock systems were shown to reduce the cost of farming inputs, increase forage quality, and with it milk production (by up to 38.5%), while nearly halving the carbon footprint of milk (Quintero et al., 2024). However, effects will depend on the prevailing climate and socio-economic conditions. For example, where milk productivity is driven by imported feed, transitions to silvopasture could temporarily reduce milk productivity. Nevertheless, potential losses in income to the dairy farmer could be offset through timber sales or governmental incentives (Apdini et al., 2024). Other co-benefits from agroforestry systems exist, which include the protection of heat stress of dairy cows due to the provision of shade while grazing, which could help maintain milk productivity and reduce mortality in an increasingly warmer climate (Brick et al., 2022; Morenz et al., 2024). For example, tree canopies in Brazilian tropical agroforestry systems reduced solar radiation exposure by 58%. However, surveys from EU nations showed that consumer interest in this practice is still relatively low, despite being the most common agricultural practice in Europe until at least the Middle Ages (King, 1987). This suggests that more research and outreach is needed to raise consumer awareness and understanding of these sustainable practices (Naspetti et al., 2021). For example, the implementation of precision agriculture using sensor technology could increase the adoption and management of agroforestry practices and raise the potential to meet increasing food demands while maintaining sustainability (Brick et al., 2022).

A more wide-reaching strategy to shift ICLS to become net zero emitters has been proposals of carbon taxes on emissions (i.e. $15–144 or 120 Euro per metric ton of CO_2-eq emitted), to include negative externalities in the milk price (Wehner & Yu, 2023). However, careful consideration in drafting these strategies need to be taken, to avoid shifting carbon emissions abroad when implemented on a national level or as part of import/export agreements (Jansson et al., 2024), or to overly burden the agricultural producer, particularly smallholder farms. For instance, studies have shown that commodity crop prices could rise with the implementation of a carbon tax scheme, leading to an increase in production costs of up to 32.6%. However, these studies primarily focused on major crops like corn, soybean, and wheat, and did not consider ICLS or pasture-based ICLS, which would be impacted differently (Dumortier & Elobeid, 2021). Other studies suggest that a carbon tax on milk production systems is less elastic compared to other commodities, because the demand for fresh milk consumption is less sensitive to price changes and income fluctuations. As such, the largest financial burden would likely be carried by upper-middle-income economies, depending on the tax rate (Wehner & Yu, 2023). Other studies find that a global carbon tax, depending on the policy, could reduce agricultural emissions by 19%, but may result in food insecurities due to consumer, rather than producer price increases, as well as a shift in diet, depending on the country (Jansson et al., 2024).

Another significant driver of GHG emissions is waste management, which can be an issue in the dairy industry, due to overproduction and limited shelf life of the product. Milk 'dumping' has often captured national and international headlines, with the COVID-19 pandemic being the most recent significant instance that caused supply shocks in the dairy industry, causing farmers to dump large volumes of milk as they were unable to distribute their goods to temporarily closed institutional buyers such as schools and restaurants (Lemke et al., 2024; Liu & Rabinowitz, 2021). Milk and other milk products are often wasted because of significant price drops due to product stockpiling. In Canada, over 6.8 billion liters of milk were wasted between 2012 and 2021, amounting to a loss of approximately 14.9 billion Canadian dollars, largely due to the country's supply management system and the overproduction incentivized by milk quotas (Elliot et al., 2025). This highlights the volatility of dairy farms in response to changing market conditions, particularly within the supply chain and processing capacity. To enhance the sustainability of dairy agriculture, it is essential to reduce overproduction and waste, which is more achievable with smaller, multiple farms rather than concentrated animal feeding operations. For example, dairy farms, particularly in nations with substantial milk production and export, prioritize consistency in milk quality and supply to maintain stability in production and market conditions. This approach often takes precedence over adapting to changes in productivity caused by seasonality, weather extremes, or other disturbances. In mature markets, where competition is high and milk quality is crucial, the pressure to maintain stable output is intensified, sometimes limiting flexibility in adapting to external environmental factors (Fiorillo & Amico, 2024). Early detection systems that assist farmers, processors, and consumers in adapting to changes in production could be a valuable approach to reducing negative externalities in the dairy supply chain (Bilotto et al., 2024), which can include sensor and data-driven technologies.

Other drivers include consumer behavior and tradition. For example, in Europe local agricultural and traditional livestock products have been an important part of the culture for centuries (Ilbery & Kneafsey, 1998), which include world renowned specialty cheeses (Merlino et al., 2022). These traditional markets often entail greater sustainability and population maintenance of rural areas, by promoting local food products (Guerrero et al., 2009), as well as more potential to adapt to local changes in environmental conditions.

Lastly, it is often stated that our increasing population demands an increased rate of production of goods and services (Patwa et al., 2021). However, evidence suggests that current caloric production levels (though not nutritionally balanced) are sufficient to sustain global population levels. The primary challenge is the unequal distribution of nutritious and healthy resources, particularly to vulnerable populations in developing countries (Béné et al., 2020; Berners-Lee et al., 2018; KC et al., 2018; Kick et al., 2017; McKenzie & Williams, 2015). For the dairy industry, research identified multifunctionality, targeting diverse services in addition to agricultural production, improving the connection between consumer and producer, as well as circular economics, including the inclusion of agroecology, integrating farm practices and natural resources with socioeconomic ecosystems (Clay et al., 2020). Research has shown that collaboration between supply chain actors during the COVID-19 dairy crisis significantly improved the adaptation and survival of production and partnerships, even superseding economic and performance goals (Lemke et al., 2024). Finally, it is worth noting that economic drivers and legislation are primary drivers of farm management, such that a farmer will prioritize their return on investment, while processors and markets will value low-cost raw ingredients (Glover et al., 2014). As a result, strategies to reduce GHG emissions will have to target a balance between economic and environmental sustainability (Garrett et al., 2017).

10 Conclusion

In this chapter we showed that to understand net zero dairy agriculture, we must (1) account for the multifunctionality of ICLS, (2) consider all influxes and effluxes across farm boundaries producing GHG emissions, (3) assess the mitigation potential of all vegetation grown on farm, especially perennial vegetation cover, (4) thoroughly assess technologies that facilitate GHG capture, and (5) understand net zero on timescales that allow for proper comparison of different processes, particularly with respect to the different global warming potentials of GHG species as functions of their atmospheric lifetimes.

The connection between socio-economic drivers, the environment, and climate has long been recognized (Ebert, 2018). A sustainable economy offers resources and their distribution in an efficient way, as well as adaptation to, for example, short-term climate changes, such as droughts, to maintain agricultural production and the distribution of resources for long periods of time. The question is if and how to marry the need for profit growth of industries with sustainable resource management. Our current economic systems often lack adaptation buffers, partly due to the centralization and intensification of production systems, as well as the consolidation of supply chains and processors, which limits resilience and reduces the ability to adapt to changes (Bojovic & McGregor, 2023). Global economic systems heavily rely on the maximization of a few cash crops and other products to sustain growth. Instead of maximum yield, we should strive for maximum energy use efficiency, which would allow for the sustainable management of our agricultural resources, allowing for environmental health, reduced GHG emissions, climate buffers, all the while being profitable. Prioritizing efficiency results in reduced pollution and more sustainable agricultural practices (Jordan, 2016). Recent research increasingly highlights the importance of circular economies. Studies demonstrate lifecycle assessments of various agricultural production systems that aim to integrate mass and energy flows in ways that balance and neutralize each other. With a rising global population, albeit reductions in fertility, this is a challenge, as the output (cash crops, milk, etc.) is sought to increase, while negative externalities are required to decline. Realistically, this can only be achieved when systems become more energy efficient, for example, by improving the efficiency of photosynthesis, all the while not increasing emissions. Further, for novel technologies to be effective, the production and supply chain resource use and GHG emissions need to be accounted for and compared with more traditional farming systems that rely on natural resources and local distribution of products. Such systems are accomplished by integrating livestock with crop production, which could also mean focusing on local distribution schemes of nutrients (i.e. manureshed), rather than energy-intensive production of chemical fertilizers, as well as diversified systems that help fight pathogens and herbivores using the natural defense mechanisms of biodiversity, rather than intensively managed treatments (Ryschawy et al., 2012). Finally, research shows that the integration of livestock, especially with mixed cropping systems (i.e. permaculture), has the potential for greater climate and economic resilience to weather or market fluctuations by integrating environmental and societal needs with a circular economy. Integrated crop-livestock systems offer better risk management than insurance schemes or subsidies, and greater food security by reducing supply chain dependencies, as well as greater environmental protection if circular nutrient management is prioritized. In conclusion, to meet net zero of the dairy industry, national and international policies must prioritize the diversification and reintegration of livestock and crop agriculture.

11 Acknowledgement

We thank Dr. Veronica Justen and Chris Wilson for their valuable inspiration during initial discussions. Susanne Wiesner acknowledges funding from the Wisconsin Dairy Innovation Hub during her appointment at the University of Wisconsin River Falls.

12 Where to look for further information

To explore practical frameworks, policy insights, and scientific guidance on crop-livestock integration and pathways toward net zero dairy production, the following resources provide comprehensive, practical and evidence-based guidance:

Key reports and initiatives

- **FAO Livestock Environmental Assessment and Performance (LEAP) Partnership:** Offers methodologies for assessing greenhouse gas emissions and mitigation options in livestock systems. https://www.fao.org/LEAP.
- **IPCC Special Report on Climate Change and Land (2019)** Summarizes mitigation pathways for agriculture and livestock within global climate frameworks. https://www.ipcc.ch/srccl.
- **FAO, The State of Food and Agriculture 2023 – Revealing the True Cost of Food:** A flagship FAO report applying true cost accounting to agrifood systems, assessing environmental, health, and social externalities across 154 countries–including smallholder crop-livestock systems. https://www.fao.org.
- **World Bank, Moving Towards Sustainability: The Livestock Sector and the World Bank:** A policy brief exploring how livestock underpins rural livelihoods in low- and middle-income countries and outlining integrated strategies to enhance environmental performance while supporting socio-economic development. https://www.worldbank.org.
- **CGIAR, Livestock & Environment Research Program:** A comprehensive initiative analyzing environmental footprints, resource-use efficiency, and governance strategies for sustainable livestock systems in developing-country contexts. https://livestock.cgiar.org.

- **UNEP/AU-IBAR, Livestock Development Strategy for Africa (LiDeSA) (2015–2035):** A policy framework addressing environmental sustainability, socio-economic growth, and pastoral resilience in Africa's integrated crop-livestock systems. https://leap.unep.org.
- **CONCITO, Climate Action in a Downstream Agricultural Emissions Trading System (2024):** Provides analysis of policy instruments for reducing agricultural emissions through market-based mechanisms, with a focus on Danish and EU contexts. https://concito.dk.

Key organizations and networks

- Sustainable Livestock Intergroup: An EU-based platform fostering dialogue on sustainable livestock production, policy development, and socio-economic impacts across Europe. https://sustainablelivestockintergroup.eu.
- International Dairy Federation (IDF): Global standards and sustainability initiatives for dairy. https://www.fil-idf.org.
- European Federation of Animal Science (EAAP): Annual conferences and research updates on livestock systems. https://www.eaap.org.

Major conferences

- IDF World Dairy Summit: Annual global event on dairy sustainability and innovation.
- EAAP Annual Meeting: Leading European conference on animal science and integrated systems.

Leading research centers

- International Livestock Research Institute (ILRI): Global research on livestock and mixed farming systems. https://www.ilri.org.
- Wageningen University & Research (The Netherlands): Expertise in sustainable livestock and crop integration. https://www.wur.nl.
- Aarhus University (Denmark): Research on climate-smart dairy systems and circular agriculture. https://www.au.dk.
- University of Reading (UK): Strong programs in dairy sustainability and agri-food systems. https://www.reading.ac.uk.

13 References

Aarons, S. R., et al. (2023). Estimating excreted nutrients to improve nutrient management for grazing system dairy farms. *Animals*, *13*(8), 1404. https://doi.org/10.3390/ani13081404.

Agga, G. E., et al. (2022). Lagoon, anaerobic digestion, and composting of animal manure treatments impact on tetracycline resistance genes. *Antibiotics*, *11*(3), 391. https://doi.org/10.3390/antibiotics11030391.

Aguirre-Villegas, H. A. & Larson, R. A. (2017). Evaluating greenhouse gas emissions from dairy manure management practices using survey data and lifecycle tools. *Journal of Cleaner Production*, *143*, 169–179. https://doi.org/10.1016/j.jclepro.2016.12.133.

Aguirre-Villegas, H. A., et al. (2014). From waste-to-worth: energy, emissions, and nutrient implications of manure processing pathways. *Biofuels, Bioproducts and Biorefining*, *8*(6), 770–793. https://doi.org/10.1002/bbb.1496.

Ajani, J. I., et al. (2013). Comprehensive carbon stock and flow accounting: a national framework to support climate change mitigation policy. *Ecological Economics*, *89*, 61–72. https://doi.org/10.1016/j.ecolecon.2013.01.010.

Apdini, T., et al. (2024). Developing sustainable dairy farms in the tropics: from policy to practice. *Agricultural Systems*, *220*, 104097. https://doi.org/10.1016/j.agsy.2024.104097.

Basarab, J., et al. (2012). Greenhouse gas emissions from calf- and yearling-fed beef production systems, with and without the use of growth promotants. *Animals*, *2*(2), 195–220. https://doi.org/10.3390/ani2020195.

Bazargani, K. & Deemyad, T. (2024). Automation's impact on agriculture: opportunities, challenges, and economic effects. *Robotics*, *13*(2), 33. https://doi.org/10.3390/robotics13020033.

Beaver, A., et al. (2021). Invited review: the welfare of dairy cattle housed in tiestalls compared to less-restrictive housing types: a systematic review. *Journal of Dairy Science*, *104*(Arch. Anim. Breed. 56 2013), 9383–9417. https://doi.org/10.3168/jds.2020-19609.

Bell, L. W. & Moore, A. D. (2012). Integrated crop–livestock systems in Australian agriculture: trends, drivers and implications. *Agricultural Systems*, *111*, 1–12. https://doi.org/10.1016/j.agsy.2012.04.003.

Bellarby, J., et al. (2013). Livestock greenhouse gas emissions and mitigation potential in Europe. *Global Change Biology*, *19*(1), 3–18. https://doi.org/10.1111/j.1365-2486.2012.02786.x.

Béné, C., et al. (2020). Global drivers of food system (un)sustainability: a multi-country correlation analysis. *PLoS One*, *15*(4), e0231071. https://doi.org/10.1371/journal.pone.0231071.

Berners-Lee, M., et al. (2018). Current global food production is sufficient to meet human nutritional needs in 2050 provided there is radical societal adaptation. *Elementa: Science of the Anthropocene*, *6*, 52. https://doi.org/10.1525/elementa.310.

Bilotto, F., et al. (2024). Towards resilient, inclusive, sustainable livestock farming systems. *Trends in Food Science & Technology*, *152*, 104668. https://doi.org/10.1016/j.tifs.2024.104668.

Birthal, P. S., et al. (2021). Effectiveness of farmers' risk management strategies in smallholder agriculture: evidence from India. *Climatic Change*, *169*(3–4), 30. https://doi.org/10.1007/s10584-021-03271-1.

Blandford, D. (2024). The Vikings have landed – Denmark's decision to tax livestock emissions as a 'game changer.' *EuroChoices*, *23*(3), 38–41. https://doi.org/10.1111/1746-692x.12440.

Bleasdale, M., et al. (2021). Ancient proteins provide evidence of dairy consumption in eastern Africa. *Nature Communications*, *12*(1), 632. https://doi.org/10.1038/s41467-020-20682-3.

Bojovic, M. & McGregor, A. (2023). A review of megatrends in the global dairy sector: what are the socioecological implications? *Agriculture and Human Values*, *40*(1), 373–394. https://doi.org/10.1007/s10460-022-10338-x.

Bolinder, M. A., et al. (2020). The effect of crop residues, cover crops, manures and nitrogen fertilization on soil organic carbon changes in agroecosystems: a synthesis of reviews. *Mitigation and Adaptation Strategies for Global Change*, *25*(6), 929–952. https://doi.org/10.1007/s11027-020-09916-3.

Bonanno, A., et al. (2018). Market power and bargaining in agrifood markets: a review of emerging topics and tools. *Agribusiness*, *34*(1), 6–23. https://doi.org/10.1002/agr.21542.

Bortolli, M. A., et al. (2024). Nutrient dynamics in integrated crop–livestock systems: effects of stocking rates and nitrogen system fertilization on litter decomposition and release. *Agronomy*, *14*(9), 2009. https://doi.org/10.3390/agronomy14092009.

Bravo-Ureta, B. E., et al. (2021). *Handbook of production economics* (pp. 1–39). https://doi.org/10.1007/978-981-10-3450-3_31-1.

Brick, E. S. R., et al. (2022). A review of agroforestry, precision agriculture, and precision livestock farming–The case for a data-driven agroforestry strategy. *Frontiers in Sensors*, *3*, 998928. https://doi.org/10.3389/fsens.2022.998928.

Brito, A. F., et al. (2022). Production performance, nutrient use efficiency, and predicted enteric methane emissions in dairy cows under confinement or grazing management system. *Translational Animal Science*, *6*(2), txac028. https://doi.org/10.1093/tas/txac028.

Brown, K. A., et al. (2022). The role of dairy in healthy and sustainable food systems: community voices from India. *BMC Public Health*, *22*(1), 806. https://doi.org/10.1186/s12889-022-13194-w.

Cameron, L., et al. (2018). A comparison of milk yields and methane production from three contrasting high-yielding dairy cattle feeding regimes: cut-and-carry, partial grazing and total mixed ration. *Grass and Forage Science*, *73*(3), 789–797. https://doi.org/10.1111/gfs.12353.

Campbell-Arvai, V. (2009). Literature review : a comparison of dairy production systems. *Environmental Science, Agricultural and Food Sciences, Economics*. https://www.semanticscholar.org/paper/Literature-Review-%3A-A-Comparison-of-Dairy-Systems/c296842a50baacc274611e6cf06fcff59517b957.

Caradus, J. R., et al. (2024). Improving human diets and welfare through using herbivore-based foods: 2. environmental consequences and mitigations. *Animals*, *14*(9), 1353. https://doi.org/10.3390/ani14091353.

Cheng, M., et al. (2022). Climate change and livestock production: a literature review. *Atmosphere*, *13*(1), 140. https://doi.org/10.3390/atmos13010140.

Clay, N., et al. (2020). Dairy intensification: drivers, impacts and alternatives. *Ambio*, *49*(1), 35–48. https://doi.org/10.1007/s13280-019-01177-y.

Cox, J. & Mason, C. (2007). Standardisation versus adaptation: geographical pressures to deviate from franchise formats. *The Service Industries Journal*, *27*(8), 1053–1072. https://doi.org/10.1080/02642060701673737.

Crippa, M., et al. (2021). Food systems are responsible for a third of global anthropogenic GHG emissions. *Nature Food*, *2*(3), 198–209. https://doi.org/10.1038/s43016-021-00225-9.

Crusciol, C. A. C., et al. (2014). Intercropping soybean and palisade grass for enhanced land use efficiency and revenue in a no till system. *European Journal of Agronomy*, *58*, 53–62. https://doi.org/10.1016/j.eja.2014.05.001.

Dasselaar, A. P., et al. (2020). Grazing of dairy cows in Europe—An in-depth analysis based on the perception of grassland experts. *Sustainability*, *12*(3), 1098. https://doi.org/10.3390/su12031098.

de Haas, Y., et al. (2021). Selective breeding as a mitigation tool for methane emissions from dairy cattle. *Animal*, 15(Suppl 1), 1–10. https://doi.org/10.1016/j.animal.2021.100294.

Dean, C., et al. (2012). Accounting for space and time in soil carbon dynamics in timbered rangelands. *Ecological Engineering*, *38*(1), 51–64. https://doi.org/10.1016/j.ecoleng.2011.10.008.

Deeken, H. F., et al. (2024). Greenhouse gas and volatile organic compound emissions of additive-treated whole-plant maize silage: part A—anaerobic fermentation period. *Chemical and Biological Technologies in Agriculture*, *11*(1), 134. https://doi.org/10.1186/s40538-024-00655-0.

Dell, C. J., et al. (2022). Challenges and opportunities for manureshed management across U.S. dairy systems: Case studies from four regions. *Journal of Environmental Quality*, *51*(4), 521–539. https://doi.org/10.1002/jeq2.20341.

Detzel, A., et al. (2022). Life cycle assessment of animal-based foods and plant-based protein-rich alternatives: an environmental perspective. *Journal of the Science of Food and Agriculture*, *102*(12), 5098–5110. https://doi.org/10.1002/jsfa.11417.

Difford, G. F., et al. (2018). Host genetics and the rumen microbiome jointly associate with methane emissions in dairy cows. *PLoS Genetics*, *14*(10), e1007580. https://doi.org/10.1371/journal.pgen.1007580.

Duan, Z., et al. (2021). Mitigation of methane and trace gas emissions through a large-scale active biofilter system at Glatved landfill, Denmark. *Waste Management*, *126*, 367–376. https://doi.org/10.1016/j.wasman.2021.03.023.

Duguma, B. (2022). Farmers' perceptions of major challenges to smallholder dairy farming in selected towns of Jimma Zone, Oromia Regional State, Ethiopia: possible influences, impacts, coping strategies and support required. *Heliyon*, *8*(6), e09581. https://doi.org/10.1016/j.heliyon.2022.e09581.

Dumortier, J. & Elobeid, A. (2021). Effects of a carbon tax in the United States on agricultural markets and carbon emissions from land-use change. *Land Use Policy*, *103*, 105320. https://doi.org/10.1016/j.landusepol.2021.105320.

Ebert, S. F. (2018). *Famines During the "Little Ice Age" (1300-1800). Socio-natural Entanglements in Premodern Societies | H-Soz-Kult. Kommunikation und Fachinformation für die Geschichtswissenschaften | Geschichte im Netz | History in the web*. (D. Collet & M. Schuh, Eds.) (Vol. 1). Heidelberg Center for the Environment (HCE): Springer Cham. https://doi.org/10.1007/978-3-319-54337-6.

Elliot, T., et al. (2025). Over 6 billion liters of Canadian milk wasted since 2012. *Ecological Economics*, *227*, 108413. https://doi.org/10.1016/j.ecolecon.2024.108413.

England, J. R., et al. (2020). Trees on farms to support natural capital: an evidence-based review for grazed dairy systems. *Science of The Total Environment*, *704*, 135345. https://doi.org/10.1016/j.scitotenv.2019.135345.

Evangelista, C., et al. (2024). Enteric methane emission in livestock sector: bibliometric research from 1986 to 2024 with text mining and topic analysis approach by machine learning algorithms. *Animals*, *14*(21), 3158. https://doi.org/10.3390/ani14213158.

FAO, IDF, & IFCN. (2014). *World mapping of animal feeding systems in the dairy sector*. Retrieved from https://openknowledge.fao.org/server/api/core/bitstreams/c6515833-4d1d-4fdf-af90-47cf5a40e15f/content.

Feyissa, A. A., et al. (2023). Enteric methane emission factors of smallholder dairy farming systems across intensification gradients in the central highlands of Ethiopia. *Carbon Balance and Management*, *18*(1), 23. https://doi.org/10.1186/s13021-023-00242-0.

Fiorillo, V. & Amico, B. M. (2024). Milk quality and economic sustainability in dairy farming: a systematic review of performance indicators. *Dairy*, *5*(3), 384–402. https://doi.org/10.3390/dairy5030031.

Flynn, K. C., et al. (2023). Manureshed management to overcome longstanding nutrient imbalances in US agriculture. *Resources, Conservation and Recycling*, *188*, 106632. https://doi.org/10.1016/j.resconrec.2022.106632.

Foley, J. A., et al. (2011). Solutions for a cultivated planet. *Nature*, *478*(7369), 337–342. https://doi.org/10.1038/nature10452.

Franzluebbers, A. J. & Hendrickson, J. R. (2024). Should we consider integrated crop–livestock systems for ecosystem services, carbon sequestration, and agricultural resilience to climate change? *Agronomy Journal*, *116*(2), 415–432. https://doi.org/10.1002/agj2.21520.

Gao, Y. & Serrenho, A. C. (2023). Greenhouse gas emissions from nitrogen fertilizers could be reduced by up to one-fifth of current levels by 2050 with combined interventions. *Nature Food*, *4*(2), 170–178. https://doi.org/10.1038/s43016-023-00698-w.

Garcia, S. N., et al. (2019). A one health perspective on dairy production and dairy food safety. *One Health*, *7*, 100086. https://doi.org/10.1016/j.onehlt.2019.100086.

Garrett, R. D., et al. (2017). Policies for reintegrating crop and livestock systems: a comparative analysis. *Sustainability*, *9*(3), 473. https://doi.org/10.3390/su9030473.

Gaucheron, F. (2011). Milk and dairy products: a unique micronutrient combination. *Journal of the American College of Nutrition*, *30*(sup5), 400S-409S. https://doi.org/10.1080/07315724.2011.10719983.

Giller, K. E., et al. (2021). The future of farming: Who will produce our food? *Food Security*, *13*(5), 1073–1099. https://doi.org/10.1007/s12571-021-01184-6.

Glover, J. L., et al. (2014). An institutional theory perspective on sustainable practices across the dairy supply chain. *International Journal of Production Economics*, *152*, 102–111. https://doi.org/10.1016/j.ijpe.2013.12.027.

Guerrero, L., et al. (2009). Consumer-driven definition of traditional food products and innovation in traditional foods. A qualitative cross-cultural study. *Appetite*, *52*(2), 345–354. https://doi.org/10.1016/j.appet.2008.11.008.

Guzmán-Luna, P., et al. (2022). Analysing the interaction between the dairy sector and climate change from a life cycle perspective: A review. *Trends in Food Science & Technology*, *126*, 168–179. https://doi.org/10.1016/j.tifs.2021.09.001.

Hand, T., et al. (2024). Net Zero by 2050: U.S. Dairy's Climate Pledge and Lessons from Idaho. Retrieved February 9, 2025, from https://www.choicesmagazine.org/choices-magazine/submitted-articles/net-zero-by-2050-us-dairys-climate-pledge-and-lessons-from-idaho.

Hannah, L., et al. (2020). The environmental consequences of climate-driven agricultural frontiers. *PLoS One*, *15*(2), e0228305. https://doi.org/10.1371/journal.pone.0228305.

Hawkins, A., et al. (2020). Effects of housing system on dairy heifer replacement cost from birth to calving: evaluating costs of confinement, dry-lot, and pasture-based systems and their impact on total rearing investment. *Frontiers in Veterinary Science*, *7*, 625. https://doi.org/10.3389/fvets.2020.00625.

Hearn, C., et al. (2021). Agronomic performance of ten perennial ryegrass varieties on commercial grassland farms. *The Journal of Agricultural Science*, *159*(7–8), 604–614. https://doi.org/10.1017/s0021859621000927.

Heinke, J., et al. (2020). Water use in global livestock production–opportunities and constraints for increasing water productivity. *Water Resources Research*, *56*(12). https://doi.org/10.1029/2019wr026995.

Henson, S. & Humphrey, J. (2010). Understanding the Complexities of Private Standards in Global Agri-Food Chains as They Impact Developing Countries. *The Journal of Development Studies*, *46*(9), 1628–1646. https://doi.org/10.1080/00220381003706494.

Herron, J., et al. (2022). Life cycle assessment of pasture-based dairy production systems: current and future performance. *Journal of Dairy Science*, *105*(7), 5849–5869. https://doi.org/10.3168/jds.2021-21499.

Hobbie, J. E. & Hobbie, E. A. (2013). Microbes in nature are limited by carbon and energy: the starving-survival lifestyle in soil and consequences for estimating microbial rates. *Frontiers in Microbiology*, *4*, 324. https://doi.org/10.3389/fmicb.2013.00324.

Hodge, I., et al. (2024). A review of potential feed additives intended for carbon footprint reduction through methane abatement in dairy cattle. *Animals*, *14*(4), 568. https://doi.org/10.3390/ani14040568.

Hoekstra, A. Y. & Mekonnen, M. M. (2012). The water footprint of humanity. *Proceedings of the National Academy of Sciences*, *109*(9), 3232–3237. https://doi.org/10.1073/pnas.1109936109.

Holly, M. A., et al. (2017). Greenhouse gas and ammonia emissions from digested and separated dairy manure during storage and after land application. *Agriculture, Ecosystems and Environment*, *239*, 410–419. https://doi.org/10.1016/j.agee.2017.02.007.

Hospers, J., et al. (2022). The evolution of the carbon footprint of Dutch raw milk production between 1990 and 2019. *Journal of Cleaner Production*, *380*, 134863. https://doi.org/10.1016/j.jclepro.2022.134863.

Hristov, A. N. (2012). Historic, pre-European settlement, and present-day contribution of wild ruminants to enteric methane emissions in the United States. *Journal of Animal Science*, *90*(4), 1371–1375. https://doi.org/10.2527/jas.2011-4539.

Hu, Q., et al. (2023). Soil organic carbon fractions in response to soil, environmental and agronomic factors under cover cropping systems: a global meta-analysis. *Agriculture, Ecosystems & Environment*, *355*, 108591. https://doi.org/10.1016/j.agee.2023.108591.

Huang, Z., et al. (2018). Reconstruction of global gridded monthly sectoral water withdrawals for 1971–2010 and analysis of their spatiotemporal patterns. *Hydrology and Earth System Sciences*, *22*(4), 2117–2133. https://doi.org/10.5194/hess-22-2117-2018.

Hyland, J. J., et al. (2018). Factors influencing dairy farmers' adoption of best management grazing practices. *Land Use Policy*, *78*, 562–571. https://doi.org/10.1016/j.landusepol.2018.07.006.

Ilbery, B. & Kneafsey, M. (1998). Niche markets and regional speciality food products in Europe: towards a research agenda. *Environment and Planning A*, *31*(12), 2207–2222. https://doi.org/10.1068/a312207.

Iliyasu, J. & Sanusi, A. R. (2024). Climate change's impact on commodity prices: a new challenge for monetary policy. *Portuguese Economic Journal*, *23*(2), 187–212. https://doi.org/10.1007/s10258-023-00237-2.

Ilyas, H. M. A., et al. (2019). The carbon footprint of energy consumption in pastoral and barn dairy farming systems: a case study from Canterbury, New Zealand. *Sustainability*, *11*(17), 4809. https://doi.org/10.3390/su11174809.

Jansson, T., et al. (2024). Carbon taxes and agriculture: the benefit of a multilateral agreement. *Climate Policy*, *24*(1), 13–25. https://doi.org/10.1080/14693062.2023.2171355.

Jian, J., et al. (2020). A meta-analysis of global cropland soil carbon changes due to cover cropping. *Soil Biology and Biochemistry*, *143*, 107735. https://doi.org/10.1016/j.soilbio.2020.107735.

Jordan, C. F. (2016). The farm as a thermodynamic system: implications of the maximum power principle. *BioPhysical Economics and Resource Quality*, *1*(2), 9. https://doi.org/10.1007/s41247-016-0010-z.

Kakade, A., et al. (2021). World eutrophic pollution of lake and river: biotreatment potential and future perspectives. *Environmental Technology & Innovation*, *23*, 101604. https://doi.org/10.1016/j.eti.2021.101604.

Kathage, J., et al. (2022). How much is policy driving the adoption of cover crops? Evidence from four EU regions. *Land Use Policy*, *116*, 106016. https://doi.org/10.1016/j.landusepol.2022.106016.

KC, K. B., et al. (2018). When too much isn't enough: does current food production meet global nutritional needs? *PLoS ONE*, *13*(10), e0205683. https://doi.org/10.1371/journal.pone.0205683.

Kelliher, F. M. & Clark, H. (2010). Methane emissions from bison–An historic herd estimate for the North American Great Plains. *Agricultural and Forest Meteorology*, *150*(3), 473–477. https://doi.org/10.1016/j.agrformet.2009.11.019.

Keyserlingk, M. A. G., et al. (2013). Invited review: sustainability of the US dairy industry. *Journal of Dairy Science*, *96*(9), 5405–5425. https://doi.org/10.3168/jds.2012-6354.

Khalafi, A., et al. (2021). Corn (Zea mays L.) growth, yield and nutritional properties affected by fertilization methods and micronutrient use. *International Journal of Plant Production*, *15*(4), 589–597. https://doi.org/10.1007/s42106-021-00148-2.

Khatri, P., et al. (2024). Understanding the intertwined nature of rising multiple risks in modern agriculture and food system. *Environment, Development and Sustainability*, *26*(9), 24107–24150. https://doi.org/10.1007/s10668-023-03638-7.

Khoshnevisan, B., et al. (2021). A critical review on livestock manure biorefinery technologies: sustainability, challenges, and future perspectives. *Renewable and Sustainable Energy Reviews*, *135*, 110033. https://doi.org/10.1016/j.rser.2020.110033.

Khoury, C. K., et al. (2014). Increasing homogeneity in global food supplies and the implications for food security. *Proceedings of the National Academy of Sciences*, *111*(11), 4001–4006. https://doi.org/10.1073/pnas.1313490111.

Kick, E. L., et al. (2017). Food production or food distribution: the key to global food security? *Perspectives on Global Development and Technology*, *16*(6), 666–682. https://doi.org/10.1163/15691497-12341455.

King, K. F. S. (1987). The history of agroforestry in Agroforestry a decade of development. In H. A. Steppler & P. K. R. Nair (Eds.), *Agroforestry a decade of development*. International Council for Research in Agroforestry ICRAF House, Nairobi, Kenya.

Kleinman, P. J. A., et al. (2022). Envisioning the manureshed: Toward comprehensive integration of modern crop and animal production. *Journal of Environmental Quality*, *51*(4), 481–493. https://doi.org/10.1002/jeq2.20382.

Klerk, J. N., & Robinson, P. A. (2022). Drivers and hazards of consumption of unpasteurised bovine milk and milk products in high-income countries. *PeerJ*, *10*, e13426. https://doi.org/10.7717/peerj.13426.

Koncz, P., et al. (2017). Extensive grazing in contrast to mowing is climate-friendly based on the farm-scale greenhouse gas balance. *Agriculture, Ecosystems & Environment*, *240*, 121–134. https://doi.org/10.1016/j.agee.2017.02.022.

Köninger, J., et al. (2021). Manure management and soil biodiversity: towards more sustainable food systems in the EU. *Agricultural Systems*, *194*, 103251. https://doi.org/10.1016/j.agsy.2021.103251.

Kozicka, K., et al. (2023). Explaining global trends in cattle population changes between 1961 and 2020 directly affecting methane emissions. *Sustainability*, *15*(13), 10533. https://doi.org/10.3390/su151310533.

Kronberg, S. L., et al. (2021). Review: closing nutrient cycles for animal production – Current and future agroecological and socio-economic issues. *Animal*, *15*, 100285. https://doi.org/10.1016/j.animal.2021.100285.

Küsters, J., et al. (2021). The effect of increased temperature and CO2 air enrichment on the nutritive value of orchard grass (Dactylis glomerata) in permanent grassland. *The Journal of Agricultural Science*, *159*(3–4), 167–176. https://doi.org/10.1017/s0021859621000460.

Lauerwald, R., et al. (2024). Carbon and greenhouse gas budgets of Europe: trends, interannual and spatial variability, and their drivers. *Global Biogeochemical Cycles*, *38*(8). https://doi.org/10.1029/2024gb008141.

Lean, I. J., et al. (2021). A meta-analysis of effects of dietary seaweed on beef and dairy cattle performance and methane yield. *PLoS One*, *16*(7), e0249053. https://doi.org/10.1371/journal.pone.0249053.

Lee, J. M., et al. (2013). Climate-change effects and adaptation options for temperate pasture-based dairy farming systems: a review. *Grass and Forage Science*, *68*(4), 485–503. https://doi.org/10.1111/gfs.12039.

Lemke, F., et al. (2024). Mitigating the effects of global disruptions on supply chains: gaining insights from the dairy industry during Covid-19. *Production Planning & Control*, *ahead-of-print*(ahead-of-print), 1–21. https://doi.org/10.1080/09537287.2024.2319736.

Lewis, T. W. & Makarewicz, J. C. (2009). Winter application of manure on an agricultural watershed and its impact on downstream nutrient fluxes. *Journal of Great Lakes Research*, *35*(sp1), 43–49. https://doi.org/10.1016/j.jglr.2008.08.003.

Liebe, D. L., et al. (2020). Contributions of dairy products to environmental impacts and nutritional supplies from United States agriculture. *Journal of Dairy Science*, *103*(11), 10867–10881. https://doi.org/10.3168/jds.2020-18570.

Liu, S., et al. (2020). Impact of manure on soil biochemical properties: a global synthesis. *Science of The Total Environment*, *745*, 141003. https://doi.org/10.1016/j.scitotenv.2020.141003.

Liu, S., et al. (2021). Rethinking methane from animal agriculture. *CABI Agriculture and Bioscience*, *2*(1), 22. https://doi.org/10.1186/s43170-021-00041-y.

Liu, Y. & Rabinowitz, A. N. (2021). The impact of the COVID-19 pandemic on retail dairy prices. *Agribusiness*, *37*(1), 108–121. https://doi.org/10.1002/agr.21687.

Lyngkhoi, D. R., et al. (2022). Trend analysis of milk production in India. *Asian Journal of Dairy and Food Research* (Of). https://doi.org/10.18805/ajdfr.dr-1789.

Månsson, H. L. (2008). Fatty acids in bovine milk fat. *Food & Nutrition Research*, *52*(0), 1821. https://doi.org/10.3402/fnr.v52i0.1821.

Manzanilla-Pech, C. I. V., et al. (2022). Selecting for feed efficient cows will help to reduce methane gas emissions. *Frontiers in Genetics*, *13*, 885932. https://doi.org/10.3389/fgene.2022.885932.

Martin, N. P., et al. (2017). Invited review: sustainable forage and grain crop production for the US dairy industry. *Journal of Dairy Science*, *100*(12), 9479–9494. https://doi.org/10.3168/jds.2017-13080.

Mashad, H. M. E., et al. (2023). Anaerobic digestion and alternative manure management technologies for methane emissions mitigation on Californian dairies. *Atmosphere*, *14*(1), 120. https://doi.org/10.3390/atmos14010120.

Mathkari, C. (2025). The cow paradox–a scoping review of dairy bovine welfare in india using the five freedoms. *Animals*, *15*(3), 454. https://doi.org/10.3390/ani15030454.

McKenzie, F. C. & Williams, J. (2015). Sustainable food production: constraints, challenges and choices by 2050. *Food Security*, *7*(2), 221–233. https://doi.org/10.1007/s12571-015-0441-1.

Mech, A., et al. (2023). Assessment of carbon footprint of milk production and identification of its major determinants in smallholder dairy farms in Karnataka, India. *Journal of Dairy Science*, *106*(12), 8847–8860. https://doi.org/10.3168/jds.2022-22153.

Melse, R. W. & Werf, A. W. van der. (2005). Biofiltration for Mitigation of Methane Emission from Animal Husbandry. *Environmental Science & Technology*, *39*(14), 5460–5468. https://doi.org/10.1021/es048048q.

Menegat, S., et al. (2022). Greenhouse gas emissions from global production and use of nitrogen synthetic fertilisers in agriculture. *Scientific Reports*, *12*(1), 14490. https://doi.org/10.1038/s41598-022-18773-w.

Merlino, V. M., et al. (2022). Are local dairy products better? Using principal component analysis to investigate consumers' perception towards quality, sustainability, and market availability. *Animals*, *12*(11), 1421. https://doi.org/10.3390/ani12111421.

Miao, R. (2020). Climate, insurance and innovation: the case of drought and innovations in drought-tolerant traits in US agriculture. *European Review of Agricultural Economics*, *47*(5), 1826–1860. https://doi.org/10.1093/erae/jbaa010.

Middelaar, C. E. V., et al. (2013). Evaluation of a feeding strategy to reduce greenhouse gas emissions from dairy farming: The level of analysis matters. *Agricultural Systems*, *121*, 9–22. https://doi.org/10.1016/j.agsy.2013.05.009.

Min, B.-R., et al. (2022). Enteric methane emissions and animal performance in dairy and beef cattle production: strategies, opportunities, and impact of reducing emissions. *Animals*, *12*(8), 948. https://doi.org/10.3390/ani12080948.

Minoli, S., et al. (2022). Global crop yields can be lifted by timely adaptation of growing periods to climate change. *Nature Communications*, *13*(1), 7079. https://doi.org/10.1038/s41467-022-34411-5.

Moraine, M., et al. (2017). A social-ecological framework for analyzing and designing integrated crop–livestock systems from farm to territory levels. *Renewable Agriculture and Food Systems*, *32*(1), 43–56. https://doi.org/10.1017/s1742170515000526.

Morenz, A. B. da S., et al. (2024). Dairy cows on integrated livestock-forestry system in the tropics. *Agroforestry Systems*, *98*(5), 1079–1090. https://doi.org/10.1007/s10457-023-00883-7.

Muhammad, I., et al. (2019). Regulation of soil CO2 and N2O emissions by cover crops: a meta-analysis. *Soil and Tillage Research*, *192*, 103–112. https://doi.org/10.1016/j.still.2019.04.020.

Mundia, C. W., et al. (2019). A regional comparison of factors affecting global sorghum production: the case of North America, Asia and Africa's Sahel. *Sustainability*, *11*(7), 2135. https://doi.org/10.3390/su11072135.

Nair, P. K. R., et al. (2010). Chapter five carbon sequestration in agroforestry systems. *Advances in Agronomy*, *108*, 237–307. https://doi.org/10.1016/s0065-2113(10)08005-3.

Naspetti, S., et al. (2021). Consumer perception of sustainable practices in dairy production. *Agricultural and Food Economics*, *9*(1), 1. https://doi.org/10.1186/s40100-020-00175-z.

Newsome, T. (2018). The coevolution of wolves and humans. *BioScience*, *68*(4), 305–306. https://doi.org/10.1093/biosci/biy017.

Niles, M. T., et al. (2019). A review of determinants for dairy farmer decision making on manure management strategies in high-income countries. *Environmental Research Letters*, *14*(5), 053004. https://doi.org/10.1088/1748-9326/ab1059.

Niles, M. T., et al. (2022). Manure management strategies are interconnected with complexity across U.S. dairy farms. *PLoS One*, *17*(6), e0267731. https://doi.org/10.1371/journal.pone.0267731.

Nouri, A., et al. (2022). When do cover crops reduce nitrate leaching? A global meta-analysis. *Global Change Biology*, *28*(15), 4736–4749. https://doi.org/10.1111/gcb.16269.

Novo, A. M., et al. (2013). Feasibility and competitiveness of intensive smallholder dairy farming in Brazil in comparison with soya and sugarcane: case study of the Balde Cheio Programme. *Agricultural Systems*, *121*, 63–72. https://doi.org/10.1016/j.agsy.2013.06.007.

Nowak, A. & Różańska-Boczula, M. (2024). A comparative view of the level of agricultural sustainability – The case of European Union member states. *Sustainable Development*, *32*(3), 2638–2652. https://doi.org/10.1002/sd.2804.

Oliveira, C. E. A., et al. (2024). Health and thermal comfort of dairy cattle in compost-bedded pack barns and other types of housing: a comparative systematic review. *AgriEngineering*, *6*(2), 1395–1416. https://doi.org/10.3390/agriengineering6020080.

Oltenacu, P. & Broom, D. (2010). The impact of genetic selection for increased milk yield on the welfare of dairy cows. *Animal Welfare*, *19*(S1), 39–49. https://doi.org/10.1017/s0962728600002220.

Ormston, S., et al. (2025). Impact of dietary forage proportion and crossbreeding on feed efficiency and methane emissions in lactating dairy cows. *Animal Nutrition*, *20*, 419–429. https://doi.org/10.1016/j.aninu.2024.08.011.

Otálora, X. D., et al. (2024). Modelling the effect of context-specific greenhouse gas and nitrogen emission mitigation options in key European dairy farming systems. *Agronomy for Sustainable Development*, *44*(1), 4. https://doi.org/10.1007/s13593-023-00940-6.

Owen, J. J. & Silver, W. L. (2017). Greenhouse gas emissions from dairy manure management in a Mediterranean environment. *Ecological Applications*, *27*(2), 545–559. https://doi.org/10.1002/eap.1465.

Paris, B., et al. (2022). Energy use in open-field agriculture in the EU: A critical review recommending energy efficiency measures and renewable energy sources adoption. *Renewable and Sustainable Energy Reviews*, *158*, 112098. https://doi.org/10.1016/j.rser.2022.112098.

Patra, A., et al. (2017). Rumen methanogens and mitigation of methane emission by anti-methanogenic compounds and substances. *Journal of Animal Science and Biotechnology*, *8*(1), 13. https://doi.org/10.1186/s40104-017-0145-9.

Patwa, N., et al. (2021). Towards a circular economy: An emerging economies context. *Journal of Business Research*, *122*, 725–735. https://doi.org/10.1016/j.jbusres.2020.05.015.

Peters, C. J., et al. (2016). Carrying capacity of U.S. agricultural land: ten diet scenarios. *Elementa*, *4*, 000116. https://doi.org/10.12952/journal.elementa.000116.

Petersen-Rockney, M., et al. (2021). Narrow and Brittle or Broad and Nimble? Comparing adaptive capacity in simplifying and diversifying farming systems. *Frontiers in Sustainable Food Systems*, *5*, 564900. https://doi.org/10.3389/fsufs.2021.564900.

Peterson, C. A., et al. (2020). Resilience of an integrated crop–livestock system to climate change: a simulation analysis of cover crop grazing in Southern Brazil. *Frontiers in Sustainable Food Systems*, *4*, 604099. https://doi.org/10.3389/fsufs.2020.604099.

Pratt, C., et al. (2012). Biofiltration of methane emissions from a dairy farm effluent pond. *Agriculture, Ecosystems & Environment*, *152*, 33–39. https://doi.org/10.1016/j.agee.2012.02.011.

Quintero, R. G., et al. (2024). A case study on enhancing dairy cattle sustainability: the impact of silvopastoral systems and improved pastures on milk carbon footprint and farm economics in Cauca department, Colombia. *Agroforestry Systems*, *98*(8), 3001–3018. https://doi.org/10.1007/s10457-024-01070-y.

Raj, A., et al. (2022). *Natural resources conservation and advances for sustainability* (pp. 289–307). https://doi.org/10.1016/b978-0-12-822976-7.00002-8.

Rayne, N. & Aula, L. (2020). Livestock manure and the impacts on soil health: a review. *Soil Systems*, *4*(4), 64. https://doi.org/10.3390/soilsystems4040064.

Ren, K., et al. (2023). Effect of replacing synthetic nitrogen fertilizer with animal manure on grain yield and nitrogen use efficiency in China: a meta-analysis. *Frontiers in Plant Science*, *14*, 1153235. https://doi.org/10.3389/fpls.2023.1153235.

Requena-i-Mora, M. & Barbeta-Viñas, M. (2024). The agrarian question in dairy farms: an analysis of dairy farms in the European Union countries. *Agriculture and Human Values*, *41*(2), 459–474. https://doi.org/10.1007/s10460-023-10488-6.

Roehe, R., et al. (2016). Bovine host genetic variation influences rumen microbial methane production with best selection criterion for low methane emitting and efficiently feed converting hosts based on metagenomic gene abundance. *PLoS Genetics*, *12*(2), e1005846. https://doi.org/10.1371/journal.pgen.1005846.

Rogelj, J., et al. (2016). Differences between carbon budget estimates unravelled. *Nature Climate Change*, *6*(3), 245–252. https://doi.org/10.1038/nclimate2868.

Röös, E., et al. (2017). Greedy or needy? Land use and climate impacts of food in 2050 under different livestock futures. *Global Environmental Change*, *47*, 1–12. https://doi.org/10.1016/j.gloenvcha.2017.09.001.

Ryschawy, J., et al. (2012). Mixed crop-livestock systems: an economic and environmental-friendly way of farming? *Animal*, *6*(10), 1722–1730. https://doi.org/10.1017/s1751731112000675.

Sánchez, I. M., et al. (2015). Systematics and evolution of the miocene three-horned palaeomerycid ruminants (Mammalia, Cetartiodactyla). *PLoS One*, *10*(12), e0143034. https://doi.org/10.1371/journal.pone.0143034.

Scholtz, M. M., et al. (2020). A balanced perspective on the importance of extensive ruminant production for human nutrition and livelihoods and its contribution to greenhouse gas emissions. *South African Journal of Science*, *116*(9/10). https://doi.org/10.17159/sajs.2020/8192.

Schütz, K. E., et al. (2011). Dairy cattle prefer shade over sprinklers: effects on behavior and physiology. *Journal of Dairy Science*, *94*(1), 273–283. https://doi.org/10.3168/jds.2010-3608.

Senapati, N., et al. (2014). Net carbon storage measured in a mowed and grazed temperate sown grassland shows potential for carbon sequestration under grazed system. *Carbon Management*, *5*(2), 131–144. https://doi.org/10.1080/17583004.2014.912863.

Serpell, J. A. (2021). Commensalism or cross-species adoption? a critical review of theories of wolf domestication. *Frontiers in Veterinary Science*, *8*, 662370. https://doi.org/10.3389/fvets.2021.662370.

Sharara, M., et al. (2017). Spatially explicit methodology for coordinated manure management in shared watersheds. *Journal of Environmental Management*, *192*, 48–56. https://doi.org/10.1016/j.jenvman.2017.01.033.

Sharara, M. A., et al. (2018). Techno-economic optimization of community-based manure processing. *Agricultural Systems*, *161*, 117–123. https://doi.org/10.1016/j.agsy.2018.01.006.

Simensen, E., et al. (2010). Housing system and herd size interactions in Norwegian dairy herds; associations with performance and disease incidence. *Acta Veterinaria Scandinavica*, *52*(1), 14. https://doi.org/10.1186/1751-0147-52-14.

Singh, B. K., et al. (2023). Climate change impacts on plant pathogens, food security and paths forward. *Nature Reviews Microbiology*, *21*(10), 640–656. https://doi.org/10.1038/s41579-023-00900-7.

Sobhi, M., et al. (2021). Nutrients recovery from fresh liquid manure through an airlift reactor to mitigate the greenhouse gas emissions of open anaerobic lagoons. *Journal of Environmental Management*, *294*, 112956. https://doi.org/10.1016/j.jenvman.2021.112956.

Soergel, B., et al. (2021). A sustainable development pathway for climate action within the UN 2030 Agenda. *Nature Climate Change*, *11*(8), 656–664. https://doi.org/10.1038/s41558-021-01098-3.

Sommer, S. G. & Knudsen, L. (2021). Impact of Danish livestock and manure management regulations on nitrogen pollution, crop production, and economy. *Frontiers in Sustainability*, *2*, 658231. https://doi.org/10.3389/frsus.2021.658231.

Soussana, J. F., et al. (2010). Mitigating the greenhouse gas balance of ruminant production systems through carbon sequestration in grasslands. *Animal*, *4*(3), 334 350. https://doi.org/10.1017/s1751731109990784.

Spiegal, S., et al. (2020). Manuresheds: advancing nutrient recycling in US agriculture. *Agricultural Systems*, *182*, 102813. https://doi.org/10.1016/j.agsy.2020.102813.

Spiekermann, U. (2011). Redefining food: the standardization of products and production in Europe and the United States, 1880–1914. *History and Technology*, *27*(1), 11–36. https://doi.org/10.1080/07341512.2011.548971.

Srinivasan, M. S., et al. (2006). Manure management and nutrient loss under winter conditions: a literature review. *Journal of Soil and Water Conservation*, *61*(4), 200–209. https://doi.org/10.1080/00224561.2006.12435884.

Stoy, P. C., et al. (2020). Methane efflux from an American bison herd. *Biogeosciences*, *18*(3), 961–975. https://doi.org/10.5194/bg-18-961-2021.

Summer, A., et al. (2018). Impact of heat stress on milk and meat production. *Animal Frontiers*, *9*(1), 39–46. https://doi.org/10.1093/af/vfy026.

Sumner, D. A. (2014). American farms keep growing: size, productivity, and policy. *Journal of Economic Perspectives*, *28*(1), 147–166. https://doi.org/10.1257/jep.28.1.147.

Tälle, M., et al. (2016). Grazing vs. mowing: a meta-analysis of biodiversity benefits for grassland management. *Agriculture, Ecosystems & Environment*, *222*, 200–212. https://doi.org/10.1016/j.agee.2016.02.008.

Teague, R. & Kreuter, U. (2020). Managing grazing to restore soil health, ecosystem function, and ecosystem services. *Frontiers in Sustainable Food Systems*, *4*, 534187. https://doi.org/10.3389/fsufs.2020.534187.

Thacharodi, A., et al. (2024). The ruminant gut microbiome vs enteric methane emission: the essential microbes may help to mitigate the global methane crisis. *Environmental Research*, *261*, 119661. https://doi.org/10.1016/j.envres.2024.119661.

Toledo, I. M., et al. (2022). Dairy cattle management and housing for warm environments. *Livestock Science*, *255*, 104802. https://doi.org/10.1016/j.livsci.2021.104802.

Tonet, R. M., et al. (2023). Typology of Brazilian dairy farms based on vulnerability characteristics. *Animal - Open Space*, *2*, 100040. https://doi.org/10.1016/j.anopes.2023.100040.

Tseten, T., et al. (2022). Strategies to mitigate enteric methane emissions from ruminant animals. *Journal of Microbiology and Biotechnology*, *32*(3), 269–277. https://doi.org/10.4014/jmb.2202.02019.

Tvarnø, C. (2022). The new era of climate law in Denmark and in the EU. *European Public Law*, *28*(Issue 1), 101–122. https://doi.org/10.54648/euro2022006.

Varma, V. S., et al. (2021). Dairy and swine manure management – Challenges and perspectives for sustainable treatment technology. *Science of The Total Environment*, *778*, 146319. https://doi.org/10.1016/j.scitotenv.2021.146319.

Vijn, S., et al. (2020). Key considerations for the use of seaweed to reduce enteric methane emissions from cattle. *Frontiers in Veterinary Science*, *7*, 597430. https://doi.org/10.3389/fvets.2020.597430.

Wang, G., et al. (2024a). A comprehensive review of building lifecycle carbon emissions and reduction approaches. *City and Built Environment*, *2*(1), 12. https://doi.org/10.1007/s44213-024-00036-1.

Wang, Y., et al. (2024b). Carbon footprint of a typical crop–livestock dairy farm in Northeast China. *Agriculture*, *14*(10), 1696. https://doi.org/10.3390/agriculture14101696.

Wangu, J., et al. (2021). Recognizing determinants to smallholders' market orientation and marketing arrangements: building on a case of dairy farming in Rural Kenya. *Land*, *10*(6), 572. https://doi.org/10.3390/land10060572.

Wankar, A. K., et al. (2021). Heat stress in dairy animals and current milk production trends, economics, and future perspectives: the global scenario. *Tropical Animal Health and Production*, *53*(1), 70. https://doi.org/10.1007/s11250-020-02541-x.

Wehner, J. & Yu, X. (2023). Carbon tax on milk products and the exact consumer welfare measure in emerging economies. *Agribusiness*, *39*(S1), 1595–1623. https://doi.org/10.1002/agr.21854.

Weisbecker, A. (2007). A legal history of raw milk in the United States. *Journal of Environmental Health*, *69*(8), 62–3.

Wiesner, S., et al. (2022). Quantifying the natural climate solution potential of agricultural systems by combining eddy covariance and remote sensing. *Journal of Geophysical Research: Biogeosciences*, *127*(9), e2024JG008391. https://doi.org/10.1029/2022jg006895.

Wilkin, S., et al. (2020). Dairy pastoralism sustained eastern Eurasian steppe populations for 5,000 years. *Nature Ecology & Evolution*, *4*(3), 346–355. https://doi.org/10.1038/s41559-020-1120-y.

Wróbel, B., et al. (2023). Challenges of pasture feeding systems–opportunities and constraints. *Agriculture*, *13*(5), 974. https://doi.org/10.3390/agriculture13050974.

Yan, X., et al. (2024). A review of mitigation technologies and management strategies for greenhouse gas and air pollutant emissions in livestock production. *Journal of Environmental Management*, *352*, 120028. https://doi.org/10.1016/j.jenvman.2024.120028.

Yang, Y., et al. (2024). Drivers of soybean-based rotations synergistically increase crop productivity and reduce GHG emissions. *Agriculture, Ecosystems & Environment*, *372*, 109094. https://doi.org/10.1016/j.agee.2024.109094.

Yenice, G. (2024). A green misconception: the real impact and role of ruminants on greenhouse gases. *Journal of Dairy & Veterinary Sciences*, *16*(5). https://doi.org/10.19080/jdvs.2024.16.555948.

Zanten, H. H. E., et al. (2023). Circularity in Europe strengthens the sustainability of the global food system. *Nature Food*, *4*(4), 320–330. https://doi.org/10.1038/s43016-023-00734-9.

Zhou, M., et al. (2023). Evaporative water loss from dairy cows in climate-controlled respiration chambers. *Journal of Dairy Science*, *106*(3), 2035–2043. https://doi.org/10.3168/jds.2022-22489.

Part 4

Conclusion

Chapter 14

Achieving net zero dairy farming: weighing the options arising from new understanding

John Webster, University of Bristol, UK

1 Introduction

The chapters in this book explore in great detail the many and various scientific principles and methods employed in the pursuit of knowledge and understanding in relation to the issue of sustainability in dairy farming, with particular emphasis on the target of net zero for greenhouse gases (GHGs). This topic may be extremely complex when examined in detail, and difficult to reconcile with other essential objectives, like the provision of good, cheap, healthy, and humane food. However, as I wrote in Chapter 1 (Webster, 2026), it may be defined by a small number of simple factors that can, in the first analysis, be addressed by a small number of simple questions (with difficult answers). Net zero for any agricultural or industrial process means what it says. It describes a perfect balance between the generation and removal of the by-products (wastes) of the process, in this case, dairy farming.

The expression 'Achieving net zero' is fashionable but scientifically sloppy shorthand for achieving sustainability and avoiding pollution of the land, the waters, and the air. The essence of good farming, throughout the ages, has been based on good husbandry: producing food and other resources from plants and animals in ways best suited to the land, water and weather, managing the

http://dx.doi.org/10.19103/AS.2025.0157.32

balance between the needs of crops and animals and sustaining the capacity of the soil, plants and animals to support these needs. The concept of net zero means that all matter is recycled (eventually). The only non-renewable element is energy from the sun and, for practical purposes, we can consider this to be inexhaustible. The history of agriculture tells of famines and other disasters attributable to acts of God and Mammon, but overall, traditional (subsistence) agriculture managed to balance the books; to maintain stable cycles of carbon, nitrogen, and other key elements.

In the times before the coming of agriculture, especially the Carboniferous period, plants, through photosynthesis, harvested the inexhaustible energy supply from the sun to store vast quantities of carbon as what we now describe as fossil fuels. Our current capacity to extract and exploit these vast reserves of primeval solar energy for our present use has self-evidently brought enormous benefits to our lives but, equally self-evidently, it is unsustainable because, in terms of energy, fossil fuels represent capital not income. Until recently this did not appear to be a problem, because there is still plenty of carbon under the ground (and sea). However, it is now clear that we have released too much of this carbon into the air, the planet is overheating, and we need to restore the equilibrium.

Industrialised farming, especially the huge increase in the intensive production of relatively cheap foods of animal origin, has not only brought great satisfaction to those of us with the resources to enjoy the riches on offer. Foods such as milk also provide a critical source of high-quality nutrients, especially protein, for those who can barely afford a subsistence diet. However, the increase in livestock production to meet the demands and desires of the human population, and the increased consumption of fossil fuels to support the livestock and dairy industries, are significant contributors to the problem of global warming. Moreover, dairy cows, like all ruminants, present the special problem of methane produced from rumen fermentation. The Global Warming Potential (GWP) of methane relative to CO_2 cannot be defined with precision since methane has a much shorter atmospheric lifetime than carbon dioxide, thus its GWP is much less over longer time periods. Thus GWP at 100 years, GWP-100 is 27.9, whereas GWP-500 is only 7.95.

Atmospheric methane has increased since pre-industrial times from 0.7 ppm to 1.9 ppm, more than doubling in concentration. This increase in methane reflects the incapacity of natural methane sinks to accommodate the increase in methane due to anthropogenic activities. From 2010 to 2019, methane emissions caused 0.5°C (about 30%) of observed global warming. Methane emissions from human activity total between 350 million tonnes and 390 million tonnes annually (Dlugokencky et al., 2011). Emissions from the agriculture and fossil fuel energy sectors were approximately equal at around 120–140 million tonnes per year, roughly twice the emissions of the waste sector (Caro et al., 2014). Over 70% of methane production from agriculture comes from the livestock sector. However, in the specific context of dairy farming, methane belched directly from cows (enteric methane production) is only part of the problem. The goal of reduced methane production relative to milk yield must be approached on the basis of a Life Cycle Analysis (LCA) that incorporates all sources (cows, wastes, fossil fuels) and sinks (plants and soils).

2 Greenhouse gas emissions from dairy farms

Chapter 3 (Rotz and Dell, 2026) presents a superb, comprehensive review and analysis of the factors contributing to the emission of GHG from dairy farms and pathways to their mitigation. I present here a simplified summary of their review to demonstrate the magnitude of the major contributors to GHG production and the potential impact of different strategies for their mitigation.

The primary GHG compounds emitted from dairy farms are methane (CH_4), nitrous oxide (N_2O), and carbon dioxide (CO_2). Sources include the animals, the manure they produce, land used to produce feed crops and bought-in resources (e.g. feed, fertiliser, and fuel). Animal emissions are predominantly CH_4, whereas emissions from cropland are N_2O. The GWP for GHGs can be summated by expressing CH_4 and N_2O relative to CO2 (CO_2e). Methane is by far the most important GHG in this context. Rotz and Dell assume 28 kg CO_2e/kg CH_4 (GWP-100). Production of GHG is expressed as kg CO_2e/kg FPCM (fat and protein corrected milk), which is, of course, the proper way to do it, since the prime target is not simply to reduce emissions but to reduce this ratio.

Thoma et al. (2013a, 2013b) report farmgate GHG emission intensities varying from 0.6 kg to 3.0 kg CO_2e/kg FPCM with a mean of 1.23 kg CO_2e/kg FPCM, which suggests that there may be considerable potential for mitigation. Rotz et al. (2010) presented a comprehensive LCA on four dairy farms in the U.S.A., an organic, pasture-based farm in Pennsylvania and three large intensive dairy units in Wisconsin (grassy), California, and Texas (dry cows mostly fed on corn and other harvested crops). Table 1 presents a simplified summary of their results. The difference between these four systems in overall production of GHG was small, 0.87–1.05, much less than the range reported by Thoma et al. (2013a). Enteric methane production contributed 57% to total output of CO_2e, kg/kg FPCM on the smaller organic farm (Pennsylvania) but only 37% on the intensive unit in California. However, emissions from the system, principally attributable to the storage and handling of manure, were much greater in both relative and absolute terms within the confinement systems. Manure disposal made a negligible contribution to emissions on the Pennsylvania unit, which practised organic grazing. The Texas and California units were especially profligate in respect to emissions associated with manure management.

Table 1 Milk production, (kg FPCM milk/cow) and major contributors to GHG emissions (kg CO_2e/kg FPCM) from four dairy farms in U.S.A. and the potential (%) of different strategies for their mitigation

	Pennsylvania grazing		Wisconsin confinement		Texas confinement		California confinement	
Milk production, kg	7640		10690		12000		12300	
GHG emissions, (CO_2e)								
Total	0.87	Total %	0.97	Total %	0.94	Total %	1.05	Total %
Enteric	0.50	57	0.48	49	0.40	42	0.39	37
Manure	0.05	6.0	0.14	14	0.34	36	0.40	38
Resources	0.21	24	0.20	20	0.18	19	0.33	30
Mitigation (%)								
Feeding	14.6		9.5		11.7		8.4	
Methane inhibitors	9.0		13.3		11.3		12.1	
Manure	32		15		25		26	
Total	25		46		53		2.1	
Potential sequestration @ 0.5 tC/ha			13		2.0			

The theoretical impact of mitigation attributable to feeding strategy mainly reflects differences in feed formulation; in particular, the contribution of fibre from grasses (grazed and conserved) and starches, from corn (maize) and other harvested crops. Increasing the ratio of starch to fibre (within limits consistent with stable fermentation and digestion) increases the metabolisable energy (ME) concentration of the diet, thereby making it possible for cows to take in more ME and so sustain higher milk yields within the constraints of appetite. It also increases the propionate:acetate ratio in the end-products of fermentation. The increase in milk yield relative to CO_2e reflects both the increased *gross* efficiency of conversion of ME with increasing intake in respect to maintenance *and* the reduction in methane production per MJ ME associated with the increase in the propionate:acetate ratio.

The 14.6% reduction in CO_2e, kg/kg FPCM *theoretically* achievable by changes in feed formulation on the Pennsylvania, grass-based unit reflects the potential combined impact of increasing intake and the starch:fibre ratio on the emission of methane per MJ ME. However, such a strategy would not only be alien to the practice of milk production from grazing cows fed on fresh and conserved pasture, but a step backwards in relation to the pursuit of net zero. Rotz and Dell estimated that carbon sequestration on this organic farm at a realistic rate of 0.5 tonnes C/ha per annum could mitigate a further 25% of CO_2e/kg FPCM.

The theoretical potential for reducing CO_2e/kg FPCM produced by the cows within the three confinement units was split approximately 45:55% between changes to feed formulation and strategies designed to inhibit enteric methane production. I made the point in my opening chapter that nearly all high-production farms, whether pastoral or confinement based, are already formulating rations designed to maximise efficiency of feed conversion to milk. Thus, while there are strong theoretical grounds for reducing CO_2e/kg FPCM through attention to feed formulation, it is likely that the best dairy farmers are already close to the limit.

3 Enteric methane production and mitigation

The analysis of contributors to methane production on the four high-production units summarised in Table 1 shows that enteric methane may contribute (only) about 40% to total emissions in confined units, the proportion rising to 60% in pastoral systems. It is likely that, in the units described by Thoma et al. (2013a) with farmgate GHG emission intensities approaching 3.0 kg CO_2e/kg FPCM, the proportion attributable to enteric methane production may be considerably lower. The estimated potential for mitigation of total CO/kg FPCM from changes in feeding and enteric methane inhibition are 8.4–14.6% and 9.0–13.3% respectively (Table 1). These are theoretical estimates. The maximum effect of increasing the concentrate:forage, (starch:fibre) ratio can only be achieved with a ration both unrealistically and unhealthily low in fibre for dairy cows.

Chapter 2 (Ramin et al., 2026) concludes that the impact of dietary carbohydrate composition exhibited only minor effects on CH_4 emission. They described this as 'surprising'. However, they expressed CH_4 emissions relative to GE intake rather than digestible energy (DE), which can lead to misleading conclusions. Increasing the concentrate:forage ratio increases digestibility, i.e. the ratio of digestible/gross energy. Thus, no change in the CH_4/GE ratio with increasing digestibility indicates a decrease in methane emissions per MJ DE and an even greater reduction in the ratio CH_4-e/ME.

Chapter 9 (Newbold, 2026) outlines four general strategies that can be used to inhibit methane production in the rumen:

1 Reduce the production of H (hydrogen) during fermentation by substituting starch and/or fat for plant cell walls in forage-based diets.
2 Inhibit methanogenesis (or methanogenic archaea) directly.
3 Provide alternatives to methanogenesis as pathways for H utilisation.
4 Metabolise methane before it leaves the rumen.

These approaches are potentially additive. Mitigation of methane production by dietary means has been considered already. The potential of approaches three and four to mitigate methane emissions would appear to be small. Thus, the major impetus has focused on the science of inhibitors of methanogenesis and their potential for incorporation into commercial feed additives.

Chapter 2 (Ramin et al., 2026) reports that the most effective feed additives based on their proven efficacy are, in decreasing order of impact:

- the synthetic compound 3-nitrooxypropanol (3-NOP), marketed as Bovaer 10®;
- red seaweed, *Asparagopsis* spp.;
- nitrate; and
- essential oils.

To date, the most promising feed additive appears to be the synthetic compound 3-NOP. A recent meta-analysis, which included 340 data from 14 experiments, reported a 33%, 31%, and 33% reduction in methane production (g/d), methane yield (g/kg dry matter intake), and methane intensity (g/kg energy corrected milk yield), respectively, when 3-NOP was fed to dairy cows (Kebreab et al., 2023). Most studies have found no reduction in milk yield at dosages between 40 mg and 200 mg 3-NOP/kg feed dry matter.

The red seaweed *Asparagopsis* spp. contains halogenated compounds, primarily bromoform, which inhibits the final step of methanogenesis in the rumen, thereby reducing methane production. Studies quoted by Ranin reported reductions in methane yield ranging from 20% to 55%. However, supplementation with *Asparagopsis* spp. significantly decreased feed intake and milk yield. Additionally, there are other concerns: damage to the rumen wall, absence of rumen papillae, haemorrhages, ulcers, and other signs of inflammation. There is also concern as to the possible transfer of bromoform into milk, since bromoform has toxic and carcinogenic properties.

Other possible methane inhibitors (e.g. nitrate, oils) have been reported to achieve 10–20% reduction in emissions without reducing feed intake or milk yield. The ionophore monensin is widely used as a feed additive for dairy cows in U.S.A. to improve milk yield, feed conversion efficiency and reduce the incidence of ketosis in early lactation, largely by increasing the ratio of glucogenic propionate to ketogenic acetate in the end-products of fermentation (Rezaei Ahvanooei et al., 2013). It is therefore, *inter alia*, a methane inhibitor but currently not licensed for use in the EU and UK.

The evidence indicates that, at present, 3-NOP appears to be the most effective of the commercially available methane inhibitors, but this does not necessarily equate with commercial success. Martins et al. (2025), assuming zero cost of 3-NOP and using US feed costs and milk prices, calculated that use of 3-NOP increased income over feed cost by only 0.14 US$/cow per day, and concluded that other financial incentives would be needed to achieve the aim of reducing methane emissions through adoption of 3-NOP on a large scale.

The inescapable conclusion to be drawn from these reports is that, while methane inhibitors, especially 3-NOP used as feed additives, may make a modest contribution to the goal of net zero (and perhaps a significant source of income for the successful developers), this, at best, is unlikely to achieve a target approaching 40% for total CO_2e emissions from large commercial dairy farms.

Chapter 6 (Van Breukelen, 2026) examines the potential for reduction in CH_4/kg FPCM through genetic selection and reviews approaches to the recording of methane production on a scale that can achieve the numbers necessary for genetic trials. First principles dictate that a reduction in CH_4e/kg FPCM can be achieved through:

1 Improving the gross efficiency of feed utilisation for milk production, through increased intake or a reduction in 'residual feed intake' (intake in excess of that 'necessary' for milk production and maintenance of body condition).
2 Anatomical or physiological differences in digestive physiology, e.g. reduced rumen volume and/or increased passage rate, that reduce the contribution of fermentation to total digestion.
3 Changes in the rumen biome attributable to changes in cow genotype.

Genetic selection for traits linked to physiological changes that reduce the contribution of rumen fermentation to the generation of ME in dairy cows could contribute to a reduction in CH_4/kg FPCM in high-performance dairy cows fed high-energy rations in confinement. However, it would be a backward step within the context of pastoral systems.

There is some evidence that the rumen microbiome composition is partially controlled by host genomics, with heritabilities ranging from 0.10 to 0.4. Genetic selection has the advantage that its effect is cumulative and permanent. However, changing populations through genetic selection can be a slow process. Selection index calculations suggest that genetic selection for CH_4 intensity (CH_4/ kg milk) could achieve a reduction of 24% by 2050 compared to 2018 levels.

The overall conclusion to be drawn from Table 1 and this summary is that we might, by 2050, reduce the contribution of GHG from enteric fermentation. However, as this is likely to constitute no more than half to the total, it is very far from being the whole story.

4 Manure management systems

The traditional, 'green' method for managing manure has, of course, been to spread it on the fields and restore essential nutrients (N and P) and sustain the quality of the soil. The separation between crops and animal production with intensive livestock systems inevitably creates an imbalance in soil nutrients, concentrated where animals are present and lacking where crops are grown. Chapter 13 (Wiesner et al., 2026) describes one approach to this problem based on the concept of a 'manureshed', which aims to balance nutrient surpluses and deficiencies by assessing nutrient needs per land area, as a function of animal head counts and manure production, thereby mimicking traditional crop-livestock agricultural systems, but at a larger scale through cooperation between multiple farmers. However, as Bittman et al. (2026) (Chapter 11) write: 'Manure is bulky and heavy, irregular and inconsistent, with an inconvenient nutrient composition that does not match crop needs. Manure nutrients are also prone to losses as gases, solutes and suspensions, and manure is a source of pathogens. Solving the manure conundrum so that it contributes value to farms, rather than merely posing a disposal issue is a process with synergies, trade-offs, paradoxes and, often, costs'. The increase in the amount of milk produced from confinement units housing hundreds or thousands of cows massively increases the problem of pollution as defined by 'too much fertiliser in the wrong place'.

Bittman et al. explore strategies designed to maximise the amount of manure that fertilises the land relative to that which pollutes the air and water sources. The most common, rapid, and least trouble-prone method of spreading slurry is by broadcasting, especially on small family farms with limited resources. However, slurry applied via broadcast is prone to substantial N losses by NH_3 volatilisation. Various low emission applicators have been developed to conserve manure N and reduce deleterious emissions, e.g. using trailing hoses, which hang loosely at or near the ground surface, and trailing shoes, which float directly on the soil surface.

The great increase in the contribution of large-scale confinement units for farm animals, especially pigs and dairy cows, together with the use of liquid-based manure handling and storage systems has inevitably led to increased absolute emissions of GHG, predominantly methane. Papers cited by Rocha (2026) (Chapter 10) reveal that manure management emissions have increased by 55.6% from 1990 to 2022 and are now responsible for 13.8% of U.S. agricultural GHG emissions. These reports confirm the conclusions summarised in Table 1 that, on large barn-based dairy units, manure-based GHG emissions make a significantly greater contribution to global warming than enteric fermentation. Manure management has therefore become the highest priority for research and development in the pursuit of reduced GHG emissions from high output, intensive dairy farms. Rocha reviews the various strategies. Most are designed to reduce reliance on liquid manure storage thereby reducing time and space in anaerobic environments that lead to methane formation.

Nitrogenous compounds, especially NH_3 and N_2O, are also potentially serious pollutants. Here again, the poison is in the dose since the nitrogen cycle is essential to life. Ammonia is produced from the reaction between urea in urine and the enzyme urease in faeces, either in animal housing facilities or in manure storage areas. While NH_3 (unlike N_2O) is not a GHG, it is no less problematic when it comes to its impact on the environment when produced in excess, deposited into water and land-based ecosystems, leading to eutrophication.

Anaerobic digesters (AD) appear to offer the most effective approach to the mitigation of GHG emissions from dairy units. Chapter 12 (Kusch-Brandt, 2026) sets out the many ways by which they can contribute to the pursuit of net zero. For dairy farms, they include:

- Reduced release of GHG through the establishment of a system of manure and slurry management that minimises the emission of GHG from anaerobic fermentation in storage.
- Substitution of fossil fuels with renewable methane generated from the AD.
- Recycling of solid and liquid residues on the land, thereby reducing the need for synthetic fertilisers and contributing to carbon sequestration.

Kusch–Brandt describes types of AD suitable for large and medium/small units. While AD may be too expensive to install for exclusive use on medium-to-small sized farms or pastoral systems, it is possible to establish joint-use AD units that augment the supply of manure and slurry from the dairy farm with fuels from multiple sites and sources: agricultural residues, food waste, organic waste from industry, biowaste from households and sewage sludge.

Other, more traditional approaches to GHG mitigation include composting, solid separators, crusting and daily spreading. These do not carry the multiple benefits of AD, but they may be more realistic options for small/medium and seasonally pastoral farms. In all but the most pastoral of dairy units, the establishment of manure management strategies and facilities that significantly reduce GHG emissions would appear to be the single most effective step towards the goal of net zero.

5 Carbon capture and sequestration

The specialist chapters in this book have identified and explored the many and various contributors to the problem of increasing concentrations of CO_2 and methane in the atmosphere and offered explanations for the wide range of variation in the estimates of their impact. There is, however, one central, inescapable truth. This is entirely a problem of the last 100 years (or so) and caused by the industrialisation of production (including farming) made possible by the expenditure of capital in the form of fossil fuels.

The very large increase in the production of milk products for human consumption cannot simply be attributed to a very large increase in the number and size of belching ruminants. I repeat the words from my introduction (Webster, 2026): 'In 2024, the total cattle population in the U.S.A. (cows, calves and finishing beef cattle) was approximately 87 million of which about 9.3 million were adult dairy cows and 28 million beef cows. In the 17th century, the bison population in the USA was approximately 60 million. The current production of GHG from ruminants in the USA has been estimated as no more than 14% greater than that when bison roamed the plains – *sustainably* (Hristov 2012)'.

It is also clear that we cannot solve the problem simply by turning the cows out into the fields. In an international comparison, use of confinement production systems normally increased the GHG emission intensity compared to systems (e.g. in New Zealand) where cows spend much of the year grazing at pasture, but often decreased the intensity of production systems in the northeastern U.S. and the Netherlands, which are much more reliant on harvested crops. The many comparisons of intensive versus grass-based systems quoted by Rotz and Dell (2026) (Chapter 3) present data to show that, when climate and soil conditions are more favourable for corn and small grain feed crops, confinement feeding systems can out-perform grass-based systems.

The route to sustainability is not grass, but grazing. With the right environmental conditions and ample moisture, grazing can recycle more carbon than systems managed by mowing. Correct grazing management can provide significant benefits to soil conditions, as well as entail a more conservative biomass removal compared to mowing and lead to some carbon sequestration and improved soil health. Sixty million bison can't have been wrong. Nevertheless, this was a system in balance. It did not involve net sequestration of carbon. While it is possible to sequester carbon in soils that have been depleted of organic matter through the production of annual crops by converting them to perennial grassland, eventually the soil will come into equilibrium with the soil C content similar to native, undisturbed soil (Soussanna et al., 2012). There is a very limited capacity for C sequestration in stable pastures with stable soils. Thus, it does not resolve our current problem.

Strategies to offset emissions include Integrated Crop-Livestock Systems (ICLS) that maximise the potential for recycling nutrients on farm, as discussed in Chapter 13 (Weisner, 2026). Other ecofriendly approaches include the coexistence of dairy cattle with perennial ('natural') vegetation, including pastures, shrublands, hedgerows, or trees. However, results from such trials tend to show limited effects on carbon sequestration, nutrient retention, and nutrient runoff. More promising approaches include silvopasture and agroforestry systems, where animals are integrated with trees and grasslands on the same land (Brick et al., 2022). I have visited silvopastoral dairy units in Brazil wherein dairy cows graze quality grass leys between rows of recently planted trees. The trees not only sequester carbon but also create a very favourable microclimate for the growth of the grass and the comfort of the cows.

Many years ago, when I was promoting, on animal welfare grounds, the development of outdoor systems for breeding pigs, I was informed by an intensive pig producer 'we couldn't possibly rear our sows outdoors, our land is too wet'. To which I replied: 'then perhaps you don't have a pig farm'. This comment has special relevance to dairy farming. I live in Somerset (U.K.), classic dairy cow country. My next-door neighbour farms his 100+ dairy cows according to an ICLS system based on grass grazed and conserved as silage, together with, in rotation, maize (ensiled) and root crops for his cows, plus barley and wheat for sale. He speaks of it as a dairy farm but it is, of course, a classic ICLS system. Moreover, as I wrote in Chapter 1, this system produces more energy and protein for human consumption from milk than the cows consume from 'competitive' feeds (food that could feed humans). This makes the obvious point that dairy farming is at its most sustainable on the land that is best suited to dairy farming.

Larger farms exhibit higher economic productivity due to economies of scale and better resource management. Nevertheless, smallholder family farms, usually operating in classic 'cow country', still make up the bulk of the global food production, contributing 50–70% of the supply. However, they often struggle with economic viability and food insecurity, particularly in lower-income countries. Despite their potential for more sustainable farming practices, these farms face challenges due to limited resources and global market dynamics and merit government support on grounds of sustainability for both the local populations and the global ecosystem.

6 Life cycle analysis of dairy systems

A plethora of recent lifecycle analyses have focused on evaluating the inputs, outputs and efficiency of agricultural systems. The most relevant of these are those based on carbon and *emergy*. When aiming for net zero, especially in regard to CO_2e, agricultural systems should be evaluated over multiple timescales, as different processes, such as nutrient cycling and decomposition, do not follow the same seasonal patterns. Chapter 3 (Rotz and Dell, 2026) provides a comprehensive review of strategies for mitigating CO_2e release. The capacity for carbon sequestration in the soil of grasslands is, I repeat, limited, although it can be improved if dairy farming is integrated with other approaches to sequestration, e.g. new growth of trees.

Dairy farming is, of course, driven by energy. The prime source of energy is the sun. Almost all the rest is sourced from carbon. *Emergy* analysis, as described in Chapter 1 (Webster, 2026), quantifies the yield of food energy for human consumption (Y) in terms of renewable (R) resources of sun, soil and water embedded in farmland, non-renewable sources (NR), loss of energy from (e.g.) soil degradation, and F, purchased goods and services such as fuel, fertiliser, labour and imported feeds. The three critical outcomes of emergy analysis are emergy yield ratio (EYR), environmental load ratio (ELR), and emergy sustainable index (ESI). These ratios are dimensionless. The emergy yield ratio describes the contribution of sun and local resources (land) to product, ELR describes the ratio of non-renewables to renewables in product. The ESI (EYR/ELR) is a measure of yield relative to environmental compatibility.

Vigne et al. (2013) compared traditional third-world dairy farms with minimal inputs from machinery and purchased feeds with modern dairy units in the pastoral regions of west France. The ESI values were >1.0 (i.e. better than net zero by this measure) for the traditional units, approximately 0.25–0.42 for the commercial farms in France. Luiz et al. (2013) compared intensive (INT), semi-intensive (SIS), and extensive (EXT) dairy units in Brazil in terms of both ESI and profitability. The INT and SIS systems ranked equal in terms of profitability, but SIS was twice as effective in terms of ESI (0.91 v. 0.56). The extensive system (EXT) was only half as profitable as the other two and, perhaps surprisingly, less effective in terms of ESI (0.66 v. 0.91).

7 Conclusion

Whether assessed in terms of CO_2e or energy, LCA makes it clear that dairy farming, considered in isolation, can never achieve net zero. However, the experts who have contributed to this book make it possible to weigh the options and set priorities for action to reduce the contribution of the dairy sector to climate change.

However, it is important first to set these options in a wider context. The increasing concentrations of atmospheric methane and CO_2 in the last 100 years are linked to the industrialisation of agriculture, and the overwhelming contributor has been the release of consumption of fossil fuels. A clear message is that the contribution of GHG emissions from enteric methane production is relatively small, 9–13% CO_2e/kg FPCM. Indeed, as noted earlier, the emission of methane from ruminants in the U.S.A is estimated to be only about 14% greater than when bison roamed the plains.

Mitigation strategies based on diet formulation and methane-inhibiting feed additives have the potential to reduce this to about 5–8% CO_2e/kg FPCM. Reports of the emission of CO_2e/kg FPCM from large, well-managed confinement units housing very high-yielding cows are remarkably similar to those from less intensive farms. The increased carbon costs for fuel, fertiliser, and purchased feeds are offset by efficiencies of scale and formulation of higher energy diets that promote increased lactation yields and reduce methane production per MJ ME intake.

On all dairy farms, especially very large confinement units, the main contributor to CO_2e is pollution from manure and other farm wastes. The first priority therefore must be to reduce (indeed, seek to eradicate) pollution from 'too much fertilizer in the wrong place'. The most effective approach to mitigation is through use of ADs.

The systems that appear to achieve the best compromise between profitability and sustainability are semi-intensive systems that incorporate seasonal grazing on permanent pastures or long leys, recycling of manure, and a high contribution of conserved, home-grown feed during the housing period. Dairy farming is best suited to 'cow country', i.e. those areas traditionally seen as suited to this form of production.

However, it is important to note permanent pastures offer little scope for carbon sequestration. More can be achieved with silvo-pastoral systems.

However, the overwhelming source of the increase in atmospheric CO_2e in the Anthropocene has not been belching cows. The most important conclusion is that the overwhelming contributor to the emission of GHG from dairy farming has been the release of CO_2e from the consumption of fossil fuels. Since it is impossible to imagine a world in which the demand for dairy products could be met by subsistence farming, the best option for dairy farming, as for all industry, must be to reduce our reliance on fossil fuels as an energy source.

9 References

Bittman, S., et al. (2026). Optimizing manure and slurry application on intensive dairy farming operations. In Webster, J. (ed.). *Achieving net zero dairy farming*, Burleigh Dodds Science Publishing, Cambridge, UK.

Brick, E. S. R., et al. (2022). A review of agroforestry, precision agriculture, and precision livestock farming–The case for a data-driven agroforestry strategy. *Frontiers in Sensors*, 3, 998928. https://doi.org/10.3389/fsens.2022.998928.

Caro, D., et al. (2014). Global and regional 731 trends in greenhouse gas emissions from livestock. *Climate Change*, 126, 203–216.

Dlugokencky, et al. (2011). Global atmospheric methane: budget, changes and dangers. *Philosophical Transactions of the Royal Society A: Mathematical, Physical and*

Engineering Sciences, 369(1943), 2058–2072. https://doi.org/10.1098/rsta.2010.0341.

Hristov, A. N. (2012). Historic, pre-European settlement, and present-day contribution of wild 955 ruminants to enteric methane emissions in the United States. *Journal of Animal Science*, 90(4), 1371–1375. https://doi.org/10.2527/jas.2011-4539.

Kebreab, E., et al. (2023). A meta-analysis of effects of 3-nitrooxypropanol on methane production, yield, and intensity in dairy cattle. *Journal of Dairy Science*, 106, 927–936.

Kusch-Brandt, S. (2026). Methane utilisation on dairy farms. In Webster, J. (ed.). *Achieving net zero dairy farming*, Burleigh Dodds Science Publishing, Cambridge, UK.

Luiz, V.T., et al. (2013). Sustainability assessment of intensification levels of Brazilian smallholder integrated dairy-crop production systems: An emergy and economic-based decision approach. *Sustainability, 15*(5), 4674.

Martins, L.F., et al. (2025). Lactational performance effects of 3-nitrooxypropanol supplementation to dairy cows: A meta-regression. *Journal of Dairy Science*, 108, 1538–1557.

Newbold, J. (2026). Development and use of methane-inhibiting feed supplements in dairy farming. In Webster, J. (ed.). *Achieving net-zero dairy farming*, Burleigh Dodds Science Publishing, Cambridge, UK.

Ramin, M., et al. (2026). Advances in understanding methane production in the dairy cow. In Webster, J. (ed.). *Achieving net zero dairy farming*, Burleigh Dodds Science Publishing, Cambridge, UK.

Rezaei Ahvanooei, M.R., et al. (2023). Effects of monensin supplementation on lactation performance of dairy cows: a systematic review and dose–response meta–analysis. *Scientific Reports*, 13, 568. https://doi.org/10.1038/s41598-023-27395-9.

Rocha, A. (2026). Manure management and processing. In Webster, J. (ed.). *Achieving net zero dairy farming*, Burleigh Dodds Science Publishing, Cambridge, UK.

Rotz, C.A. and Dell, C. (2006). Assessing greenhouse gas emissions from dairy farms. In Webster, J. (ed.). *Achieving net zero dairy farming*, Burleigh Dodds Science Publishing, Cambridge, UK.

Rotz C.A., et al. (2010). The carbon footprint of dairy production systems through partial life cycle assessment. *Journal of Dairy Science*, 93, 1266–1282.

Soussanna, J.F., et al. (2010). Mitigating the greenhouse gas balance of ruminant production systems through carbon sequestration in grasslands. *Animal*, 4, 334–340.

Thoma, G., et al. (2013a). Greenhouse gas emissions from milk production and consumption in the United States: a cradle-to-grave life cycle assessment circa 2008. *International Dairy Journal*, 31, S3–S14. https://doi.org/10.1016/j.idairyj.2012.09.010.

Thoma, G., et al. (2013b). Regional analysis of greenhouse gas emissions from USA dairy farms: a cradle to farm-gate assessment of the American dairy industry circa 2008. *International Dairy Journal*, 31, S29–S40. https://doi.org/10.1016/j.idairyj.2012.09.010.

van Breukelen, A.E. (2026). Genetics and breeding for reduced methane emissions. In Webster, J. (ed.). *Achieving net zero dairy farming*, Burleigh Dodds Science Publishing, Cambridge, UK.

Vigne, M., et al. (2013). Emergy evaluation of contrasting dairy systems at multiple levels. *Journal of Environmental Management*, 129, 44–53.

Wang, J. and He, Q. P. (2023). Methane removal from air: challenges and opportunities. *Methane*, 2(4), 404–414. https://doi.org/10.3390/methane2040027.

Webster, J. (2026). Sustainability, nutrient and energy flows in the dairy cow: an overview. In Webster, J. (ed.). *Achieving net zero dairy farming*, Burleigh Dodds Science Publishing, Cambridge, UK.

Wiesner, S., et al. (2026). Integrating dairy farming and crop production. In Webster, J. (ed.). *Achieving net zero dairy farming*, Burleigh Dodds Science Publishing, Cambridge, UK.

Index

www.ingramcontent.com/pod-product-compliance
Ingram Content Group UK Ltd.
Pitfield, Milton Keynes, MK11 3LW, UK
UKHW021825100926
8556IPUK00003B/163

9 781835 450185